Solar Radiation, Modelling and Remote Sensing

Solar Radiation, Modelling and Remote Sensing

Special Issue Editors

Dimitris Kaskaoutis
Jesús Polo

MDPI • Basel • Beijing • Wuhan • Barcelona • Belgrade

MDPI

Special Issue Editors
Dimitris Kaskaoutis
National Observatory of Athens
Greece

Jesús Polo
Avda Complutense
Spain

Editorial Office
MDPI
St. Alban-Anlage 66
4052 Basel, Switzerland

This is a reprint of articles from the Special Issue published online in the open access journal *Remote Sensing* (ISSN 2072-4292) from 2018 to 2019 (available at: https://www.mdpi.com/journal/remotesensing/special_issues/solar_RS)

For citation purposes, cite each article independently as indicated on the article page online and as indicated below:

LastName, A.A.; LastName, B.B.; LastName, C.C. Article Title. *Journal Name* **Year**, *Article Number*, Page Range.

ISBN 978-3-03921-004-6 (Pbk)
ISBN 978-3-03921-005-3 (PDF)

Cover image courtesy of Jesús Polo.

Contents

About the Special Issue Editors

Dimitris Kaskaoutis was awarded his PhD from the University of Ioannina, Greece, in 2009. After serving for 4 years as Professor in Indian universities, he then worked at the National Observatory of Athens as a Research Assistant. He has been working in Atmospheric Physics with an emphasis on solar radiation transfer into the atmosphere; modification of the solar spectrum due to aerosols; aerosol optical, physical, and chemical properties; climate implications; and meteorology and dust dynamics over climate-sensitive areas, like the Eastern Mediterranean, the Middle East, and South Asia. Dr. Kaskaoutis has published extensively in scientific journals and at national and international conferences, and has authored 2 book chapters and served as a reviewer for numerous scientific journals.

Jesús Polo is a Senior Researcher at the Photovoltaic Solar laboratory at CIEMAT (Center of Research in Energy, Environment and Technology). His main interests are focused on solar resource assessment, solar radiation modeling, and solar system (PV and CSP) performance modeling. Dr. Polo's research interests include radiative transfer, atmospheric physics and optics, remote sensing and satellite-derived retrievals for energy meteorology, and performance studies of PV and CSP plants. Dr. Polo has been actively collaborating in expert groups of energy meteorology for several programs of the International Energy Agency (IEA-SHC, IEA-PVPS, and IEA—SolarPACES). He is the main author of numerous journal articles and conference communications on these topics and has supervised numerous MSc theses in collaboration with other universities.

Preface to "Solar Radiation, Modelling and Remote Sensing"

Surface solar radiation is of vital importance for life on Earth, radiation–energy balance, photosynthesis and photochemical reactions, meteorological and climatic conditions, and the water cycle. Solar radiation is the most abundant renewable energy resource and, therefore, the demand for environmentally clean energy solutions and the reduction of greenhouse gas emissions have shifted the global interest towards solar energy exploitation for sustainable development and electricity demands. Solar radiation measurements are necessary in the assessment of potential solar energy resources, while the scarce spatial coverage renders solar radiation modeling and remote sensing necessary for atmospheric and energy applications. This Special Issue aims to review techniques for solar radiation measurements and modeling (solar radiation networks, historical developments, technique comparisons, and standard comparisons between models) and remote sensing using satellite and advanced statistical techniques, like artificial neural networks for solar radiation and energy mapping from regional to global scales. Satellite remote sensing of solar radiation provides better spatial coverage, and various methods have been developed on this theme, with the main disadvantages being the increased uncertainties and the required validations against ground-based measurements or modeling data. An accurate knowledge of global solar radiation and its direct and diffuse components at the ground are very important for the design of solar power systems, as well as for climatological and agricultural issues.

Dimitris Kaskaoutis, Jesús Polo
Special Issue Editors

remote sensing

MDPI

Editorial

Editorial for the Special Issue "Solar Radiation, Modeling, and Remote Sensing"

Dimitris Kaskaoutis [1] and Jesús Polo [2],*

[1]   Atmospheric Research Team, Institute of Environmental Research and Sustainable Development,
      National Observatory of Athens, 11810 Athens, Greece; dkask@noa.gr
[2]   Photovoltaic Solar Energy Unit, Renewable Energy Division (Department of Energy), CIEMAT,
      40 28040 Madrid, Spain
*   Correspondence: jesus.polo@ciemat.es

Received: 16 May 2019; Accepted: 17 May 2019; Published: 20 May 2019

Abstract: Surface-solar radiation is of vital importance for life on Earth, radiation–energy balance, photosynthesis, and photochemical reactions, meteorological and climatic conditions, and the water cycle. Solar radiation measurements are growing in quality and density but they are still scarce enough to properly explain the spatial and temporal variability. As a consequence, great efforts are still being devoted to improving modeling and retrievals of solar radiation data. This Special Issue reviews techniques for solar radiation modeling and remote sensing using satellite and advanced statistical techniques for solar radiation. Satellite remote sensing of solar radiation provides better spatial coverage, and various methods have been presented on this issue covering several aspects: updated models for solar radiation modeling under clear sky conditions, new approaches for retrieving solar radiation from satellite imagery and validation against ground data, forecasting solar radiation, and modeling photosynthetically active radiation.

Keywords: solar radiation; radiative transfer; solar energy systems; solar radiation forecasting

---

1. Overview of the Issue: Solar Radiation Modeling and Remote Sensing

Accurate solar radiation knowledge and its characterization on the Earth's surface are of high interest in many aspects of environment and engineering sciences. Radiative transfer that takes place in the atmosphere leads to the high spatial and temporal variability of solar radiation along the planet. This variability, which has important consequences in both climatic studies and in solar system performance analysis, has promoted the effort in improving the modeling capabilities for the determination of the solar irradiance components [1]. In the absence of clouds, the models for estimating solar irradiance are denoted as clear sky models, in which the accuracy mostly resides on the detail and quality knowledge of the atmospheric components that act as attenuators of the solar radiation traversing the atmosphere [2,3]. The evolution of clear sky models has brought significant improvements in parallel to the advanced reached in atmospheric retrievals; in this issue, the first contribution is precisely a paper containing the update of a well-known and widely used clear sky model named SOLIS [4].

Modeling solar irradiance from satellite imagery has become the most widely used models for retrieving solar irradiance under total sky conditions, particularly in the solar energy community [5]. Satellite-based models can be divided into different criteria. Thus, according to the kind of satellite, models using polar orbiting satellite imagery and models for geostationary satellites can be found, each with different applicability. On the other hand, depending on the approaches and assumptions used in the algorithms the classification divides the models into physical models and semi-empirical (or cloud index based) models [6]. There has been enormous growth in the amount of work, authors, research, developments, and improvements in satellite-derived solar radiation through the last 30 years and this growth is still ongoing. This Special Issue is witness to this assessment, since contributions 2 to 6 deal with different aspects of solar radiation modeling and validation using satellite information [7–11]. Moreover, this is not the only evidence of the ongoing progress in satellite-derived solar radiation since there are many thorough reviews elsewhere and groups of experts working on improving the modeling capabilities and accuracy (task 16 IEA-PVPS, http://www.iea-pvps.org/).

Contributions 7 and 8 make use of remote sensing techniques and solar radiation for a different approach [12,13]. The former is focused on the canopy and forest information obtained from LIDAR and Solar Analyst model from ESRI's ArcGIS to estimate solar radiation at forest stands in various scales. Contribution 8 focuses on statistical techniques for improving the photosynthetically active radiation (PAR) derived from satellite Kato bands CM-SAF product.

Satellite-based models can be also used for forecasting purposes and this is a broad area of research and development during the last decade. The impact of solar radiation forecasting tools in solar energy penetration is clearly highlighted by the number of scientific publications in recent years [14]. Contribution 9 of this Special Issue is one example; it deals with the methods for nowcasting solar irradiance using Meteosat Second Generation images [15]. The next contribution, paper 10, focuses on forecasting aerosols using remote sensing and its impact on solar energy plants performance in sites with high aerosol loading climatology [16].

Finally, the last contribution of this Special Issue is aimed at studying solar radiation variability at different temporal scales and the connection with satellite-derived cloud properties using GOES images [17].

2. Conclusions

Solar radiation modeling, forecasting, and characterization have been and still are broad areas of study, research, and development in the scientific community. This Special Issue contains a small sample of current activities in this field. Both the environmental and climatology community, as part of the solar energy world, share a high interest in improving the modeling tools and capabilities for more reliable and accurate knowledge of solar irradiance components worldwide. The work presented in this Special Issue also remarks on the significant role that remote sensing technologies play in retrieving and forecasting solar radiation information.

Author Contributions: The two authors contributed equally to all aspects of this editorial.

Acknowledgments: The authors would like to thank the authors who contributed to this Special Issue and to the reviewers who dedicated their time for providing the authors with valuable and constructive recommendations.

Conflicts of Interest: The authors declare no conflict of interest.

References

1. Polo, J.; Martín-Pomares, L.; Gueymard, C.A.; Balenzategui, J.L.; Fabero, F.; Silva, J.P. Fundamentals: Quantities, Definitions, and Units. In *Solar Resource Mapping—Fundamentals and Applications*; Green Energy and Technology; Polo, J., Martín-Pomares, L., Sanfilippo, A., Eds.; Springer: Zurich, Switzerland, 2019; pp. 1–14.

2. Antonanzas-Torres, F.; Urraca, R.; Polo, J.; Perpiñán-Lamigueiro, O.; Escobar, R. Clear sky solar irradiance models: A review of seventy models. *Rene. Sus. Energ. Rev.* **2019**, *107*, 374–387. [CrossRef]

3. Ruiz-Arias, J.A.; Gueymard, C.A. Worldwide inter-comparison of clear-sky solar radiation models: Consensus-based review of direct and global irradiance components simulated at the earth surface. *Sol. Energy* **2018**, *168*, 10–29. [CrossRef]

4. Ineichen, P. High turbidity Solis clear sky model: Development and validation. *Remote Sens.* **2018**, *10*, 435. [CrossRef]

5. Sengupta, M.; Habte, A.; Kurtz, S.; Dobos, A.; Wilbert, S.; Lorenz, E.; Stoffel, T.; Renne, D.; Myers, D.; Wilcox, S.; et al. *Best Practices Handbook for the Collection and Use of Solar Resource Data for Solar Energy Applications*; NREL/TP-5D00-63112: Golden, CO, USA, 2015.

6. Polo, J.; Perez, R. Solar radiation modeling from satellite imagery. In *Solar Resource Mapping—Fundamentals and Applications*; Green Energy and Technology; Polo, J., Martín-Pomares, L., Sanfilippo, A., Eds.; Springer: Zurich, Switzerland, 2019; pp. 183–197.
7. Anderson, M.; Diak, G.; Gao, F.; Knipper, K.; Hain, C.; Eichelmann, E.; Hemes, K.S.; Baldocchi, D.; Kustas, W.; Yang, Y. Impact of Insolation Data Source on Remote Sensing Retrievals of Evapotranspiration over the California Delta. *Remote Sens.* **2019**, *11*, 216. [CrossRef]
8. Ameen, B.; Balzter, H.; Jarvis, C.; Wey, E.; Thomas, C.; Marchand, M. Validation of Hourly Global Horizontal Irradiance for Two Satellite-Derived Datasets in Northeast Iraq. *Remote Sens.* **2018**, *10*, 1651. [CrossRef]
9. Romano, F.; Cimini, D.; Cersosimo, A.; Di Paola, F.; Gallucci, D.; Gentile, S.; Geraldi, E.; Larosa, S.T.; Nilo, S.; Ricciardelli, E.; et al. Improvement in Surface Solar Irradiance Estimation Using HRV/MSG Data. *Remote Sens.* **2018**, *10*, 1288. [CrossRef]
10. Riihelä, A.; Kallio, V.; Devraj, S.; Sharma, A.; Lindfors, A. V Validation of the SARAH-E Satellite-Based Surface Solar Radiation Estimates over India. *Remote Sens.* **2018**, *10*, 392. [CrossRef]
11. Lee, S.-H.; Kim, B.-Y.; Lee, K.-T.; Zo, I.-S.; Jung, H.-S.; Rim, S.-H.; Riihelä, A.; Kallio, V.; Devraj, S.; Sharma, A.; et al. Retrieval of Reflected Shortwave Radiation at the Top of the Atmosphere Using Himawari-8/AHI Data. *Remote Sens.* **2018**, *10*, 213. [CrossRef]
12. Olpenda, A.S.; Stereńczak, K.; Będkowski, K. Modeling Solar Radiation in the Forest Using Remote Sensing Data: A Review of Approaches and Opportunities. *Remote Sens.* **2018**, *10*, 694. [CrossRef]
13. Vindel, J.M.; Valenzuela, R.X.; Navarro, A.A.; Zarzalejo, L.F.; Paz-Gallardo, A.; Souto, J.A.; Méndez-Gómez, R.; Cartelle, D.; Casares, J.J. Modeling Photosynthetically Active Radiation from Satellite-Derived Estimations over Mainland Spain. *Remote Sens.* **2018**, *10*, 849. [CrossRef]
14. Gueymard, C.A.; Pedro, H.T.C.; Coimbra, C.F.M. History and trends in solar irradiance and PV power forecasting: A preliminary assessment and review using text mining. *Sol. Energy* **2018**, *168*, 60–101.
15. Gallucci, D.; Romano, F.; Cersosimo, A.; Cimini, D.; Di Paola, F.; Gentile, S.; Geraldi, E.; Larosa, S.; Nilo, S.T.; Ricciardelli, E.; et al. Nowcasting Surface Solar Irradiance with AMESIS via Motion Vector Fields of MSG-SEVIRI Data. *Remote Sens.* **2018**, *10*, 845. [CrossRef]
16. Kosmopoulos, P.G.; Kazadzis, S.; El-Askary, H.; Taylor, M.; Gkikas, A.; Proestakis, E.; Kontoes, C.; El-Khayat, M.M. Earth-Observation-Based Estimation and Forecasting of Particulate Matter Impact on Solar Energy in Egypt. *Remote Sens.* **2018**, *10*, 1870. [CrossRef]
17. Xia, S.; Mestas-Nuñez, A.M.; Xie, H.; Tang, J.; Vega, R. Characterizing Variability of Solar Irradiance in San Antonio, Texas Using Satellite Observations of Cloudiness. *Remote Sens.* **2018**, *10*, 2016. [CrossRef]

remote sensing

MDPI

Article

High Turbidity Solis Clear Sky Model: Development and Validation

Pierre Ineichen

Department F.-A. Forel for Environmental and Aquatic Sciences, Institute for Environmental Sciences, University of Geneva, 1205 Genève, Switzerland; pierre.ineichen@unige.ch; Tel.: +41-22-379-0640

Received: 8 January 2018; Accepted: 8 March 2018; Published: 10 March 2018

Abstract: The Solis clear sky model is a spectral scheme based on radiative transfer calculations and the Lambert–Beer relation. Its broadband version is a simplified fast analytical version; it is limited to broadband aerosol optical depths lower than 0.45, which is a weakness when applied in countries with very high turbidity such as China or India. In order to extend the use of the original simplified version of the model for high turbidity values, we developed a new version of the broadband Solis model based on radiative transfer calculations, valid for turbidity values up to 7, for the three components, global, beam, and diffuse, and for the four aerosol types defined by Shettle and Fenn. A validation of low turbidity data acquired in Geneva shows slightly better results than the previous version. On data acquired at sites presenting higher turbidity data, the bias stays within ±4% for the beam and the global irradiances, and the standard deviation around 5% for clean and stable condition data and around 12% for questionable data and variable sky conditions.

Keywords: Solis scheme; clear sky; radiation model; radiative transfer; high turbidity; water vapor

---

1. Introduction

Anthropogenic activities have become an important factor in climate change and continuous monitoring of the solar irradiance reaching the ground is essential to understand the impact of such changes on the environment. Unfortunately, the density of the ground measurement network is insufficient, especially on continents like Africa or Asia. To circumvent this lack of measured data, meteorological satellites are of great help and models converting the satellite images into the different radiation components are becoming increasingly useful.

The first step in evaluating the solar irradiance from satellite images is a good knowledge of the highest possible radiation transmitted by a cloudless atmosphere. This can be done with the help of clear sky model taking into account the aerosol and water vapor content of the atmosphere. Multiple clear sky models can be found in the literature, but none of them is able to evaluate solar radiation for high turbidities resulting from heating and transportation activities and encountered in countries like India or China. For example, the Gueymard CPCR2 model [1] is limited to β turbidity values lower than 0.4 (which correspond to an aerosol optical depth at 700 nm aod_{700} of 0.64 for rural aerosol type, Angström exponent α = 1.3), Gueymard REST2 [2] is limited to β = 1 (aod_{700} = 1.6, rural aerosol), Bird's model [3] is defined for visibility values up to 23 km (which corresponds to an aod_{700} = 0.27), and the ESRA clear sky scheme [4–6] was developed for Linke turbidity values T_L [7] not exceeding a value of 7 (i.e., an aerosol optical depth aod_{700} = 0.44 for a 2 cm water vapor column w).

The spectral Solis clear sky scheme was first developed within the Mesor European program, whose subject was the management and exploitation of solar resource knowledge. Due to the spatial and temporal range of the satellite images, the on line process to evaluate irradiance maps implies fast analytical models. In [8], Ineichen published a broadband fast analytical version of the Solis model for rural aerosol type, and in 2010 a version in the form of an Excel tool for the four types of aerosols was defined by Shettle and Fenn [9] and Shettle [10]. These versions were limited to broadband aerosol optical depth values aod_{700} lower than 0.45; they diverge for higher turbidity, as shown by Zhang [11]. This is illustrated in Figure 1, where the normal beam irradiance I_{bn} is represented versus the aerosol optical depth for the Solis 2008 model and different solar elevation angles.

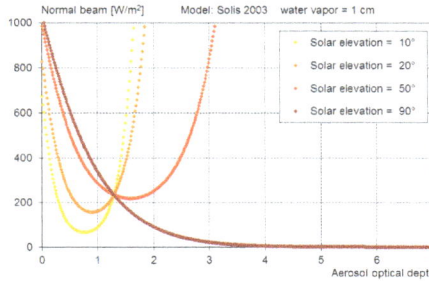

Figure 1. Behavior of the normal beam irradiance with the aerosol optical depth for the 2008 Solis version and different solar elevation angles.

This paper presents the development and validation of a new version of the analytical Solis scheme valid for the three radiation components, the global, the beam, and the diffuse, and the four aerosol types, urban, rural, maritime, and tropospheric.

2. Basis of the Solis Scheme

The original Solis clear sky model [12] was first developed within the European program Joule in 2004 in the frame of the Heliosat project, whose aim was to derive and assess the evaluation of the solar irradiance components from the geostationary meteorological satellite MSG (Meteosat Second Generation). The scheme is based on the spectral Radiative Transfer Model (RTM) LibRadTran [13,14] and the Lambert–Beer attenuation relationship:

$$I_{bn} = I_0 \cdot e^{-M\tau}, \tag{1}$$

where I_0 is the extraterrestrial irradiance, I_{bn} the normal beam irradiance reaching the ground, M the optical air mass, and τ the extinction coefficient. In order to adapt this relation to broadband irradiance, it has to be modified slightly. Three distinct extinction coefficients, one for each of the irradiance components, are defined, as well as three exponent coefficients in the Lambert–Beer attenuation relation. The diffuse irradiance is the result of the scattering process of the beam component during its crossing of the atmosphere. In order to minimize the number of coefficients, a common modified extraterrestrial irradiance I_0' expressed in W/m² is defined.

The three broadband relations then take the following form:

$$I_{bn} = I_0' \cdot e^{(-\tau_b / \sin^b h)} \quad I_{gh} = I_0' \cdot e^{(-\tau_g / \sin^g h)} \cdot \sin h \quad I_{dh} = I_0' \cdot e^{(-\tau_d / \sin^d h)}, \tag{2}$$

where I_{bn}, I_{gh} and I_{dh} represent the normal beam irradiance, the horizontal global, and the horizontal diffuse components, respectively, and h is the solar elevation angle.

The model is then driven by seven parameters: the modified extraterrestrial irradiance I_0', the three extinction coefficients τ_b, τ_g and τ_d for the beam, global, and diffuse irradiance, respectively, and b, g, and d are the corresponding sine exponent coefficients.

The justification of these equations can be found in [1,12].

3. Solis Extension to High Turbidity

To circumvent the limitation of the previous Solis scheme, we developed a new version that is applicable for aerosol optical depth values *aod* up to 7. This large range of turbidity values is needed to avoid unrealistic irradiance values resulting from extreme input parameters to automatic processes. However, one has to keep in mind that above *aod* = 2, it is not clear that the turbidity can be considered an aerosol optical depth, but a "turbidity" due to bigger particles like sand, or thin, high-altitude clouds. Nevertheless, it is important that in on line automatic production processes, the model does not diverge and still derives coherent values, even with some discrepancies from ground measurements.

3.1. Model Development

The main parameters driving the determination of the solar radiation reaching the ground for a cloudless atmosphere are the aerosol optical depth *aod*, the total atmospheric precipitable water vapor column *w* expressed in centimeters, the altitude of the considered site, and the aerosol type. Other atmospheric content such as O_3, NO_x, etc. have a very low influence of the atmospheric transmittivity and will be considered constant in the development of the model.

In the first step, we made spectral calculations with LibRadTran for each combination of the following parameters.

These RTM calculations permit us to generate the seven coefficients that drive the model: I_o', τ_b, b, τ_g, g, τ_d, and d. The next step is to develop an analytical formulation for these coefficients with the four input parameters given in Table 1.

Table 1. Aerosol types, altitudes, optical depths aod_{550}, and water vapor columns *w* values for the RTM calculations with LibRadTran.

Aerosol Type	Altitude	Aod_{550}	W
urban	sea level	0.01	0.01
rural	500 m	0.03	0.03
maritim	1000 m	0.05	0.05
tropospheric	2000 m	0.1	1
	3000 m	0.15	0.15
	4000 m	0.2	0.2
	5000 m	0.4	0.3
	6000 m	0.7	0.5
	7000 m	1	1
		1.5	1.5
		2	2
		3	3
		4	4
		5	6
		6	8
		7	10

The analysis of I_o', τ_b, τ_g dependence with the *aod* shows a similar behavior for these three parameters. A third-order polynomial model of the form is applicable:

$$\begin{aligned}
I_o'/I_0 &= a_i \cdot aod^3 + b_i \cdot aod^2 + c_i \cdot aod + d_i \\
\tau_b &= a_b \cdot aod^3 + b_b \cdot aod^2 + c_b \cdot aod + d_b \\
\tau_g &= a_g \cdot aod^3 + b_g \cdot aod^2 + c_g \cdot aod + d_g.
\end{aligned} \tag{3}$$

The behavior of I_o'/I_0 and τ_g is represented in Figure 2.

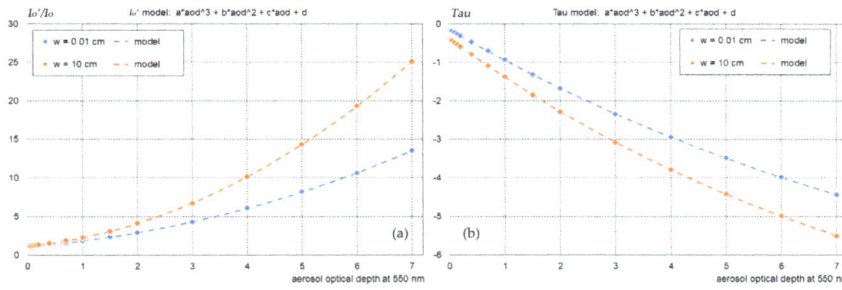

Figure 2. Behavior of the enhanced solar constant I_0'/I_0 (**a**) and the extinction coefficient τ (**b**) with the spectral aerosol optical depth at 550 nm.

We then analyzed the dependence of each of the a, b, c and d coefficients with the atmospheric water vapor content w. The behavior of the four coefficients shows a common dependence of the form:

$$n = n_1 \cdot w^{0.5} + n_2 \cdot \ln w + n_3, \tag{4}$$

where n replaces the a, b, c, and d coefficients for I_0', τ_b and τ_g.

The best fit for these coefficients is illustrated in Figure 3.

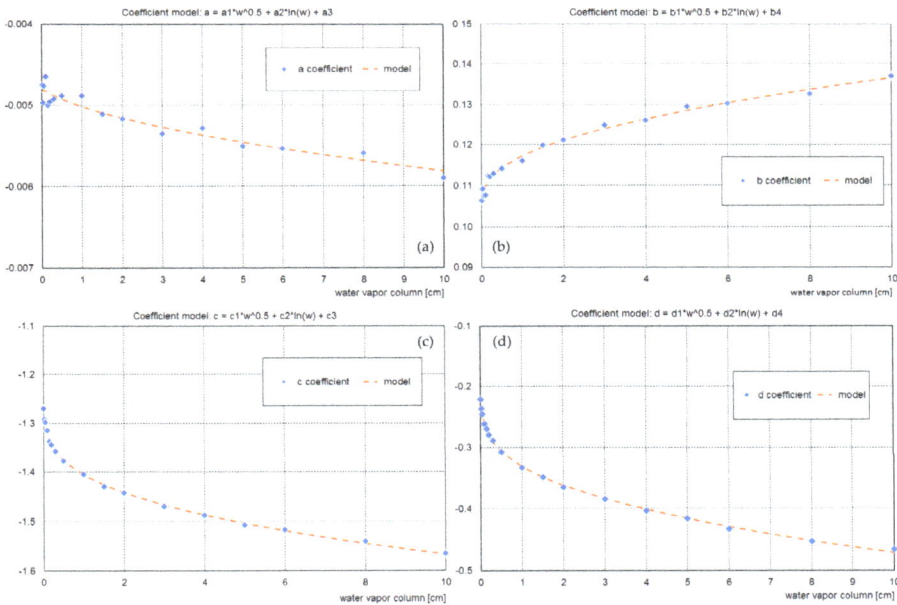

Figure 3. Behavior of the cubic equation coefficients a, b, c and d, respectively graph (**a**–**d**), with the water vapor column w.

Finally, the dependence of the three above coefficients on the altitude and the normalized atmospheric pressure p/p_0 (the pressure p at a given altitude normalized by the corresponding sea level pressure p_0) is analyzed. It appears that a linear regression gives good results.

The model has the following form (for a_i, b_i, c_i, d_i, a_b, b_b, c_b, d_b, a_g, b_g, c_g, and d_g):

$$
\begin{aligned}
a &= a_1 \cdot w^{0.5} + a_2 \cdot \ln w + a_3 \\
b &= b_1 \cdot w^{0.5} + b_2 \cdot \ln w + b_3 \\
c &= c_1 \cdot w^{0.5} + c_2 \cdot \ln w + c_3 \\
d &= d_1 \cdot w^{0.5} + d_2 \cdot \ln w + d_3
\end{aligned}
\tag{5}
$$

and the general form n_i of a, b, c and d coefficient is obtained by a linear function of p/p_o:

$$
n_i = n_{i1} \cdot p/p_o + n_{i2}. \tag{6}
$$

Finally, the inputs for the model for I_o'/I_o, τ_b, and τ_g, consist of the aerosol optical depth at 550 nm, aod_{550}; the water vapor content of the atmosphere, w; and the relative atmospheric pressure, p/p_o. The corresponding coefficients are given in a $4 \times 3 \times 2$ matrix (see Appendix A).

For τ_d, the best correlation we found is a relation with τ_b and τ_g that has the form:

$$
\tau_d = c_{td1} + c_{td2} \cdot \tau_g + c_{td3}/\tau_b + c_{td4} \cdot \tau_g^2 + c_{td5}/\tau_b^2 + c_{td6} \cdot \tau_g/\tau_b. \tag{7}
$$

The exponents of $\sin h$ in the Lambert–Beer function are best fitted as follows:

$$
\begin{aligned}
g &= ca_1 + ca_2 \cdot \ln w + ca_3 \cdot \ln aod + ca_4 \ln^2 w + ca_5 \cdot \ln^2 aod + ca_6 \cdot \ln w \cdot \ln aod \\
b &= cb_1 + cb_2 \cdot w + cb_3 \cdot aod \\
d &= cd_1 + cd_2 \cdot \ln w + cd_3 \cdot aod + cd_4 \cdot aod^2 + cd_5 \cdot aod^3 + cd_6 \cdot aod^4 + cd_7 \cdot aod^5.
\end{aligned}
\tag{8}
$$

All coefficients for the four aerosol types are given in Appendix A, as well as in the form of an Excel tool [15].

3.2. Solis Analytical Model Versus RTM Calculated Parameter Comparison

The comparison of the seven parameters of the model is expressed as scatter plots between the Solis analytical model parameters and the corresponding LibRadTran calculated parameters. The validation points should be aligned on the 1:1 diagonal. On the graphs are also given the average parameter value, the mean bias difference *mbd*, the standard deviation *sd* and the correlation coefficient R^2. An illustration is given in Figure 4 for I_o' and the g coefficient.

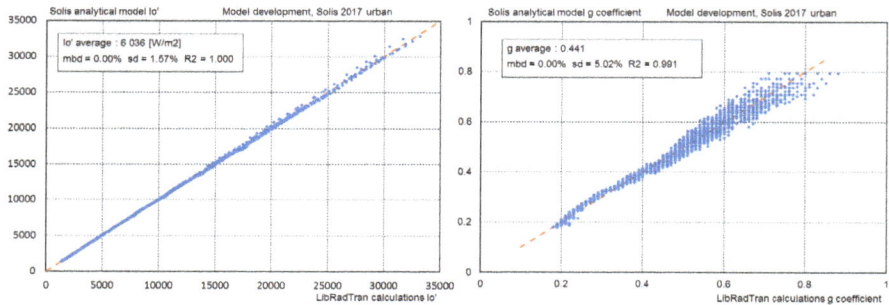

Figure 4. Modeled I_o' coefficient (**a**) and modeled g coefficient (**b**) versus the corresponding RTM calculated value.

The average *mbd* and *sd* values cannot be taken as an absolute validation of the precision as all the values calculated from the input matrix (altitude, *aod*, and *w*) have the same weight. Nevertheless, it gives a good idea of the roughness of the parameter best fit. These values are given in Table 2.

Table 2. Average, *mbd*, *sd* and R^2 for the I_o', τ_b, b, τ_g, g, τ_d, and d model.

aerosol type	I_o'				τ_g				a			
	average	mbd	sd	R^2	average	mbd	sd	R^2	average	mbd	sd	R^2
rural	6284	0	93	0.998	−1.49	0.00	0.01	1.000	0.41	0.00	0.02	0.993
urban	6036	0	95	0.997	−1.68	0.00	0.02	1.000	0.44	0.00	0.02	0.991
tropospheric	5380	0	40	0.999	−1.40	0.00	0.01	1.000	0.41	0.00	0.02	0.992
maritime	12,210	−8	134	0.998	−1.69	0.00	0.01	1.000	0.39	0.00	0.02	0.994
					τ_b				b			
rural					−2.21	0.00	0.01	1.000	0.42	0.00	0.02	0.965
urban					−2.21	0.00	0.01	1.000	0.44	0.00	0.03	0.941
tropospheric					−2.08	0.00	0.01	1.000	0.40	0.00	0.02	0.976
maritime					−2.68	0.00	0.01	1.000	0.45	0.00	0.02	0.912
					τ_d				c			
rural					−2.75	0.01	0.04	0.998	0.32	0.00	0.01	0.995
urban					−3.08	0.01	0.05	0.999	0.29	0.00	0.01	0.995
tropospheric					−2.67	0.01	0.04	0.998	0.33	0.00	0.01	0.995
maritime					−2.86	0.01	0.04	0.999	0.31	0.00	0.01	0.995

3.3. Modeled versus RTM Calculated Irradiance Comparison

In the same way, the modeled irradiances are compared to the corresponding irradiances estimated from the coefficients calculated with LibRadTran. The validation is expressed as scatter plots and the usual first-order statistics. Here, again due to the unweighted input matrix, the validation values obtained do not give absolute precision. The scatter plots given in Figure 5 illustrate the behavior of the model. They represent the modeled values plotted against the corresponding values evaluated from RTM calculations, for the three components and urban type aerosols.

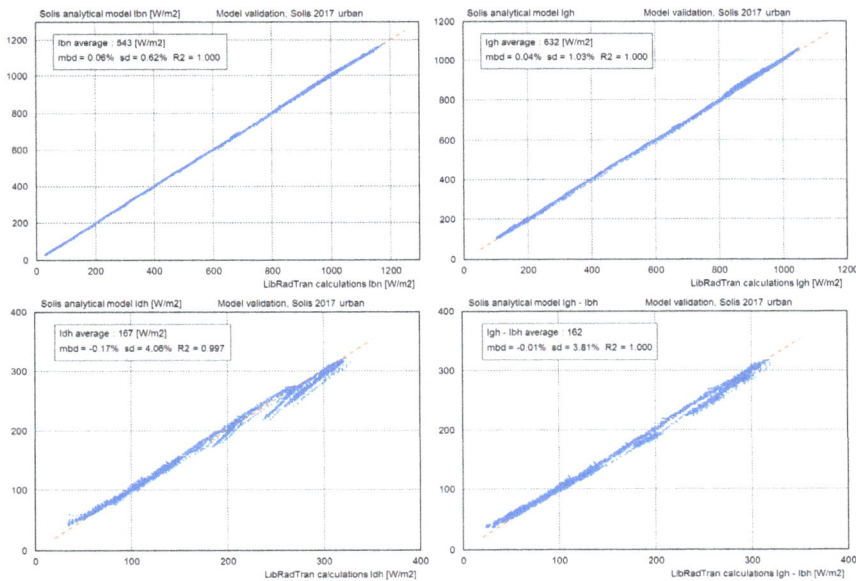

Figure 5. Modeled against RTM calculated irradiance comparison for the beam (**a**), the global (**b**), the diffuse (**c**), and the closure diffuse (**d**) components.

To obtain the diffuse component, there are two possibilities: the use of the model coefficients I'_o, τ_d and d, or to evaluate it from the global and the beam components with the use of the closure equation $I_{dh} = I_{gh} - I_{bn} \cdot \sin h$. The diffuse component obtained from the closure equation looks slightly better in Figure 4, but it presents negative values for very low solar elevations.

The comparison statistics are given in Table 3.

Table 3. Average, *mbd*, *sd*, and R^2 for the irradiance components.

	I_{gh}				I_{bn}				I_{dh}			
aerosol type	average	*mbd*	*sd*	R^2	average	*mbd*	*sd*	R2	average	*mbd*	*sd*	R^2
rural	727	0	4	1.000	549	0	5	1.000	223	0	7	0.998
urban	632	0	7	1.000	543	0	3	1.000	167	0	7	0.997
tropospheric	740	0	5	1.000	559	0	7	1.000	227	0	7	0.999
maritime	753	−1	17	0.997	519	−3	18	0.999	266	2	11	0.998

All the development and validation graphs for the rural aerosol type are given in Appendix B.

4. Model Validation against Ground Measurements

The first step in the model validation process is to apply stringent quality control to the collected ground data. Then, only data acquired under clear sky conditions should be selected; the validation is then applied on these data and the validity of the model is quantified using classical first-order statistical indicators: the mean bias difference (*mbd*) and the standard deviation (*sd*).

4.1. Quality Control of the Ground Data

The validity of the results obtained from the use of measured data is highly correlated to the quality of the data bank used as a reference. Controlling data quality is therefore the first step in the process of validating models against ground data. This should be done properly during the acquisition process and automated in order to rapidly detect significant instrumental problems like sensor failure, errors in calibration, orientation, leveling, tracking, consistency, etc. Normally, this quality control process should be done by the institution responsible for the measurements. Unfortunately, this is not always the case. Even if some quality control procedures have been implemented, they might not be sufficient to catch all errors, or the data points might not be flagged to indicate the source of the problem. A stringent control quality procedure must therefore be adopted in the present context, and its various elements are given in what follows.

If the three solar irradiance components, beam, diffuse, and global, are available, a consistency test can be applied, based on the closure equation that links them:

$$I_{gh} = I_{dh} + I_{bn} \cdot \sin h. \tag{9}$$

A posteriori automatic quality control cannot detect all acquisition problems, however. The remaining elements to be assessed are threefold:

- The measurement's time stamp (needed to compute the solar geometry);
- The sensors' calibration coefficient used to convert the acquired data into physical values;
- The coherence between the parameters.

It has to be noted that, for the data of Solar Village, even if the closure equation satisfies the selection conditions, as can be seen in Figure 6, a clear pattern appears due to calibration differences of up to 6–8% between the winter months and the summer months. The bias between the beam irradiance measurements and the closure equation values are negative in the summer months; it is not clear what the source of the difference is. This can be an issue of the temperature and/or solar altitude dependence of the sensors; it is difficult to find out a posteriori what component is questionable and why.

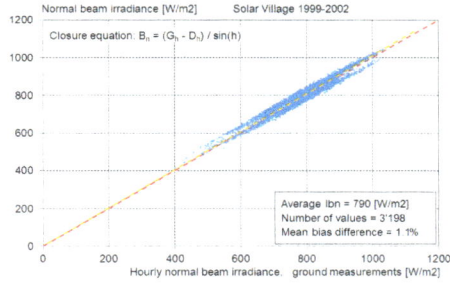

Figure 6. Illustration of the calibration uncertainty for the site of Solar Village.

A complete description of the quality control procedure is given by Ineichen [16].

4.2. Clear Conditions Selection

To perform the selection, the criteria defined in Ineichen [17] are applied:

- If three components are available, the closure equation must be satisfied within -50 W/m^2 -5% and $+50$ W/m^2 $+5\%$; if only two components are available, all the values are kept.
- The modified global clearness K_t' (as defined by Perez et al. in [18]) of the measurements is higher than 0.65.
- The stability of the global clearness index $\Delta K_t'$ is better than 0.01 ($\Delta K_t'$ is evaluated by the difference of the considered hour and the average of the considered hour, the preceding one, and the following one)

This selection minimizes the cloud contamination. It is restrictive, but for the purposes of the comparison, it ensures that only clear and stable conditions are selected. It shows its importance particularly with data from Xianghe, where often the beam component is very low or missing in the middle of the day, while the corresponding global component shows clear conditions (due to saturation?). This is illustrated in Figure 7.

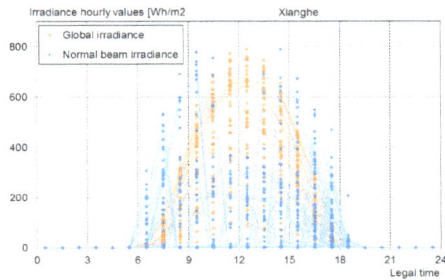

Figure 7. Illustration of the questionable data for the site of Xianghe.

4.3. Statistical Indicators and Graphical Representation

The comparison is done on an hourly basis, the model–measurements difference is computed, so that a positive value of the mean bias difference represents an overestimation of the model. The following indicators are used to quantify model performance:

- First-order statistics for a given site: the mean bias difference (*mbd*) and the standard deviation (*sd*). In addition, qualitative visualization is made with the help of model versus measurement scatterplots.

- The seasonal dependence of the bias and its dependence with the aerosol optical depth (*aod*);
- The frequency distribution of the model–measurements differences and the corresponding cumulated frequencies.

4.4. Ground Measurements for the Validation

To assess the validity of the model for low turbidity values, we first made a validation of the same data as in [18] acquired in Geneva for the years 2004 to 2011, where the *aod* is pretty low.

The average aerosol optical depth in Geneva is on the order of 0.17, with a maximum of 0.5 during polluted episodes. As no specific aerosol optical depth values are measured in Geneva, we used one of the best state-of-the-art models, REST2 [5], to retrieve by retrofit the *aod* following the method described in [19].

The next validation step is to do a comparison with ground data acquired at a site with higher turbidity values. To conduct the validation, sites that jointly acquire the irradiance components and the aerosol optical depth are needed. They are very scarce, so that not all conditions could be analyzed. Irradiance data from three Baseline Surface Radiation Network (BSRN) sites and from one Indian National Institute of Wind Energy (NIWE) site are used for the validation. The aerosol optical depths are retrieved from the Aerosol Robotic Network (AERONET). The latitude, longitude, altitude, networks, and acquired parameters are given in Table 4.

Table 4. Latitude, longitude, altitude, networks, and acquired parameters for the considered sites.

Site	Network	Latitude	Longitude	Altitude	Global	Beam	Diffuse	T_a	Hr	Aod
Ilorin	BSRN/AERONET	8.53	4.57	350	x		x	x	x	x
Jaipur	NIWE/AERONET	26.81	75.86	403	x	x	x	x	x	x
Solar Village	BSRN/AERONET	24.91	46.41	650	x	x	x	x	x	x
Xianghe	BSRN/AERONET	39.75	116.69	22	x	x	x			x

The median values of the *aod* for the four sites are situated between 0.3 and 0.55, with maximum values around 1.5 to 4. The cumulated frequencies of occurrence of the aerosol optical depth and the atmospheric water vapor content are given in Figure 8.

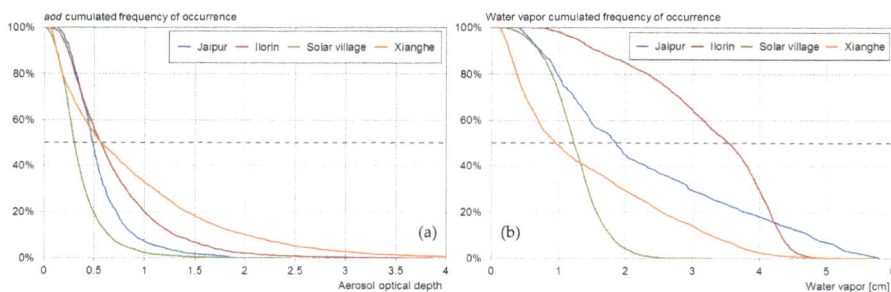

Figure 8. Cumulated frequency of occurrence of the aerosol optical depth (**a**) and the atmospheric water vapor content (**b**) for the four sites.

5. Validation Results and Discussion

The results of the validation against the ground data acquired in Geneva are illustrated in Figure 9 for the normal beam and the global horizontal irradiance. Done on the same dataset as in [17], it shows that the first-order statistics are slightly better than for the previous version of Solis.

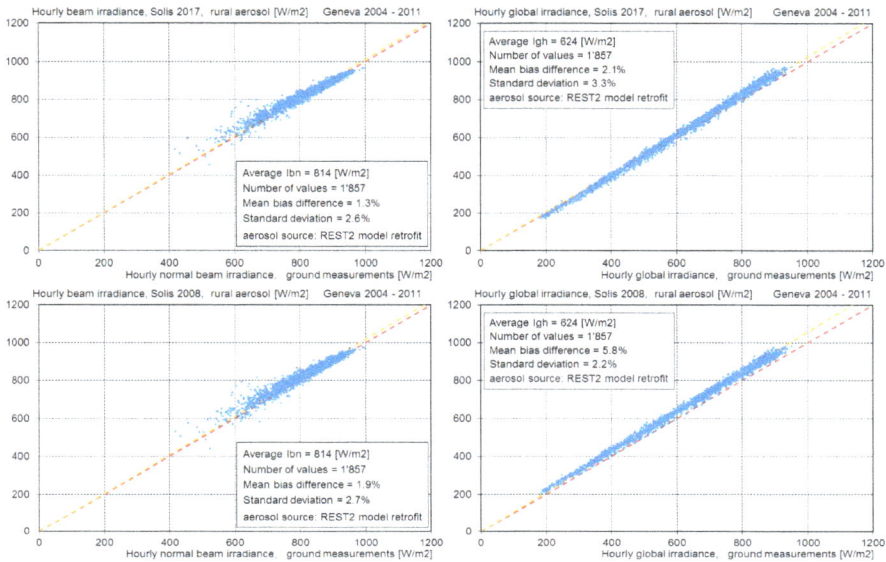

Figure 9. Validation against ground data acquired in Geneva for the beam irradiance (**a**) and the global component (**b**). The *aod* is retrofitted with REST2. For comparison, the same graphs are given for the previous version of Solis (**c,d**).

The first-order statistics of the model validation against data acquired at high turbidity sites are given in Table 5 for the rural aerosol type, and in Table 6 for the urban aerosol type. The tables report the number of *nb* values kept for the validation, the average measured irradiance and irradiation, the relative mean bias difference, *mbd*, and the standard deviation, *sd*. Results are provided for the three irradiance components and for all sites. Due to the different climates and quality of the data, the number of selected data can be very different from one site to the other.

Table 5. First-order statistics for the four sites and rural aerosol type. The irradiance is expressed in $(W \cdot h/m^2 \cdot h)$ and the irradiation in $(kWh/m^2 \cdot day)$.

Site	Hourly Values									
	Normal Beam				Horizontal Global			Horizontal Diffuse		
	nb	Average	mbd	sd	Average	mbd	sd	Average	mbd	sd
Ilorin 1998–2005	289	646	−2%	12%	845	1%	5%	277	2%	22%
Jaipur 2016–2017	387	670	4%	5%	715	3%	3%	198	4%	13%
Solar Village 1999–2002	3198	790	−2%	5%	784	−1%	4%	165	4%	19%
Xianghe 2008–2015	1161	762	−2%	13%	612	−5%	6%	132	−10%	33%
	"daily values"									
	nb	average	mbd	sd	average	mbd	sd	average	mbd	sd
Ilorin 1998–2005	39	2.3	0%	8%	3.0	1%	4%	0.9	−1%	20%
Jaipur 2016–2017	69	3.6	4%	4%	3.8	3%	2%	1.0	5%	10%
Solar Village 1999–2002	453	5.5	−2%	3%	5.5	−1%	3%	1.2	4%	12%
Xianghe 2008–2015	178	4.9	−2%	14%	3.9	−5%	6%	0.8	−10%	33%

Table 6. First-order statistics for the four sites and urban aerosol type. The irradiance is expressed in (W·h/m^2·h) and the irradiation in (kWh/m^2·day).

Site	Hourly Values									
	Normal Beam				Horizontal Global			Horizontal Diffuse		
	nb	Average	Mbd	Sd	Average	Mbd	Sd	Average	Mbd	Sd
Ilorin 1998—2005	289	646	−3%	12%	845	−8%	7%	277	−20%	20%
Jaipur 2016—2017	387	670	3%	6%	715	−4%	4%	198	−16%	11%
Solar Village	3198	790	−3%	5%	784	−5%	4%	165	−14%	19%
Xianghe 2008—2015	1161	762	−2%	13%	612	−8%	8%	132	−24%	30%
	Daily Values									
	nb	Average	Mbd	sd	Average	Mbd	Sd	Average	Mbd	Sd
Ilorin 1998—2005	39	2.3	−2%	9%	3.0	−7%	6%	0.9	−22%	19%
Jaipur 2016—2017	69	3.6	3%	4%	3.8	−3%	3%	1.0	−15%	8%
Solar Village	453	5.5	−3%	3%	5.5	−5%	3%	1.2	−13%	13%
Xianghe 2008—2015	178	4.9	−2%	14%	3.9	−8%	9%	0.8	−24%	26%

The main results drawn from the table are the following:

- For all sites, choosing the rural aerosol type gives the best results in terms of mean bias and standard deviation, especially for the global component.
- For Ilorin, the high standard deviation is due to the high variability of the sky conditions; it can also be a result of the beam retrieval from the global and the diffuse components. For Xianghe, the poor quality of the data induces this higher standard deviation.
- The Jaipur and Solar Village datasets show the overall best results in term of mean bias and standard deviation.
- The "daily values" represent the sum of only the hourly values kept for the validation: it is not the daily integral (the sum of all the values of the day). This is particularly the case during partially cloudy days where only a few hourly values are acquired during cloudless conditions. The consequence is that, for example, at Ilorin, where the sky conditions are highly variable and clear days from sunrise to sunset are very scarce, the average "daily value" is low.

A deeper analysis of the results leads to the following general observations:

- Ground measurements and modeled values can be represented on the same graph in different colors. If the model faithfully reproduces the measurements, the envelopes of the data clouds should be similar. This is illustrated for the site of Solar Village in Figure 10 where the beam clearness index is plotted against the global clearness index on the left graph (a) and the diffuse fraction against the global clearness index on the right graph (b); the measurements are represented in green and the modeled values in blue.
- The trends of the bias as a function of the aerosol optical depth show no particular pattern for all the sites, except for Xianghe, where a non-significant trend is due to the dispersion. These dependences are illustrated in Figure 11 in hourly values, for the site of Solar Village and rural aerosol type.
- The seasonal dependence of the bias for both the global and the beam irradiance components shows no significant pattern. An illustration is given in Figure 12 for data from Solar Village in hourly values.
- With the exception of the site of Ilorin, the distributions of the bias around the 1:1 model–measurements axis are near normal; for this reason, the first-order statistics represented by the mean bias and the standard deviation can be considered reliable. An example is given in Figure 13 for the site of Solar Village. The figure displays the hourly bias frequency distribution for both I_{bn} and I_{gh} irradiance components. The cumulated frequencies (red curve) are also represented in the graphs.

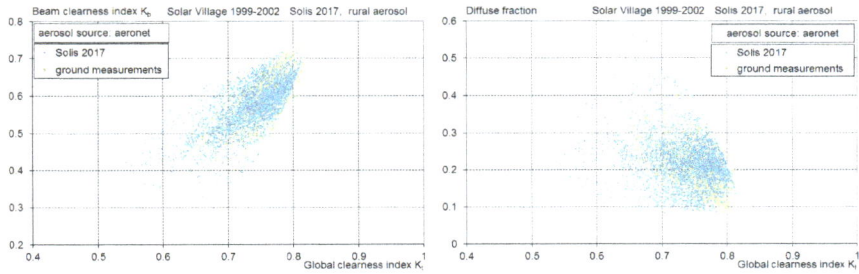

Figure 10. Beam clearness index (**a**) and diffuse fraction (**b**) versus the global clearness index.

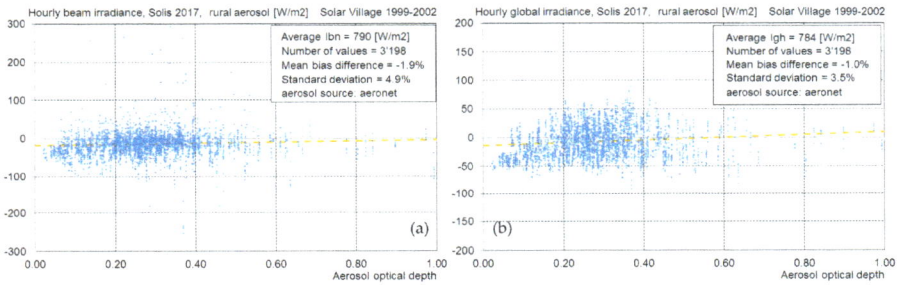

Figure 11. Bias versus the aerosol optical depth for normal beam irradiance (**a**) and global irradiance (**b**) for the site of Solar Village.

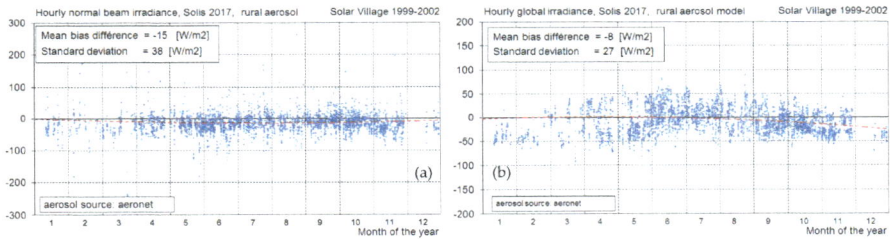

Figure 12. Seasonal dependence of the bias for the normal beam irradiance (**a**) and the global irradiance (**b**) irradiance components for the site of Solar Village.

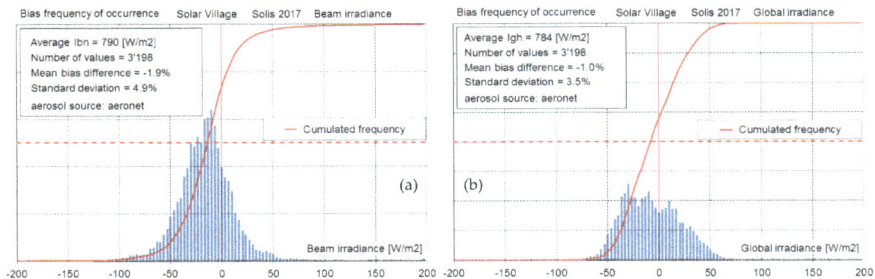

Figure 13. Frequency of occurrence of the bias around the 1:1 axis for the beam (**a**) and the global (**b**) irradiance components.

6. Conclusions

When dealing with satellite images to derive the irradiance components worldwide and every 15 min, the calculation time should be as short as possible. The analytical Solis clear sky scheme was developed to fulfill this requirement but had the weakness of being limited to aerosol optical depth values lower than 0.45. The new analytical Solis scheme is valid for aod_{550} values up to seven, even if very high values are not realistic as an optical depth; it is probably due to bigger particles like sand, or thin, high-altitude clouds. Nevertheless, in contrary to other clear sky models, this permits us to produce coherent irradiance values, even with questionable input data. Indeed, for example, the *aod*, due to the difficulty of retrieving it by automatic processes, can take on out-of-range or unrealistic values.

The new Solis clear sky scheme is developed for aerosol optical depths values aod_{550} from 0.02 to 7, for atmospheric water vapor w content from 0.01 to 10 cm, for altitude from sea level to 7000 m, and for four aerosol types, as defined by Shettle and Fenn [2]. For the urban, rural, and tropospheric aerosol types, the validation against the original LibRadTran RTM calculations presents no bias and a standard deviation lower than 1% for the global and the beam component, and 3% for the diffuse. When dealing with maritime aerosol type, the standard deviation is 2.2%, 3.4%, and 4%, respectively, for the global, the beam, and the diffuse components.

The validation against ground measurements acquired in Geneva, Ilorin, Jaipur, Solar Village, and Xianghe gives a mean bias difference less than ±4%, a standard deviation of 3–6% for the global component, and a mean bias difference less than ±3% with a standard deviation of 5–13% for the beam component. The 12–13% standard deviations are due to either the high variability of the sky conditions or the poor quality of the ground data.

As it is very difficult to find ground measurements covering the complete range of validity of the new Solis analytical scheme, it cannot be completely validated. However, the new version faithfully reproduces the RTM LibRadTran calculations.

Nomenclature

Symbol	Description	Symbol	Description
I_{bn} or B_n	normal beam irradiance	M	optical air mass
I_{gh} or G_h	global horizontal irradiance	h	solar elevation
I_{dh} or D_h	diffuse horizontal irradiance	p	atmospheric pressure
I_0	solar constant	p_0	atm. pressure at sea level
I_0'	modified solar constant		
aod	aerosol optical depth	*nb*	number of involved values
w	atmospheric water vapor column	*mbd*	mean bias difference
τ	extinction coefficient	*sd*	standard deviation
RTM	Radiation Transfer Model	R^2	correlation coefficient
BSRN	Baseline Surface Radiation Network		
NIWE	National Institute of Wind Energy		
AERONET	AERosol Robotic NETwork		

Appendix A Model Coefficients

Solis 2017 Rural Aerosol Type

I_0'/I_0

	11	12	21	22	31	32
a_i	0.000261031	−0.001810295	−0.000334819	−0.001533816	−0.000118343	0.000100560
b_i	0.019908235	0.195865498	0.019649806	0.030513908	0.003246244	0.010001152
c_i	0.035765704	0.737572097	0.017074308	0.021712639	0.004844209	0.005917312
d_i	0.072875947	1.003594313	0.009537179	−0.000136749	0.001883135	0.000378418

Solis 2017 Rural Aerosol Type

τ_g

	11	12	21	22	31	32
a_g	0.000193733	−0.002501003	−0.000156219	−0.000277670	−0.000059040	0.000006851
b_g	−0.001159980	0.058524158	0.002548229	0.004829749	0.000788522	0.000522673
c_g	−0.001449466	−0.833692155	−0.012697614	−0.030033430	−0.004681306	−0.011682439
d_g	−0.141809777	−0.085671381	−0.015070590	−0.030811277	−0.007315151	−0.008232404

coef a

ca_1	ca_2	ca_3	ca_4	ca_5	ca_6
0.393927007	−0.014924316	−0.092174236	−0.001048172	−0.009163093	0.006108964

τ_b

	11	12	21	22	31	32
a_b	0.000771802	−0.007478508	−0.000180242	−0.000368091	−0.000049938	0.000049572
b_b	−0.009565174	0.148774990	0.002336539	0.007087530	0.001102680	0.000284776
c_b	0.024325289	−1.461925559	−0.009964879	−0.043313859	−0.007992277	−0.015281853
d_b	−0.191282122	−0.099096810	−0.022005054	−0.027680914	−0.006793639	−0.008818264

coef b

cb_1	cb_2	cb_3
0.482261237	−0.016678672	0.914171831

τ_d

c_{td1}	c_{td2}	c_{td3}	c_{td4}	c_{td5}	c_{td6}
1.626089946	0.581443300	0.602607926	−0.041935067	0.048133609	−3.902240255

coef d

cd_1	cd_2	cd_3	cd_4	cd_5	cd_6	cd_7
0.113144991	−0.004165229	0.621087873	−0.359248954	0.092564432	−0.011283985	0.000525184

Solis 2017 Urban Aerosol Type

$I_0{}'/I_0$

	11	12	21	22	31	32
a_i	0.000518011	0.002795295	−0.000216902	−0.001116789	−0.000071507	0.000376597
b_i	0.018999224	0.180668837	0.020488032	0.028605398	0.002968967	0.008819303
c_i	0.035015670	0.599675484	0.008512409	0.015089145	0.004338910	0.004266916
d_i	0.072017631	1.005299407	0.009572912	0.000381392	0.001904339	0.000343349

τ_g

	11	12	21	22	31	32
a_g	0.000428378	−0.000564884	−0.000375237	−0.000072798	0.000128450	−0.000046323
b_g	−0.002733005	0.036214885	0.004374967	0.003084955	−0.001012500	0.000930531
c_g	0.003834122	−0.883655173	−0.015857854	−0.028437392	−0.001164850	−0.013464145
d_g	−0.141741963	−0.079752002	−0.014920981	−0.030293926	−0.007709979	−0.007819176

coef a

ca_1	ca_2	ca_3	ca_4	ca_5	ca_6
0.436328716	−0.015982197	−0.099473876	−0.001106107	−0.013273397	0.006225334

τ_b

	11	12	21	22	31	32
a_b	0.000565740	−0.005280564	−0.000003576	−0.000404517	−0.000044866	0.000073274
b_b	−0.007458342	0.118092608	0.000640665	0.007424899	0.000959090	−0.000036289
c_b	0.020375757	−1.380477103	−0.006939622	−0.043687713	−0.007132214	−0.014709261
d_b	−0.192470443	−0.093585418	−0.021132099	−0.027442728	−0.007179027	−0.008451631

coef b

cb_1	cb_2	cb_3
0.498959654	−0.017636916	0.926155270

τ_d

c_{td1}	c_{td2}	c_{td3}	c_{td4}	c_{td5}	c_{td6}
1.886930729	0.541083141	0.605281729	−0.041322484	0.048799518	−4.249335048

coef d

cd_1	cd_2	cd_3	cd_4	cd_5	cd_6	cd_7
0.112629823	−0.003088767	0.569910110	−0.337879216	0.087849430	−0.010751282	0.000501473

Solis 2017 Tropospheric Aerosol Type

$I_o{}'/I_o$

	11	12	21	22	31	32
a_i	−0.000147911	−0.008122292	−0.001047420	−0.001931206	−0.000132858	−0.000370950
b_i	0.022292028	0.158574811	0.020947936	0.025683688	0.002656382	0.010480509
c_i	0.032885091	0.757904643	0.014309659	0.030928721	0.005870070	0.005967166
d_i	0.073428621	0.999900199	0.009230660	−0.001121132	0.001854948	0.000311441

τ_g

	11	12	21	22	31	32
a_g	0.000217441	−0.003126729	−0.000132644	−0.000291300	−0.000002894	−0.000029457
b_g	−0.001809258	0.068030215	0.002161751	0.005160292	0.000310547	0.000821525
c_g	−0.004470877	−0.811312676	−0.010715753	−0.031925980	−0.004215436	−0.011986451
d_g	−0.138496207	−0.088561634	−0.016862218	−0.029763820	−0.006341627	−0.008895780

coef a

ca_1	ca_2	ca_3	ca_4	ca_5	ca_6
0.395322938	−0.015327403	−0.090285920	−0.001118039	−0.008727568	0.006173611

τ_b

	11	12	21	22	31	32
a_b	0.000953948	−0.008496386	−0.000088009	−0.000432762	−0.000056725	0.000028537
b_b	−0.010786534	0.163275231	0.001512516	0.007763434	0.001126151	0.000593655
c_b	0.022706050	−1.435580065	−0.008631610	−0.044554649	−0.007857069	−0.016684930
d_b	−0.192368280	−0.099465438	−0.020840853	−0.028520351	−0.007159173	−0.008504057

coef b

cb_1	cb_2	cb_3
0.481658656	−0.016555667	0.891497733

τ_d

c_{td1}	c_{td2}	c_{td3}	c_{td4}	c_{td5}	c_{td6}
1.580452405	0.558086449	0.586357700	−0.045663645	0.045790439	−3.883558603

coef d

cd_1	cd_2	cd_3	cd_4	cd_5	cd_6	cd_7
0.113480292	−0.004518770	0.624749275	−0.358574830	0.092385725	−0.011277025	0.000525609

Solis 2017 Maritim Aerosol Type

$I_o{}'/I_o$

	11	12	21	22	31	32
a_i	0.014787172	0.081267371	0.007611657	0.005687211	0.001057216	0.004652006
b_i	−0.029098068	0.099939972	−0.004466220	0.017123333	0.000502026	−0.005634380
c_i	0.135010702	1.016360983	0.043578839	0.033117203	0.009447832	0.024152082
d_i	0.062039742	0.972152257	0.005356544	−0.001177970	0.001368473	−0.002323682

τ_g

	11	12	21	22	31	32
a_g	−0.000258280	−0.000991598	0.000087922	−0.000450453	−0.000103283	0.000036369
b_g	0.003832840	0.038507962	−0.000539919	0.006663150	0.001347514	0.000076692
c_g	−0.020444433	−0.882340428	−0.002557914	−0.034648986	−0.006330049	−0.009504023
d_g	−0.138105804	−0.087198021	−0.019195997	−0.028342978	−0.006284626	−0.009009817

coef a

ca_1	ca_2	ca_3	ca_4	ca_5	ca_6
0.370255687	−0.013481692	−0.096888113	−0.000980585	−0.008966197	0.006430510

τ_b

	11	12	21	22	31	32
a_b	−0.000015662	−0.004194881	0.000107149	−0.000489497	−0.000118976	0.000094402
b_b	−0.003898058	0.107002592	−0.000186923	0.007563674	0.001566986	−0.000369907
c_b	0.010411340	−1.597417220	−0.006692200	−0.040428247	−0.007628685	−0.012973060
d_b	−0.198484327	−0.092013892	−0.019285874	−0.029172125	−0.007407717	−0.008863425

coef b

cb_1	cb_2	cb_3
0.487380474	−0.016088575	0.952963502

τ_d

c_{td1}	c_{td2}	c_{td3}	c_{td4}	c_{td5}	c_{td6}
1.633065156	0.676496607	0.651637829	−0.033919132	0.053967101	−3.761466993

coef d

cd_1	cd_2	cd_3	cd_4	cd_5	cd_6	cd_7
0.109205787	−0.003827579	0.635588699	−0.372637706	0.095930104	−0.011645247	0.000539553

Appendix B   Validation Graphs

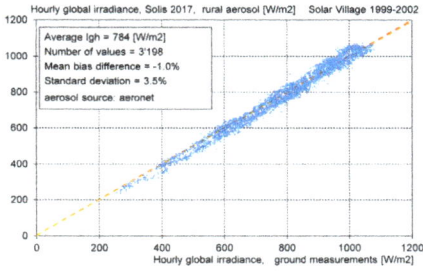

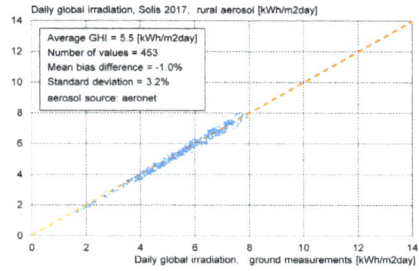

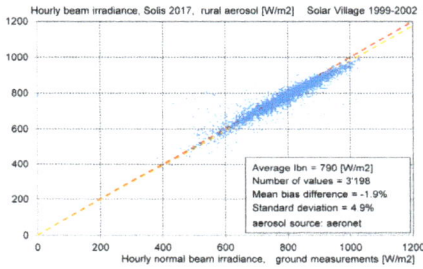

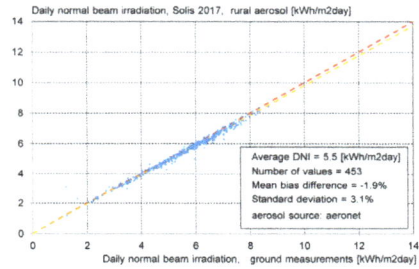

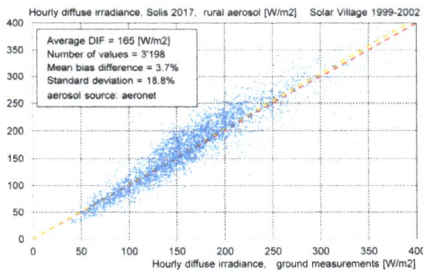

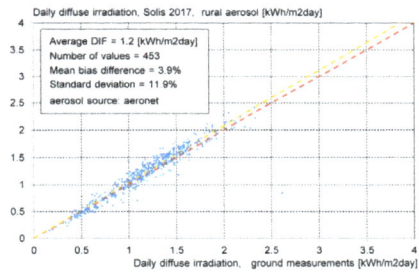

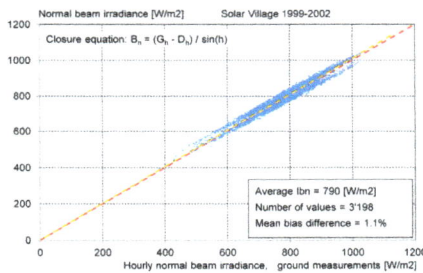

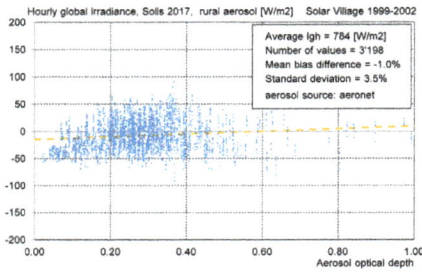

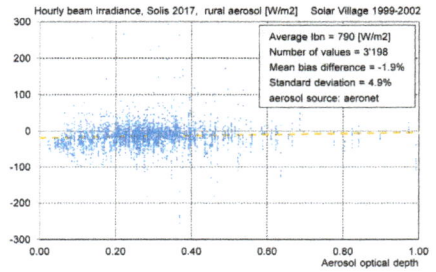

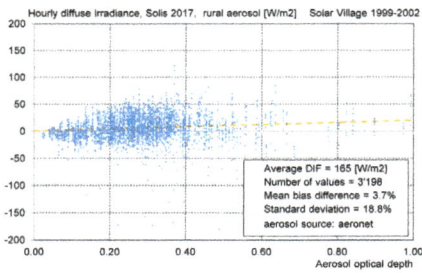

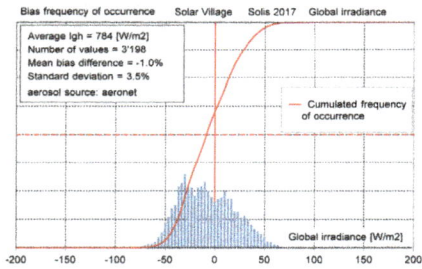

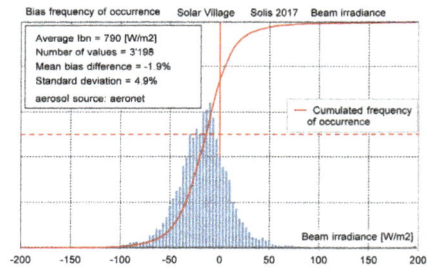

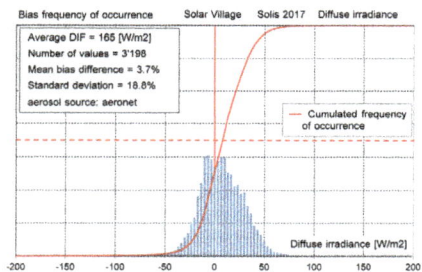

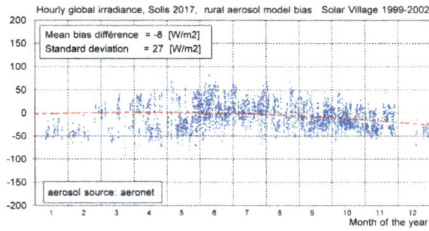

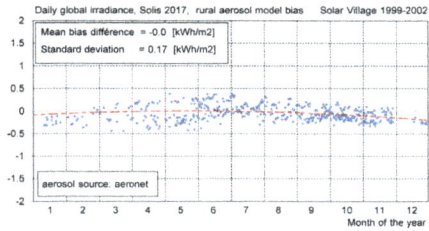

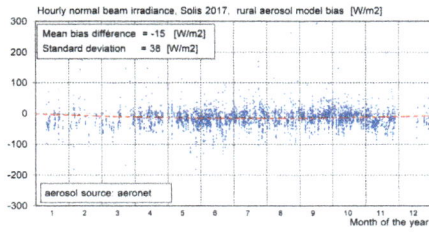

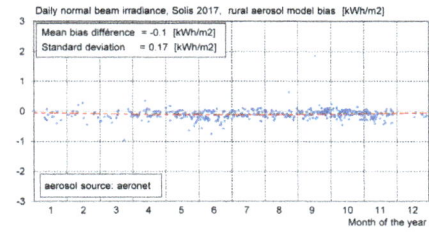

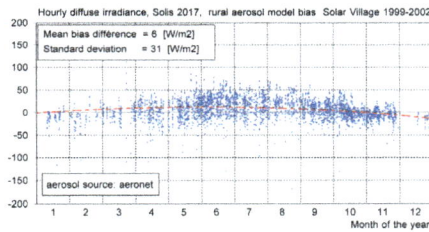

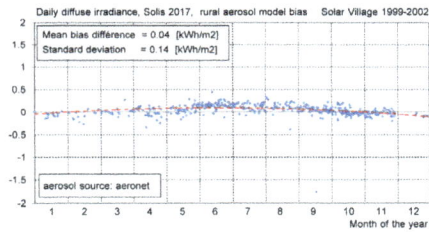

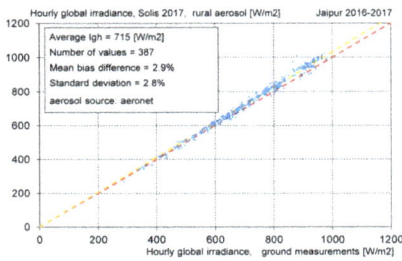

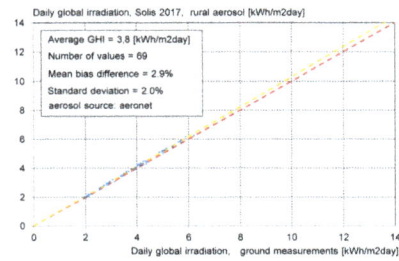

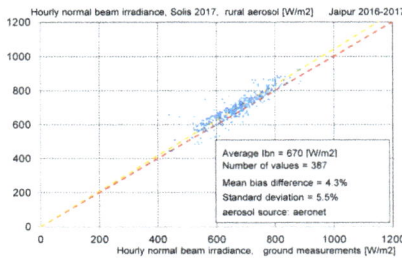

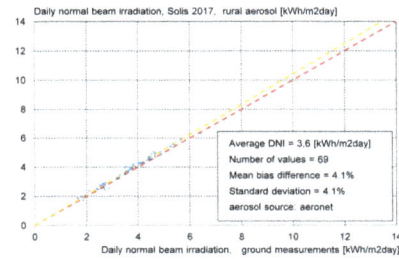

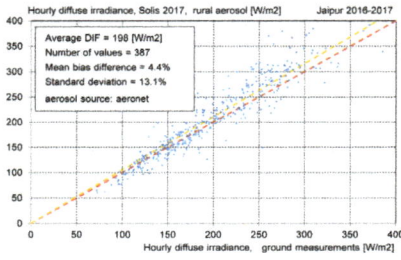

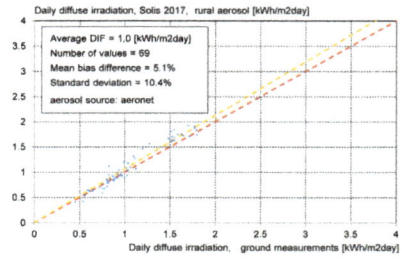

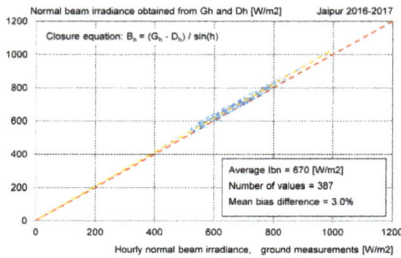

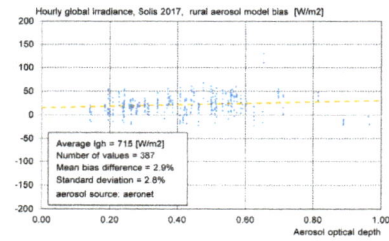

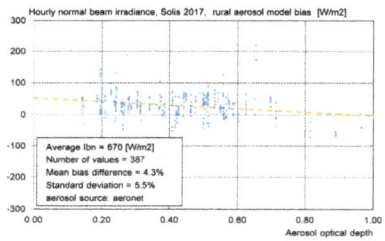

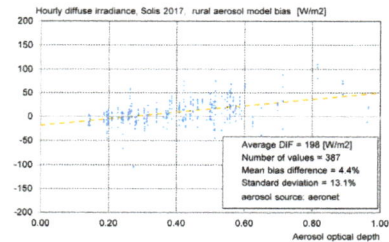

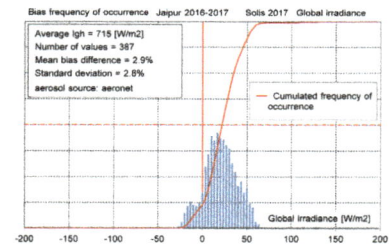

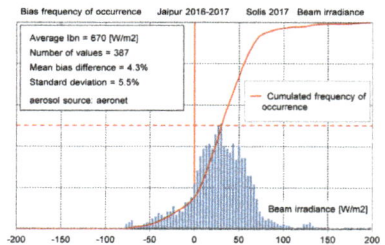

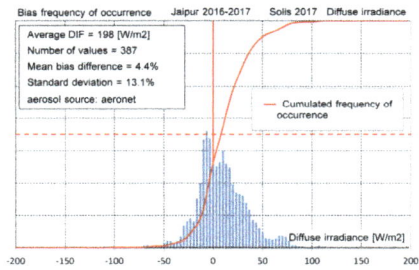

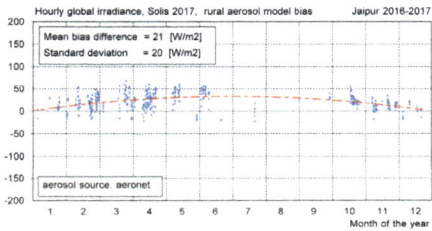

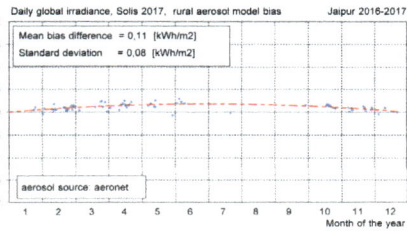

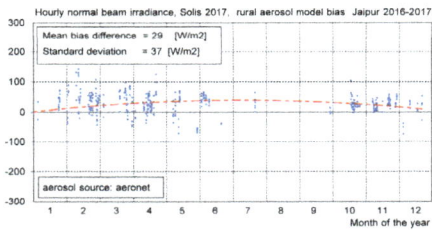

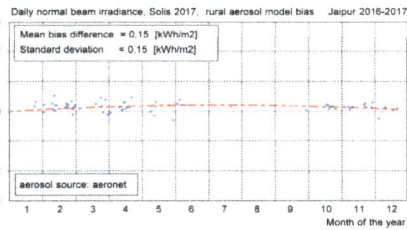

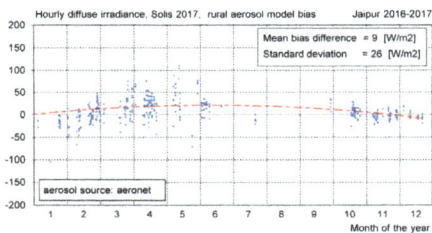

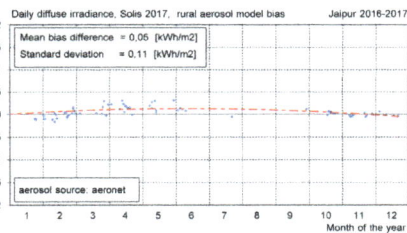

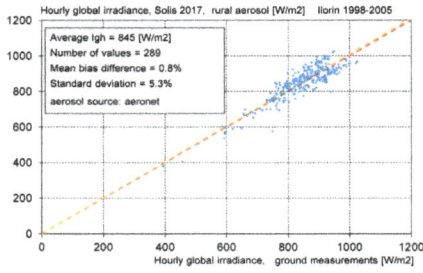

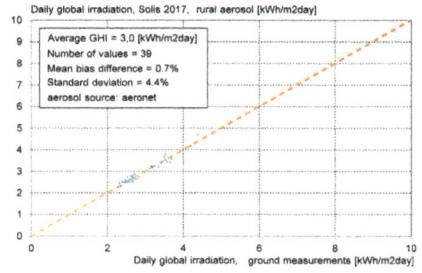

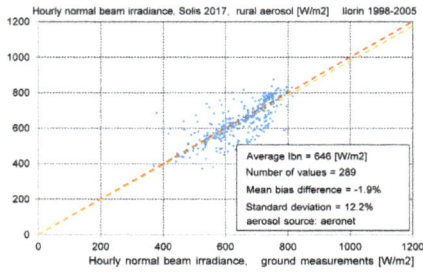

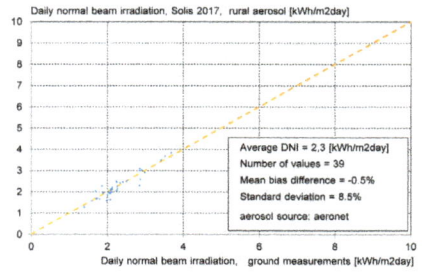

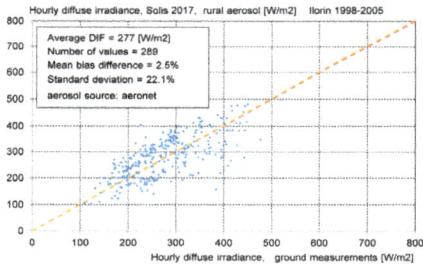

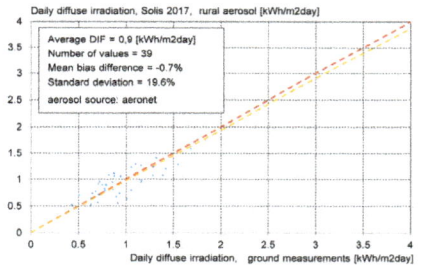

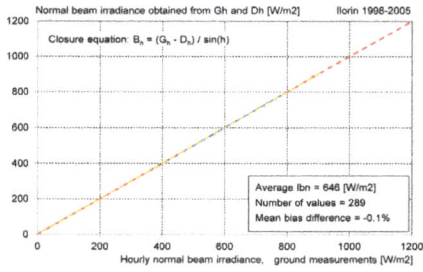

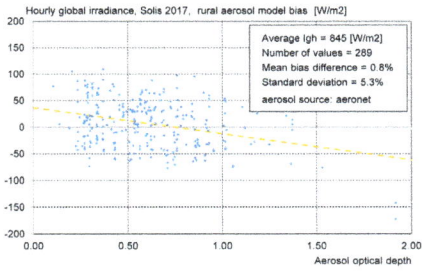

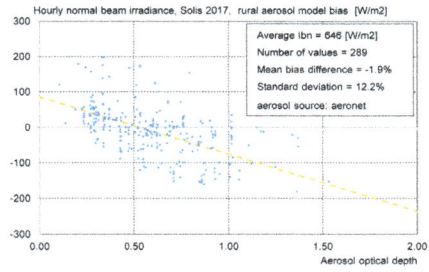

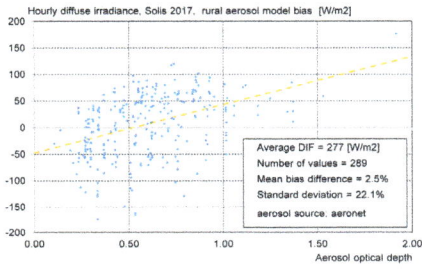

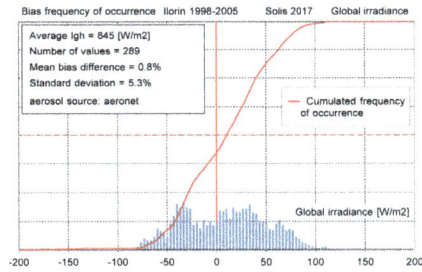

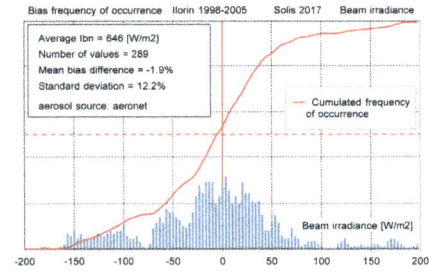

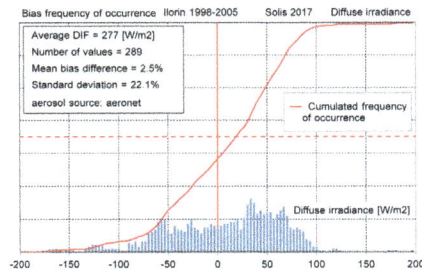

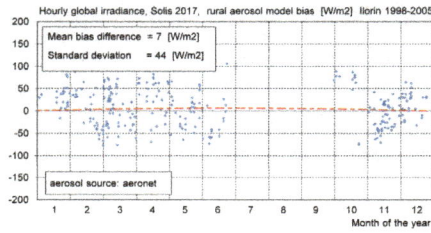

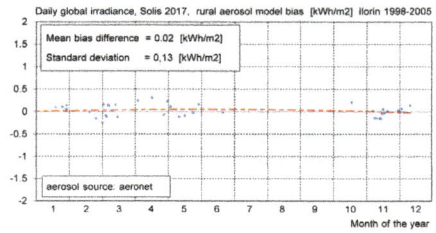

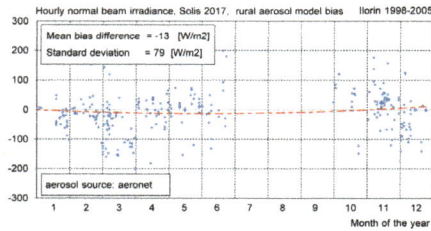

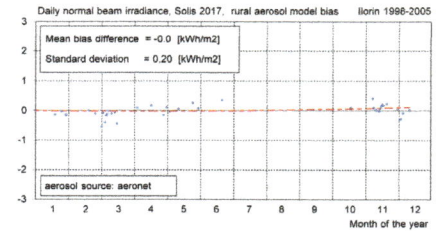

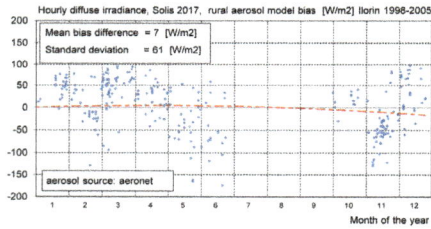

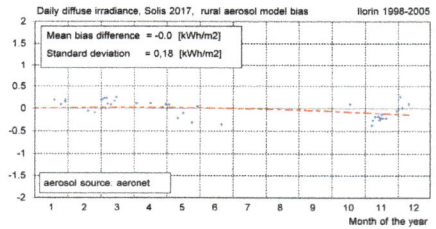

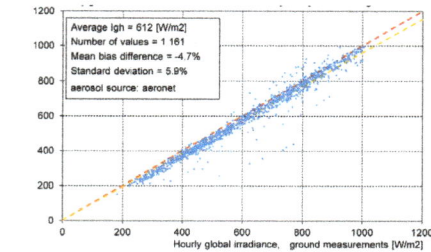

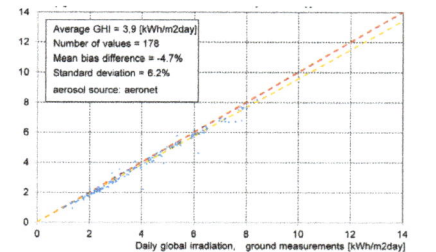

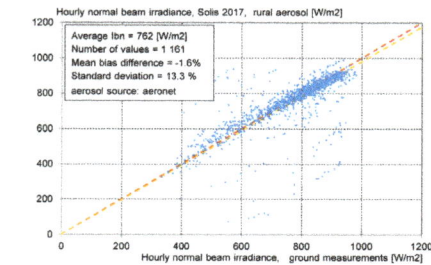

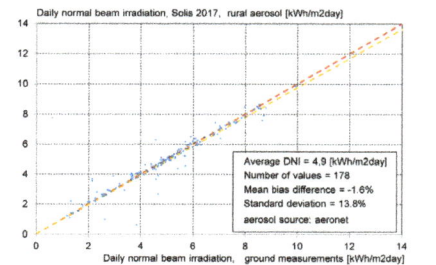

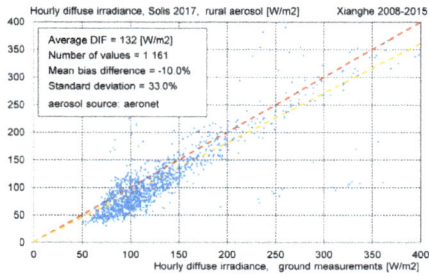

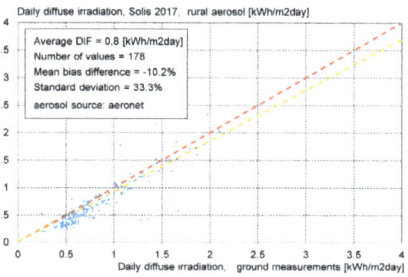

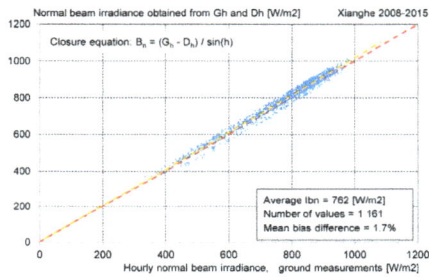

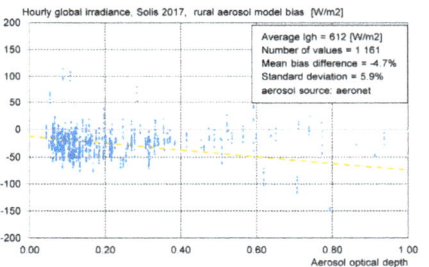

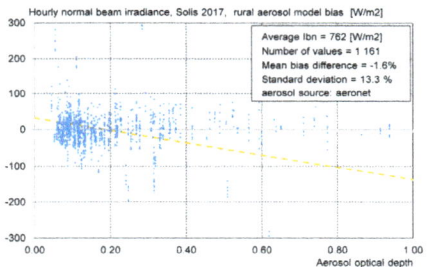

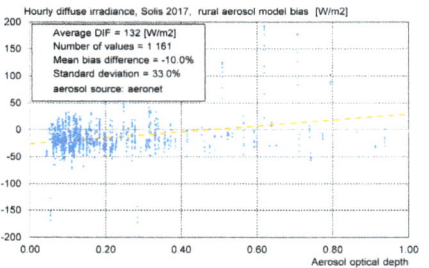

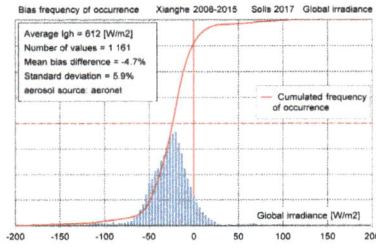

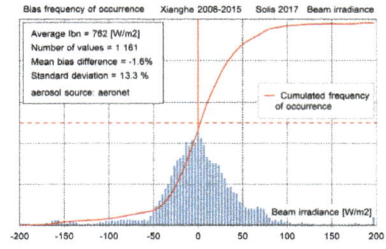

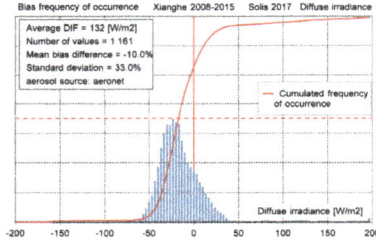

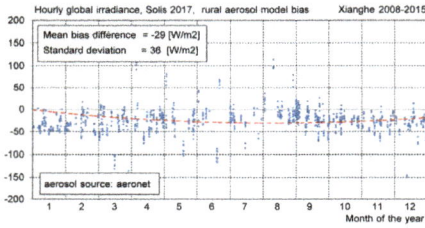

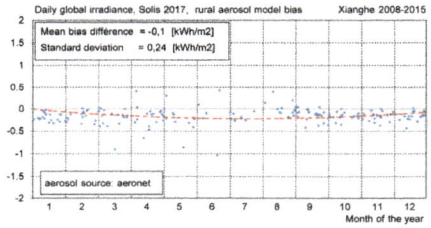

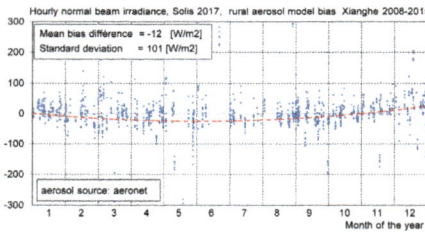

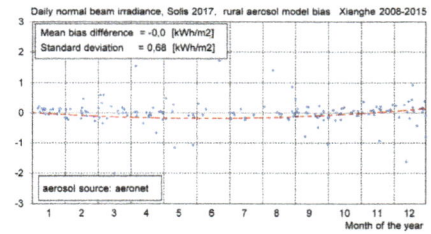

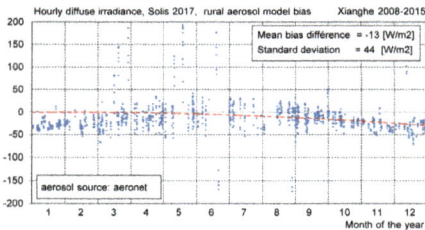

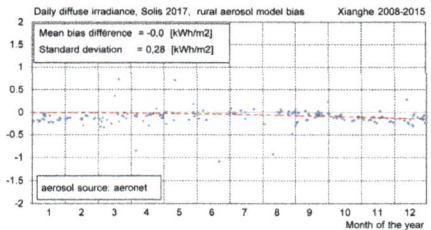

References

1. Gueymard, C. A two-band model for the calculation of clear Sky Solar Irradiance, Illuminance, and Photosynthetically Active Radiation at the Earth Surface. *Sol. Energy* **1989**, *43*, 253–265. [CrossRef]
2. Gueymard, C. Direct solar transmittance and irradiance predictions with broadband models. Part 1: Detailed theoretical performance assessment. *Sol. Energy* **2003**, *74*, 355–379. [CrossRef]
3. Bird, R.E.; Huldstrom, R.L. Direct insolation models. *Trans. ASME J. Sol. Energy Eng.* **1980**, *103*, 182–192. [CrossRef]
4. Wald, L. *European Solar Radiation Atlas*; Commission of the European Communities Presses de l'Ecole, Ecole des Mines de Paris: Paris, France, 2000.
5. Rigollier, C.; Bauer, O.; Wald, L. On the Clear Sky Model of the ESRA—European Solar Radiation Atlas—With Respect to the Heliosat Method. *Sol. Energy* **2000**, *68*, 33–48. [CrossRef]
6. Geiger, M.; Diabaté, L.; Ménard, L.; Wald, L. A web Service for Controlling the Quality of Measurements of Global Solar Irradiation. *Sol. Energy* **2002**, *73*, 475–480. [CrossRef]
7. Linke, F. Transmissions-Koeffizient und Trübungsfaktor. *Beiträge zur Physik der freien Atmosphäre* **1992**, *10*, 91–103.
8. Ineichen, P. A broadband simplified version of the Solis clear sky model. *Sol. Energy* **2008**, *82*, 758–762. [CrossRef]
9. Shettle, E.P.; Fenn, R.W. Models Fort the Aerosol of Lower Atmosphere and the Effect of Humidity Variations on Their Optical Properties. 1979. Available online: http://www.dtic.mil/docs/citations/ADA085951 (accessed on 2 March 2018).
10. Shettle, E.P. Models of aerosols, clouds and precipitation for atmospheric propagation studies. In Proceedings of the AGARD, Atmospheric Propagation in the UV, Visible, IR, and MM-Wave Region and Related Systems Aspects Conference, Copenhagen, Denmark, 9–13 October 1989.
11. Zhang, T.; Stackhouse, P.W., Jr.; Chandler, W.S.; Westberg, D.J. Application of a global-to-beam irradiance model to the NASA GEWEX SRB dataset: An extension of the NASA Surface meteorology and Solar Energy datasets. *Sol. Energy* **2014**, *110*, 117–131. [CrossRef]
12. Mueller, R.W.; Dagestad, K.F.; Ineichen, P.; Schroedter-Homscheidt, M.; Cros, S.; Dumortier, D.; Kuhlemann, R.; Olseth, J.A.; Piernavieja, G.; Reise, C.; et al. Rethinking satellite based solar irradiance modelling—The SOLIS clear-sky module. *Remote Sens. Environ.* **2004**, *91*, 160–174. [CrossRef]
13. Mayer, B.; Kylling, A. Technical note: The libRadtran software package for radiative transfer calculations—Description and examples of use. *Atmos. Phys.* **2005**, *5*, 1855–1877. [CrossRef]
14. Mayer, B.; Kylling, A.; Emde, C.; Buras, R.; Hamann, U. LibRadTran: Library for Radiative Transfer Calculations, Edition 1.0 for Libradtran Version 1.5-beta. 2010. Available online: https://clouds.eos.ubc.ca/~phil/courses/eosc582_2010/textfiles/libRadtran.pdf (accessed on 3 March 2018).
15. Ineichen, P. Solis 2017 Excel Tool. Password: Solis2017. Available online: http://www.adpi.ch/Solis2017/Solis2017-tool.xlsx (accessed on 28 February 2017).
16. Ineichen, P. Satellite Irradiance Based on MACC Aerosols: Helioclim 4 and SolarGIS, Global and Beam Components Validation. In Proceedings of the EuroSun 2014, Aix-les-Bains, France, 16–19 Septembre 2014.
17. Ineichen, P.; Barroso, C.S.; Geiger, B.; Hollmann, R.; Marsouin, A.; Mueller, R. Satellite Application Facilities irradiance products: hourly time step comparison and validation over Europe. *Int. J. Remote Sens.* **2009**, *30*, 5549–5571. [CrossRef]
18. Perez, R.; Ineichen, P.; Seals, R.; Zelenka, A. Making full use of the clearness index for parameterizing hourly insolation conditions. *Sol. Energy* **1990**, *45*, 111–114. [CrossRef]
19. Ineichen, P. Validation of models that estimate the clear sky global and beam solar irradiance. *Sol. Energy* **2016**, *132*, 332–344. [CrossRef]

remote sensing

MDPI

Article

Validation of Hourly Global Horizontal Irradiance for Two Satellite-Derived Datasets in Northeast Iraq

Bikhtiyar Ameen [1,2,*], Heiko Balzter [1,3], Etienne Wey [4], Claire Thomas [4] and Mathilde Marchand [4]

[1] Centre for Landscape and Climate Research, School of Geography, Geology and Environment, University of Leicester, University Road, Leicester LE1 7RH, UK; hb91@leicester.ac.uk (H.B.); chj2@leicester.ac.uk (C.J.)

[2] Department of Geography, College of Humanities, University of Sulaimani, Kirkuk Road, Sulaimani, Iraq-Kurdistan Region 46001, Iraq

[3] National Centre for Earth Observation (NCEO), University of Leicester, University Road, Leicester LE1 7RH, UK

[4] Transvalor, 06255 Mougins, France; etienne.wey@transvalor.com (E.W.); claire.thomas@transvalor.com (C.T.); mathilde.marchand@transvalor.com (M.M.)

* Correspondence: bma17@le.ac.uk or Bikhtiyar.84@gmail.com; Tel.: +44-116-223-1018

Received: 27 August 2018; Accepted: 15 October 2018; Published: 17 October 2018

Abstract: Several sectors need global horizontal irradiance (GHI) data for various purposes. However, the availability of a long-term time series of high quality in situ GHI measurements is limited. Therefore, several studies have tried to estimate GHI by re-analysing climate data or satellite images. Validation is essential for the later use of GHI data in the regions with a scarcity of ground-recorded data. This study contributes to previous studies that have been carried out in the past to validate HelioClim-3 version 5 (HC3v5) and the Copernicus Atmosphere Monitoring Service, using radiation service version 3 (CRSv3) data of hourly GHI from satellite-derived datasets (SDD) with nine ground stations in northeast Iraq, which have not been used previously. The validation is carried out with station data at the pixel locations and two other data points in the vicinity of each station, which is something that is rarely seen in the literature. The temporal and spatial trends of the ground data are well captured by the two SDDs. Correlation ranges from 0.94 to 0.97 in all-sky and clear-sky conditions in most cases, while for cloudy-sky conditions, it is between 0.51–0.72 and 0.82–0.89 for the clearness index. The bias is negative for most of the cases, except for three positive cases. It ranges from −7% to 4%, and −8% to 3% for the all-sky and clear-sky conditions, respectively. For cloudy-sky conditions, the bias is positive, and differs from one station to another, from 16% to 85%. The root mean square error (RMSE) ranges between 12–20% and 8–12% for all-sky and clear-sky conditions, respectively. In contrast, the RMSE range is significantly higher in cloudy-sky conditions: above 56%. The bias and RMSE for the clearness index are nearly the same as those for the GHI for all-sky conditions. The spatial variability of hourly GHI SDD differs only by 2%, depending on the station location compared to the data points around each station. The variability of two SDDs is quite similar to the ground data, based on the mean and standard deviation of hourly GHI in a month. Having station data at different timescales and the small number of stations with GHI records in the region are the main limitations of this analysis.

Keywords: solar radiation; global horizontal irradiance; satellite-derived dataset; validation

---

1. Introduction

Global horizontal irradiance (GHI), both in the atmosphere and on the earth's surface, is a crucial parameter in the fields of atmosphere interaction, solar energy, architecture, and agriculture.

High-resolution GHI data are required for studying those fields. Therefore, several studies have tried to estimate solar radiation (SR) and its components from either ground measurements or satellite images using several models [1–6]. The ground measurements of GHI have high accuracy and high temporal availability, whereas the high spatial resolution of recorded data and the number of stations with SR data are limited in most geographical areas. The reasons are the purchase and high maintenance costs of pyranometers. Satellite images have been analysed to estimate GHI in order to cover the scarcity of ground measurement data. Most of the affordable satellite images for that purpose are the geostationary satellite images, namely Meteosat First Generation (MFG) and Meteosat Second Generation (MSG)/Spinning Enhanced Visible and Infrared Imager (SEVIRI), The Japanese Geostationary Meteorological Satellite (GMS), and the Geostationary Operational Environmental Satellite system (GOES) [7]. Hence, others such as the Moderate Resolution Imaging Spectroradiometer (MODIS) [8] and Landsat images have been used [9], but their temporal resolution is not acceptable.

The basic idea of estimating GHI from satellite images is to find the relationship between satellite images and ground measurements, either with statistical or physical approaches [10]. The popular method of Heliosat-2 (H2), which is based on developing Heliosat-1 to be more physical than empirical, can be, applied to large time-series data of meteorological satellites. The H2 principle is that the radiance of a cloud pixel is high in the visible band. It tests the difference of the reflectance between the cloud pixel and the clear-sky pixel; this is called a cloud index. This data and the data of the Linke turbidity factor are used to measure GHI [11]. The H2 has been developed by changing some inputs to the model in several studies [12–14]. Other studies also used satellite imagery with different techniques for GHI estimations [15–17].

There are several satellite-derived datasets (SDDs) for establishing, measuring, modelling and estimating GHI, which can be found in [18,19]. An SDD from MSG has been used to create a solar map [20]. Studies which merged ground data with the SDD for the same above purpose reveal that the merging technique for producing a solar map is better than interpolating ground data [21,22]. SDDs have also been combined with meteorological data to calibrate a GHI model [23]. The same data combinations have also been analysed to create GHI datasets for crop modelling over Europe [24]. SDDs have been utilised to assess long-term trends of a GHI time series [25]. SDDs are quite useful because of the limitations of ground data for GHI applications.

SDDs are necessary for many fields because they provide GHI for many areas and countries. Therefore, the validation of SDDs is crucial to investigate their reliability by using various methods in different geographical and climate areas, and several prior studies achieved that. For instance, the Satellite Application Facility on Climate Monitoring (CM SAF) dataset based on predicting GHI from MFG and MSG has been validated with the ground data at several stations in Europe in the period 1983–1985 [26]. The data from the same dataset with the H2 method for converting satellite images to GHI and others (SolarGis and Solemi) have also been validated in 22 cities in Europe [27]. Similarly, the CM SAF dataset of GHI has been compared against ground data at 20 stations in Sweden and Norway, and the result reveals good agreement with an accuracy of 15 W/m^2, corresponding to an error of roughly 8% [28]. In addition, GHI retrievals from different CM SAF products have been validated against ground measurements at eight sites in Europe, under various sky conditions. [29]. More broadly, Zhang et al. [19] have evaluated the result of the six re-analysed datasets for obtaining GHI with the ground data from measurement networks such as the Baseline Surface Radiation Network (BSRN) and others for 674 cities around the globe, with an overall bias found to be from 11–50 W/m^2. Hence, they used a large volume of data in various climate regions and countries; however, the results are shown according to the datasets and measurement networks, rather than for each station. This is useful for comparison between the SDDs and measurement networks, but it does not reflect a real situation for the individual stations. Moradi et al. [30] also estimated daily GHI with the H2 method in Iran, evaluating the result of the model with four stations in the country, which revealed a good agreement with 12% RMSE and 2% bias. Schillings et al. [31] validated direct normal irradiance (DNI) data at weather stations in eight cities in Saudi Arabia with Meteosat-7 data using the H2 method. The results indicate a good agreement with a mean bias of 4.3% from hourly data. Similarly, AL-Jumaily et al. [32] evaluated the GHI data of two Iraqi weather stations with the same method usingMeteosat-8 data for the year 2005. Positive biases of 0.024 KWh/m^2 and 0.012 KWh/m^2 GHI for daily mean values were found for both cities. The authors indicate that further research comparing Meteosat-8 data with other areas of Iraq is needed. It is necessary for studies to validate more than one SDD for comparison between them, and then select the most accurate one. Recently, GHI data from the MFG and MSG have been evaluated over India, which show an overestimation bias of 10–20% of daily mean [33]. Some other studies have evaluated SDDs over the United States in different climate regions against several ground-based measurements [34–37].

The most popular SDD is that arranged by the Solar Radiation Data (SoDa) portal [38], which contains several projects; one of them is HelioClim-3 version 4 and version 5 (HC3v4-5), which are based on the H2 method for converting satellite images of MSG to GHI. Another is the Copernicus Atmosphere Monitoring Service (CAMS) Radiation Service (CRS), which is based on Heliosat-4 (H4) for the same purpose.

Those SDDs have been validated by several studies in various areas. For example, Thomas et al. [39] have validated the hourly GHI from SDDs such as HC3v4-5 and CRS for 42 stations in Brazil. The result reveals a high correlation (an average of 96%) between HC3v4-5 and ground measurements, whereas that with CRS is lower by 2%. Similarly, r values above 0.92 for 15 min and 0.98 for daily GHI with a bias of roughly 5% were found when comparing HC3v4-5 and CRS to ground data at 14 stations over the world [40]. Hourly GHI and DNI from HC3v4-5 for all-sky conditions, and using the McClear dataset for clear-sky conditions, have been validated with ground data in seven stations over Egypt, with RMSE ranges from 6–22% [41]. Marchand et al. [42] have validated hourly GHI from HC3v4-5 and CRS with ground data at five stations in United Arab Emirates and Oman. The overall validation result is nearly 15% of the RMSE on average.

This study aims to validate the hourly GHI from HC3v5 and CRSv3 against ground measurements at nine stations in the northeast of Iraq, being the first study validating those SDDs in that region. One objective of this study is to evaluate the spatial-temporal performance of those datasets in all-sky, clear-sky, and cloudy-sky conditions and with the clearness index. Another objective is to use a new approach for validation, which is limited in the literature, comparing the GHI from ground measurements at a station against the GHI from SDDs at each station location and the two points around it, at a spatial resolution of 5 km (corresponding to that of MSG imagery), and with each point collected from a different pixel.

The study is organised as follows. The study location, ground data and SDDs are described in Section 2. The validation results are shown in Section 3. The discussion is set out in Section 4, and finally a conclusion is provided.

2. Materials and Methods

2.1. Study Site and Ground Data

The study area is located in the northeast of Iraq (latitudes [34°08′20″–37°22′36″], and longitudes [42°32′00″–46°14′29″]). It has a complex topography (mountains, hills and plain areas). According to the Koppen classification [43], two climate regions are seen in the region, which are the Mediterranean Sea and semi-arid climates (Figure 1).

The hourly GHI data with some other climate parameters were collected from two station types. First, the data are from tower stations. The pyranometer used for recording data in these stations is the Kipp and Zonen CMP6 Pyranometer. The data were collected for the period 2011–2014 from five stations, which lacked some years, from the Ministry of Electricity-Kurdistan Regional Government (KRG) (Table 1). Others are automatic stations equipped with an Vaisala QMS101 Pyranometer. The data were collected from the General Directorate of Meteorology and Seismology-KRG from four stations (2013–2016), which lacked some years (Table 2).

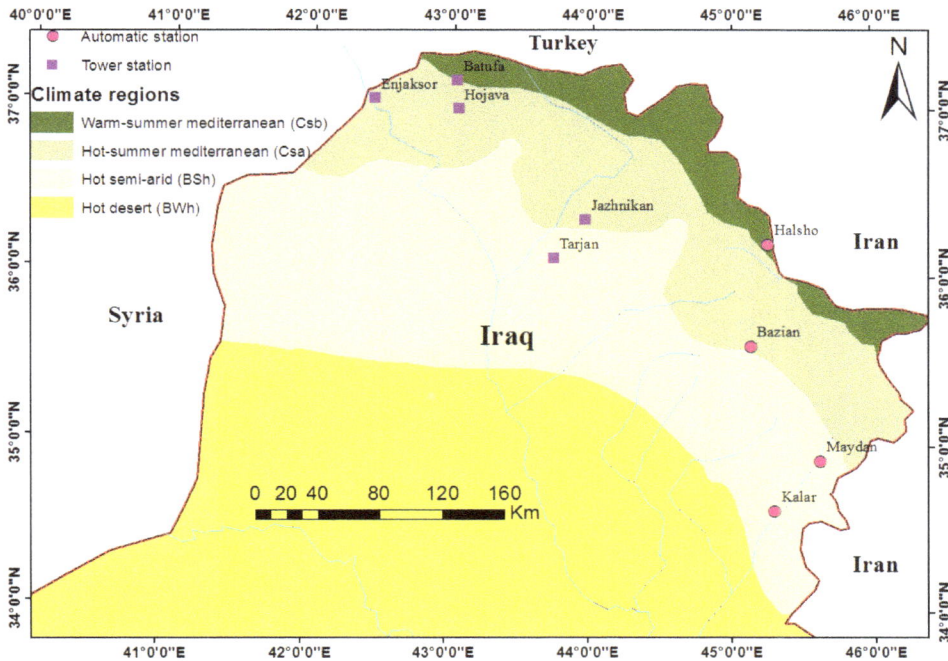

Figure 1. Climate regions, station types, and their distribution in the study area [43,44].

Table 1. Tower stations with hourly GHI from Kipp and Zonen CMP6 Pyranometer.

Station	Coordinates (Degrees)		Elevation a.s.l (m)	Period
Batufa	37.1764 N	43.0236 E	947	01/01/2011–31/12/2013
P1-Batufa	37.1952 N	42.9478 E	854	
P2-Batufa	37.1689 N	43.1042 E	885	
Enjaksor	37.0603 N	42.4353 E	509	01/01/2011–31/12/2014
P1-Enjaksor	37.0642 N	42.3544 E	433	
P2-Enjaksor	37.0533 N	42.4936 E	520	
Hojava	37.0075 N	43.0369 E	933	01/01/2011–31/12/2013
P1-Hojava	37.0331 N	42.9803 E	856	
P2-Hojava	37.0061 N	43.0883 E	940	
Jazhnikan	36.3564 N	43.9556 E	430	01/01/2011–31/10/2013
P1-Jazhnikan	36.3672 N	43.8936 E	376	
P2-Jazhnikan	36.3347 N	44.0294 E	467	
Tarjan	36.1258 N	43.7353 E	276	01/01/2011–31/12/2013
P1W-Tarjan	36.1297 N	43.6686 E	263	
P2-Tarjan	36.1208 N	43.7931 E	308	

Note: The periods in the table are available for the ground measurements; the SDDs for the same periods have been collected for each station location and points around the stations, which were used for validation.

2.2. Quality Controal of GHI Measurments

Data normalising and cleaning were done by setting the solar elevation angle above 15°. Missing values were found and set as not applicable (NA). The two datasets were harmonised for true local solar time. All of the GHI ground data were tested with the BSRN tests [45] and other quality control tests [46,47]. Full information about the quality control of the ground data in this study can be found in [46]. Systematic errors were removed from the data, and some questionable values of data according to various tests were not used in the validation process.

Table 2. Automatic stations with hourly GHI Vaisala QMS101 Pyranometer.

Station	Coordinates (Degrees)		Elevation a.s.l (m)	Period
Halsho	36.2097 N	45.2598 E	1105	01/01/2013–31/12/2016
P1-Halsho	36.2201 N	45.2235 E	1119	
P2-Halsho	36.2058 N	45.3000 E	1395	
Bazian	35.6021 N	45.1376 E	892	01/04/2014–30/12/2016
P1-W Bazian	35.6059 N	45.0689 E	872	
P2-E Bazian	35.5796 N	45.1817 E	828	
Maydan	34.9194 N	45.6224 E	330	01/01/2014–31/12/2016
P1-Maydan	34.9203 N	45.5656 E	388	
P2-Maydan	34.9182 N	45.6716 E	396	
Kalar	34.6244 N	45.3049 E	218	01/01/2014–31/12/2016
P1-Kalar	34.6220 N	45.1768 E	230	
P2-Kalar	34.6237 N	45.4103 E	210	

Note: The periods in the table are available for the ground measurements; the SDDs for the same periods have been collected for each station location and points around the stations, which were used for validation.

2.3. Satellite-Derived Datasets

The SoDa portal [38] is owned by MINES ParisTech and Transvalor. It provides a dataset of solar radiation components, which are based on converting satellite images of MSG in the field view of the SEVIRI instrument covering Europe, Africa, the Middle East and part of South America (Figure 2) by the HC3 and CRSv3 datasets. The hourly GHI data from HC3v5 and CRSv3 for each station location and for points around each station have been collected from the SoDa website, based on the available period of ground data.

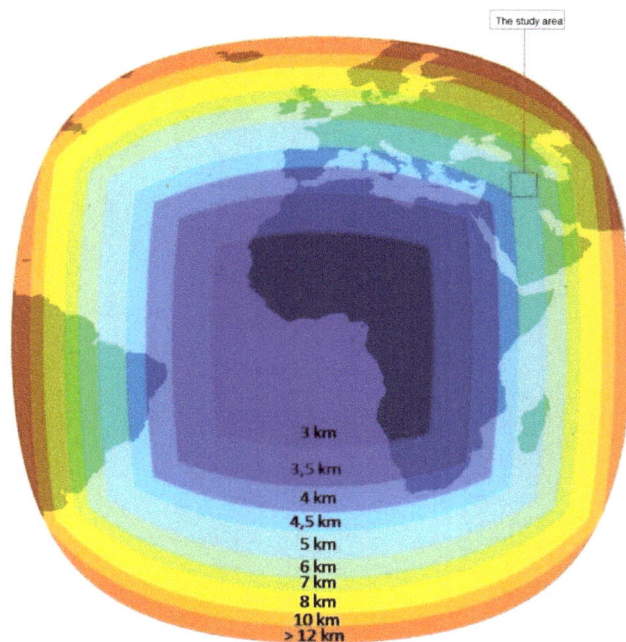

Figure 2. Spatial coverage of the SEVIRI instrument for MSG images [38].

2.3.1. HelioClim-3 (HC3)

The HC3 dataset has been created by converting MSG images to estimate the GHI for every 15 min since February 2001 using the original H2 method. The principle of H2 is to calculate solar radiation statistically by the cloud cover index, which is created by the reflectance in the visible image of MSG and ground albedo [30]. The method has been modified several times by various inputs. It initially refers to Cano et al. [10], and a new method was published in [11]. The MSG image processing in this model gives GHI. Then, a DNI and diffuse horizontal irradiance are estimated [41].

The most common version of HC3 is v4 and v5. V4 inputs are the clear-sky model of the European Solar Radiation Atlas (ESRA) and the Linke turbidity factor. One limitation of this release is that it is not detecting a local effect on the Linke turbidity factor [48]. The clear-sky model gives solar radiation globally every three hours as in the free cloudy-sky [49]. HC3v5 works largely on the same principle as HC3v4, but is different because it uses the McClear model [42]. McClear is also a model for providing solar radiation under clear-sky conditions. It counts the optical depth of atmosphere as a column, which contains aerosol, water vapour and ozone. It is provided by the Copernicus atmosphere monitoring service [50]. The data within those datasets are Available online [38] for MSG coverage for free from 2004 to 2006, and with payment from 2007 upwards.

2.3.2. Copernicus Atmosphere Monitoring Service (CAMS), Radiation Service (CRS)

CRS is a dataset of solar radiation components, which provides Heliosat-4 data using the satellite images of MSG. H4 is a modified method of their previous version. Ground albedo from MODIS and the McClear model are used in this method [39,48,49]. The data are available for free from 2004 until two days before for the areas covered by MSG images. The third version of CRS is available after bias correction [38]. This study has used CRS version 3 (CRSv3). Further information about the HC3v4-5 and CRSv3 projects can be found at SoDa [38] and [39–42,48,49,51].

2.4. Validation Criteria

The validation approach is illustrated in Figure 3. Most of the previous studies for validation of GHI SDD against ground data have separated data into all-sky and clear-sky conditions [27,31,41]. The division also depends on the clearness index (Kt). The Kt is calculated by dividing hourly GHI ground data to the top-of-atmosphere radiation on the horizontal surface (TOA). The TOA was collected from SoDa [38]. For calculating the Kt of SDDs see Figure 3. The Kt was used for validation and setting limits among the various sky conditions [27,52] as below:

- Clear-sky conditions: $0.65 < Kt \leq 1$
- Intermediate sky conditions: $0.3 < Kt \leq 0.65$
- Cloudy-sky conditions: $0 < Kt \leq 0.3$

This study separates the ground data into all-sky, clear-sky and cloudy-sky conditions based on the above Kt limits. This is to test the SDDs in various situations and to demonstrate under which situations the SDDs are the most accurate.

The approach uses the ground data of a station to assess the SDDs with data from the station location pixel and with another two points of SDD pixel data. One pixel data point is selected to the east and another is selected to the west of a station at a distance of 6–10 km, (Tables 1 and 2, Figure 4). This is to select a different pixel from the station location pixel; given the spatial resolution of MSG imagery is 5 km in the case study region (Figure 2). Hereafter, P1 is called the west point for each station, and P2 is called the east point. This is for further investigation into the validation of SDD for more than one-pixel around the station and to address whether the SDD values from neighbouring pixels are the same or different. This is because the solar radiation intensity may be the same in an area of 25 km^2 [23,53].

The validation performance between ground data and SDDs, namely HC3v5 and CRSv3, have been evaluated by statistical indicators, being the correlation coefficient (r) in Equation (1), the bias in Equation (2), and the relative bias in Equation (3), the root mean square error (RMSE) in Equation (4), and the relative RMSE (rRMSE) in Equation (5) [48,54] for the all-sky conditions for hourly GHI for the stations and points around the stations and clearness index for all-sky conditions at stations, and the GHI for the clear-sky and cloudy conditions at the stations.

$$r = \frac{\sum_{i=1}^{n} (Xi - \overline{X})(Yi - \overline{Y})}{\sqrt{\sum_{i=1}^{n} (Xi - \overline{X})^2} \sqrt{\sum_{i=1}^{n} (Yi - \overline{Y})^2}} \tag{1}$$

$$Bias = \frac{\sum_{i=1}^{n} (Yi - Xi)}{n} \tag{2}$$

$$rBias = \frac{Bias}{Mean\ Xi} * 100 \tag{3}$$

$$RMSE = \sqrt{\frac{\sum_{i=1}^{n} (Yi - Xi)^2}{n}} \tag{4}$$

$$rRMSE = \frac{RMSE}{Mean\ Xi} * 100 \tag{5}$$

where n = the number of observations, Xi = the GHI of ground data and Yi = the GHI of a SDD.

The performance of two SDDs against the ground data have also been assessed in all-sky conditions to demonstrate the variability within reproducing the ground data by SDDs by using the hourly mean and standard deviation of GHI in a month. The monthly mean and standard deviation of hourly GHI were calculated for ground data and SDDs for each month in the selected period of a station.

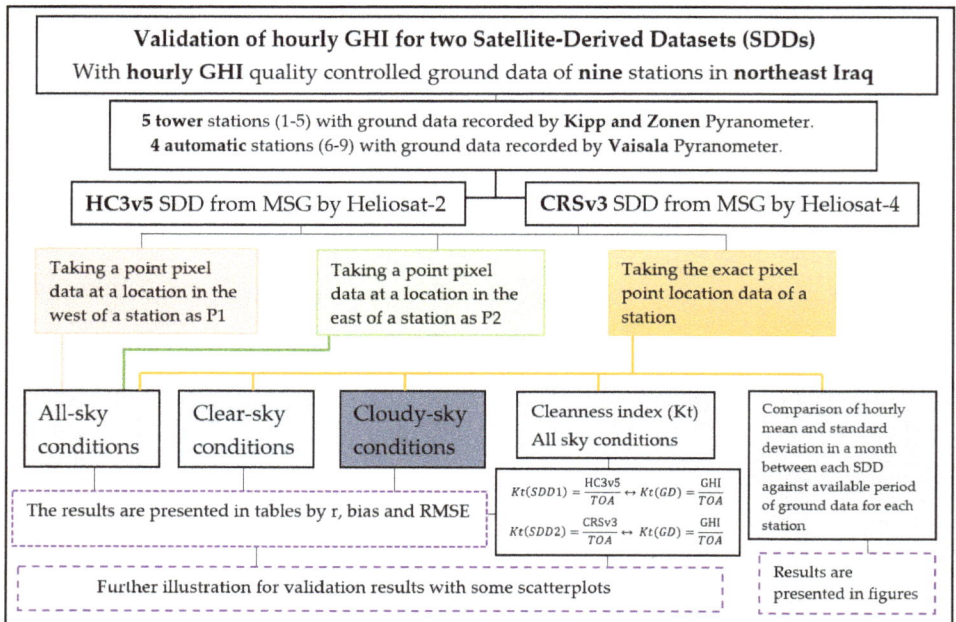

Figure 3. The flowchart of the approach.

Figure 4. Example of point pixel selection of SDD around Bazian station.

3. Results

The results of validating SDDs against ground measurement at nine stations in the northeast of Iraq are shown as follows. Table 3 represents the results of the hourly GHI in all-sky conditions for the stations and the points around them; Table 4 represents the results of the clearness index in all-sky conditions for the stations; and Table 5 represents the results of the hourly GHI in clear-sky and cloudy-sky conditions for the stations. Figures 5 and 6 show the results of the hourly mean and standard deviation in a month for each SDD with ground data. Figures 7 and 8 give further results between the stations with the points around them, and between SDDs in all-sky, clear-sky and cloudy-sky conditions for the results of the validation percentages of the bias and the RMSE, respectively. The results of some stations as examples with scatterplots are shown in Figures 9–11 for the GHI and clearness index in all-sky conditions, and the GHI in clear-sky and cloudy-sky conditions.

Overall, the study is focused on all-sky conditions to show the results in different ways such as within the clearness index, the mean and standard deviation in a month and other statistical indicators, which have been used in all three sky conditions. This is to avoid complex results when presenting all of the above data in a variety of sky conditions.

Table 3. Validation of hourly GHI under all-sky conditions for stations and points around them. Mean, bias and RMSE units are W/m^2.

Stations	Number of Data	Mean	HC3v5					CRSv3				
			r	Bias	%	RMSE	%	r	Bias	%	RMSE	%
Batufa	10,218	511	0.96	−6	−1.2	74	14	0.95	−27	−5.3	92	18
P1	10,218	511	0.95	−15	−2.9	72	14	0.94	−28	−5.5	100	20
P2	10,218	511	0.96	−16	−3.1	77	15	0.95	−29	−5.7	93	18
Enjkasor	13,622	518	0.97	−4	−0.8	64	12	0.95	−20	−3.9	85	16
P1	13,622	518	0.96	−9	−1.7	75	14	0.94	−19	−3.7	90	17
P2	13,622	518	0.97	−6	−1.2	64	12	0.95	−21	−4.1	86	17
Hojava	10,195	503	0.96	0	0	74	15	0.95	−19	−3.8	89	18
P1	10,195	503	0.95	0	0	83	17	0.94	−20	−4	96	19
P2	10,195	503	0.96	3	0.6	78	16	0.94	−16	−3.2	91	18
Jazhnikan	9856	518	0.96	−13	−2.5	73	14	0.95	−20	−3.9	82	16
P1	9856	518	0.96	−14	−2.7	77	15	0.95	−21	−4.1	84	16
P2	9856	518	0.96	−12	−2.3	73	14	0.95	−21	−4.1	83	16

Table 3. *Cont.*

Stations	Number of Data	Mean	HC3v5					CRSv3				
			r	Bias	%	RMSE	%	r	Bias	%	RMSE	%
Tarjan	10,261	521	0.96	−20	−3.8	74	14	0.95	−26	−5	81	16
P1	10,261	521	0.96	−21	−4	78	15	0.95	−26	−5	83	16
P2	10,261	521	0.96	−20	−3.8	73	14	0.95	−26	−5	80	15
Halsho	13,183	503	0.96	1	0.2	81	16	0.95	7	1.4	89	18
P1	13,183	503	0.95	6	1.2	84	17	0.95	7	1.4	90	18
P2	13,183	503	0.94	1	0.2	93	18	0.95	5	1	90	18
Bazian	8884	515	0.96	−8	−1.6	69	13	0.96	−2	−0.4	76	15
P1	8884	515	0.96	−6	−1.2	74	14	0.95	−4	−0.8	79	15
P2	8884	515	0.97	−8	−1.6	68	13	0.96	−2	−0.4	76	15
Maydan	9089	514	0.97	−5	−1	68	13	0.96	3	0.6	73	14
P1	9089	514	0.96	−6	−1.2	72	14	0.96	1	0.2	75	15
P2	9089	514	0.97	−3	−0.6	65	13	0.96	6	1.2	71	14
Kalar	7979	474	0.95	20	4.2	84	18	0.94	19	4	84	18
P1	7979	474	0.94	19	4	88	19	0.93	19	4	88	19
P2	7979	474	0.95	21	4.4	84	18	0.92	25	5.3	97	20

Table 4. Validation of hourly GHI under all-sky conditions for the clearness index. Mean, bias and RMSE units are W/m^2.

Stations	Number of Data	Mean	HC3v5 Clearness Index					CRSv3 Clearness Index				
			r	Bias	%	RMSE	%	r	Bias	%	RMSE	%
Batufa	10,218	0.602	0.89	−0.009	−1.5	0.102	16.94	0.85	−0.038	−6.3	0.121	20.1
Enjkasor	13,622	0.611	0.89	−0.01	−1.64	0.089	14.57	0.83	−0.032	−5.2	0.117	19.1
Hojava	10,195	0.592	0.87	−0.004	−0.68	0.1	16.89	0.85	−0.03	−5.0	0.116	19.5
Jazhnikan	9856	0.602	0.86	−0.024	−3.99	0.1	16.61	0.85	−0.031	−5.1	0.107	17.7
Tarjan	10,261	0.612	0.85	−0.034	−5.56	0.103	16.83	0.84	−0.04	−6.5	0.109	17.8
Halsho	13,183	0.584	0.87	−0.003	−0.51	0.109	18.66	0.84	0.008	1.3	0.12	20.5
Bazian	8884	0.593	0.87	−0.014	−2.36	0.093	15.68	0.85	−0.006	−1.0	0.098	16.5
Maydan	9089	0.594	0.87	−0.016	−2.69	0.091	15.32	0.86	−0.003	−0.5	0.092	15.4
Kalar	7979	0.565	0.83	0.013	2.3	0.099	17.52	0.82	0.017	3.0	0.097	17.1

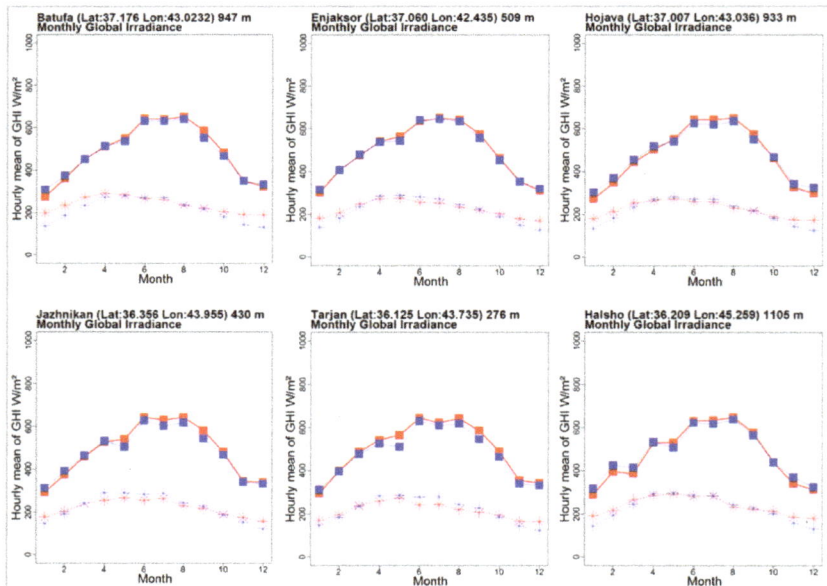

Figure 5. *Cont.*

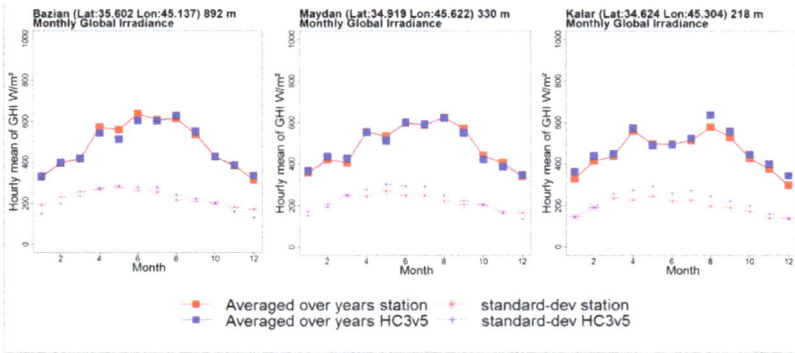

Figure 5. Monthly mean and standard deviation of hourly GHI data in each month aggregated over the data availability for each station with HC3v5. The difference between dots reveals the errors in a month and vice versa. If the dot of the SDD in a month is above the dot of the ground data, it denotes overestimation; otherwise, it denotes underestimation.

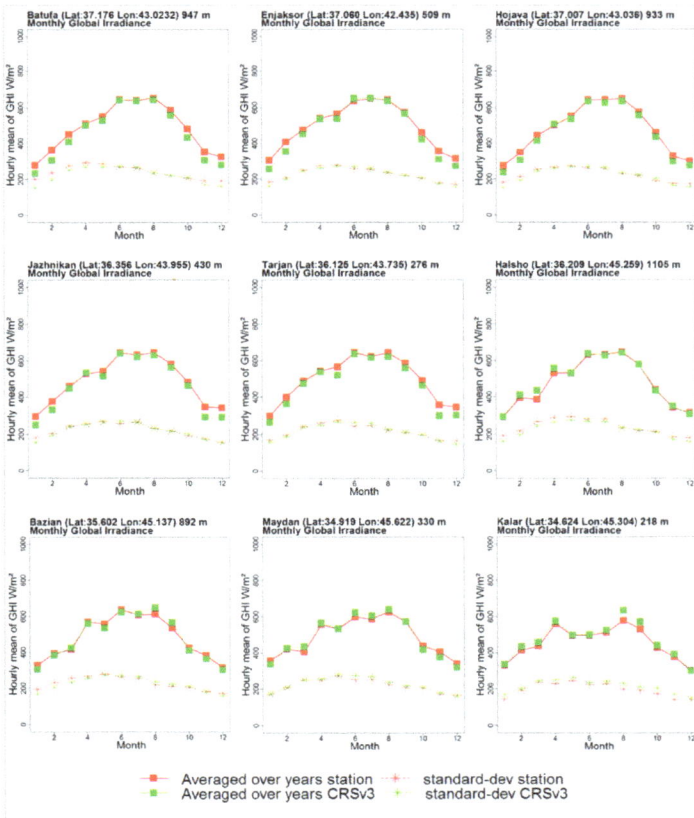

Figure 6. Monthly mean and standard deviation of hourly GHI data in each month aggregated over data availability for each station with CRSv5. The difference between dots reveals errors in a month and vice versa. If the dot of the SDD in a month is above the dot of the ground data, it denotes overestimation; otherwise, it denotes underestimation.

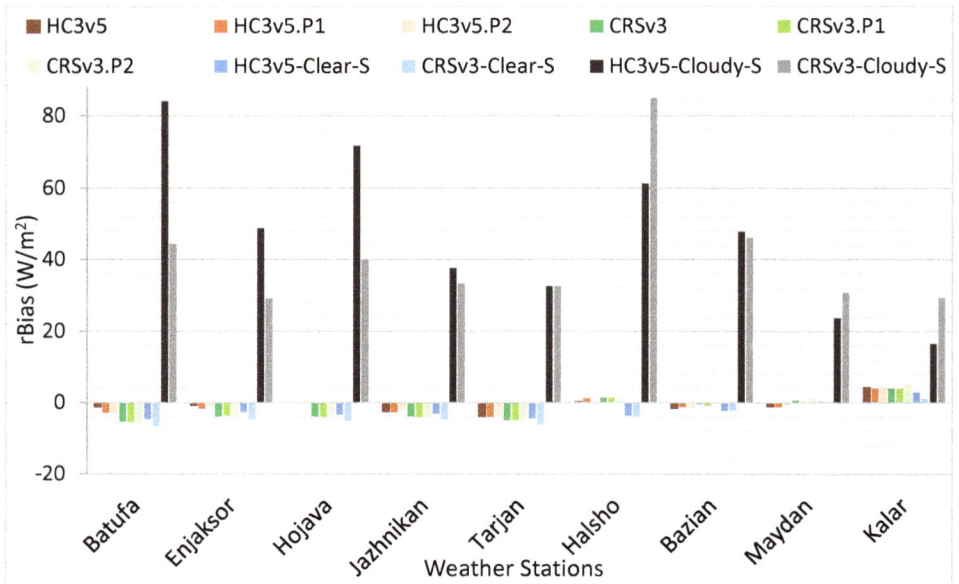

Figure 7. Comparison of rBias for the hourly GHI for all-sky conditions among stations with points around them for HC3v5 and CRSv3. Clear-skies and cloudy-skies at stations are represented by blue, light blue, black and grey colours, respectively.

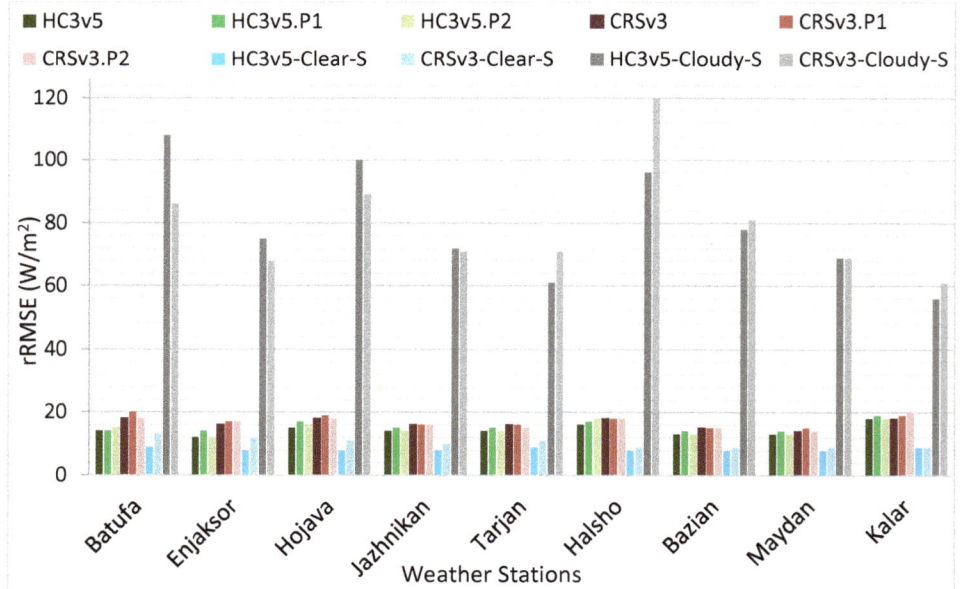

Figure 8. As in Figure 7, but for rRMSE.

Table 5. Validation of hourly GHI under clear-sky and cloudy-sky conditions. Mean, bias and RMSE units are W/m^2.

Station	Condition	Number of Data	Mean	HC3v5					CRSv3				
				r	Bias	%	RMSE	%	r	Bias	%	RMSE	%
Batufa	Clear-sky	5937	679	0.97	−32	−4.7	58	9	0.95	−45	−6.6	86	13
	Cloudy-sky	1448	106	0.68	89	84	114	108	0.63	47	44.3	91	86
Enjkasor	Clear-sky	7955	672	0.97	−18	−2.7	53	8	0.95	−33	−4.9	78	12
	Cloudy-sky	1507	113	0.68	55	48.7	85	75	0.64	33	29.2	77	68
Hojava	Clear-sky	5669	672	0.97	−23	−3.4	56	8	0.95	−34	−5.1	77	11
	Cloudy-sky	1365	113	0.64	81	71.7	113	99	0.59	45	39.8	101	89
Jazhnikan	Clear-sky	5513	681	0.97	−21	−3.1	54	8	0.96	−33	−4.8	69	10
	Cloudy-sky	1018	117	0.63	44	37.6	84	72	0.69	39	33.3	83	71
Tarjan	Clear-sky	5983	666	0.97	−29	−4.4	62	9	0.96	−41	−6.2	72	11
	Cloudy-sky	965	120	0.68	39	32.5	73	61	0.68	39	32.5	85	71
Halsho	Clear-sky	7498	677	0.97	−25	−3.7	56	8	0.97	−26	−3.8	61	9
	Cloudy-sky	2083	106	0.61	65	61.3	102	96	0.6	90	84.9	127	120
Bazian	Clear-sky	4673	690	0.97	−16	−2.3	54	8	0.96	−15	−2.2	64	9
	Cloudy-sky	937	115	0.64	55	47.8	90	78	0.65	53	46.1	93	81
Maydan	Clear-sky	4729	678	0.97	2	0.3	55	8	0.96	2	0.3	62	9
	Cloudy-sky	813	127	0.51	30	23.6	88	69	0.65	39	30.7	88	69
Kalar	Clear-sky	2755	659	0.96	19	2.9	57	9	0.95	7	1.1	60	9
	Cloudy-sky	686	133	0.64	22	16.5	74	56	0.72	39	29.3	81	61

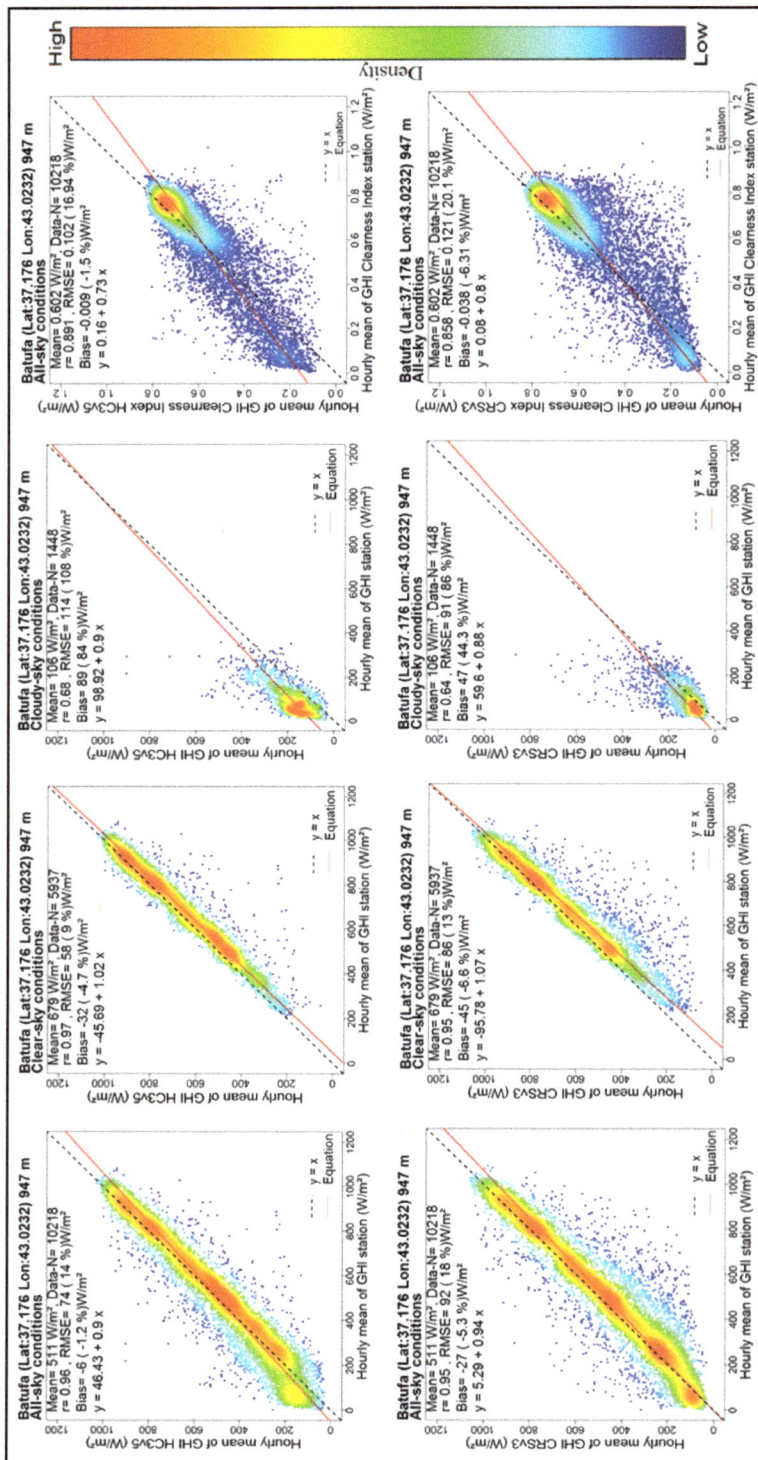

Figure 9. Scatter plot between hourly GHI ground measurements and SDDs (HC3v5 and CRSv3) for Batufa station for all-sky, clear-sky and cloudy-sky conditions and the clearness index.

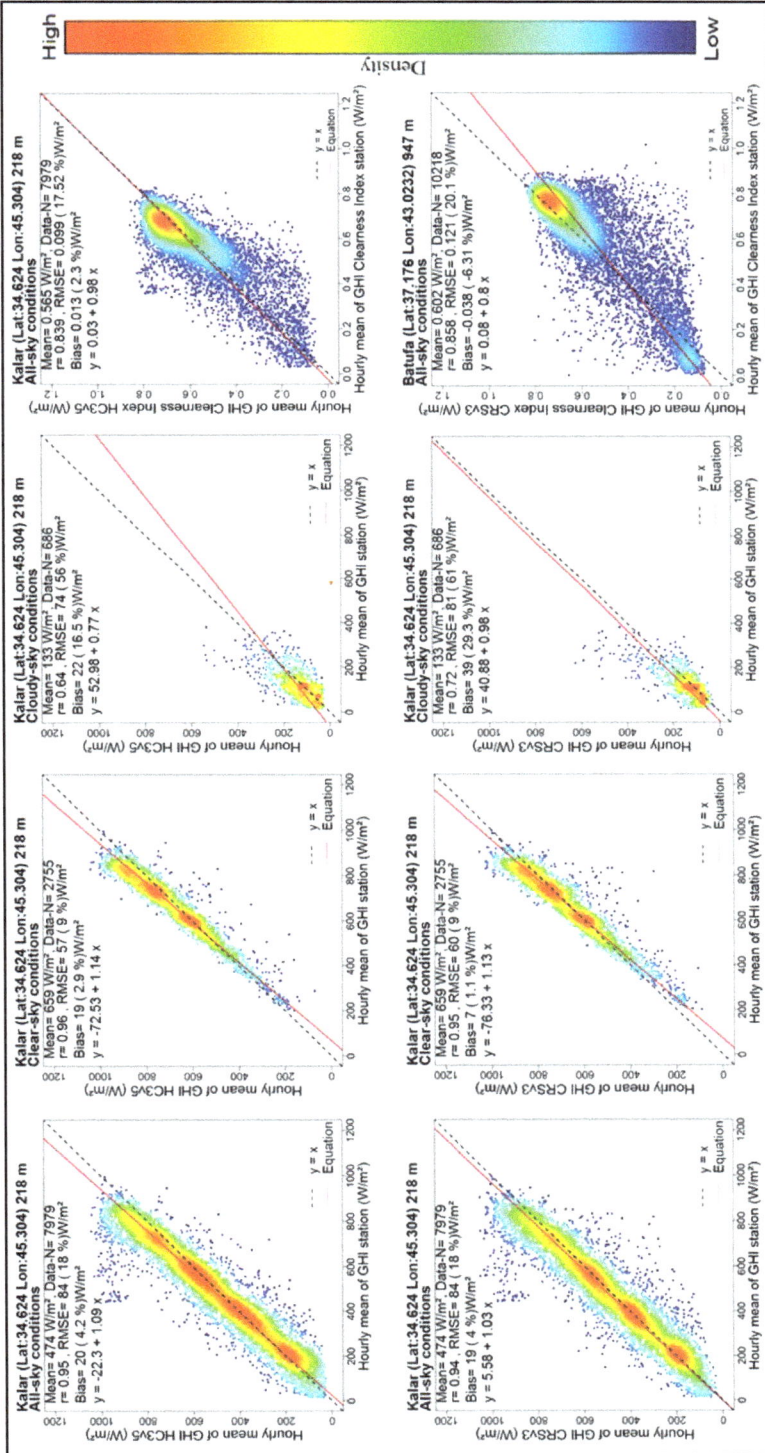

Figure 10. Scatter plot between hourly GHI ground measurements and SDDs (HC3v5 and CRSv3) for Kalar station for all-sky, clear-sky and cloudy-sky conditions and the clearness index.

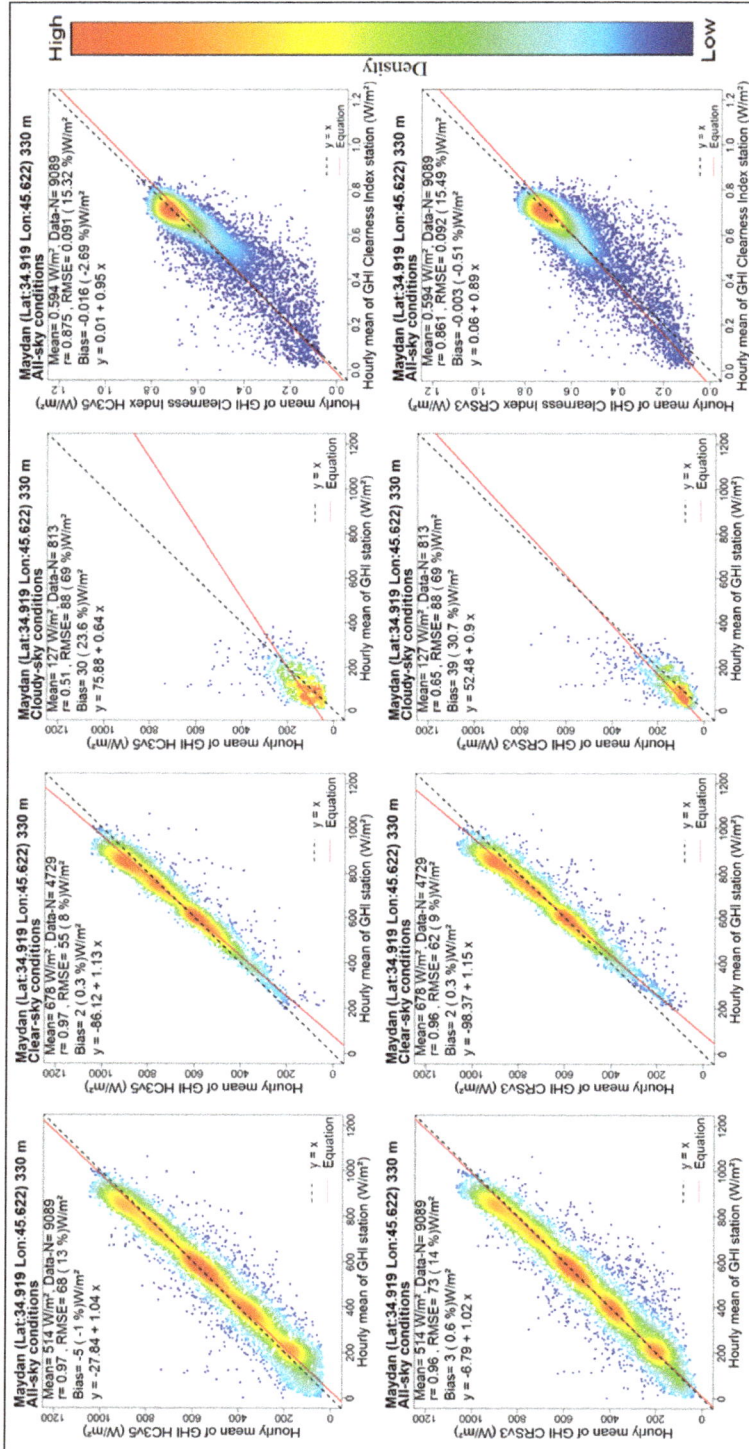

Figure 11. Scatter plot between hourly GHI ground measurements and SDDs (HC3v5 and CRSv3) for Maydan station for all-sky, clear-sky and cloudy-sky conditions and the clearness index.

3.1. All-Sky Conditions

The results of the validation for the all-sky conditions are presented in Table 3. The correlation is from 0.94 to 0.97 in all of the stations and 0.92–0.97 in the points around them. Interestingly, zero bias was recorded at Hojava station, and it was near zero in several other cases (Halsho, Maydan and Bazian stations) for both SDDs. Negative (underestimation) bias was recorded in several cases. It ranges from -21 W/m^2 (-4%) to -3 W/m^2 (-0.6%) for HC3v5, which is lower than CRSv3 in most cases, which ranges from -27 W/m^2 (-5.3%) to -2 W/m^2 (-0.4%). Moreover, it was lower than -6% for all of the stations. However, a positive (overestimation) bias was recorded in some of the cases. The highest case of bias in the study was recorded at Kalar station, which was 25 W/m^2 (5.3%) for CRSv3 and 21 W/m^2 (4.4%) for HC3v5. Some other positive rates were recorded at Halsho and Maydan stations, which were lower than 2% for both SDDs (Figure 7, Table 3).

The bias at each station was compared to the points around the station, and was nearly the same with no more than 4% difference for each station. Overall, the rate of bias in HC3v5 was less than that in CRSv3 (Figure 7, Table 3).

The RMSE was under 21% in all of the cases. Its lowest range, at Enjaksor, was 64 W/m^2 (12%), and increased to the highest value of 88 W/m^2 (19%) at Kalar station for HC3v5. It was generally high for CRSv3 ranging from 71 W/m^2 (14%) to 100 W/m^2 (20%). Most of the other rates of RMSE were between $14-18\%$ for both SDDs.

The RMSE for the points around the stations compared to the station location are nearly the same (Figure 8). Overall, the RMSE for HC3v5 was less than that for CRSv3 (Figure 8).

The smooth scatter density plot illustrates the residual and correlation between ground data and SDDs in some cases. For example, Figure 9 for Batufa station shows that the density of observations were mostly under the 1:1 line, which indicated a recorded negative bias, and the RMSE was acceptable for HC3v5 and high for CRSv3, while some of the other values above the 1:1 line are under 200 W/m^2. However, Figure 10 (Kalar station) shows that the majority of observations are above the 1:1 line and some points are far from the line. This corresponds to positive bias and high RMSE in the station.

Low rates of bias and RMSE were recorded at Maydan station, which is shown in Figure 11 for HC3v5 and CRSv3 respectively. The best-fit line in red is nearly the same as the 1:1 line, more so for HC3v5 than CRSv3 (at the same station).

The results of the clearness index in the all-sky conditions are represented in Table 4. The percentages of bias and RMSE were quite similar to the GHI. Hence, the r values at all of the stations were lower than the GHI, which ranged from 0.82–0.89. The r values were higher in HC3v3 than CRSv3 when comparing each station for both. The scatter density plot shows the highest density of observations at around 0.7 W/m^2 for all of the stations. Negative bias at Batufa, Maydan and positive bias at Kalar can be seen, although the values were low. Several values are far from the 1:1 line, resulting in RMSE to be from 15–21% at all of the stations (Figures 9–11).

Results of the mean and standard deviation of the GHI in months are represented in Figure 5 for HC3v5 and in Figure 6 for CRSv3. The two figures demonstrate the distribution of the two SDDs with ground data in each month, expressed by the standard deviation. However, some differences were recorded in the winter months at Batufa, Hojava, Halsho and Bazian stations, whereas for summer months differences were recorded at Maydan and Kalar stations for both SDDs.

3.2. Clear-Sky and Cloudy-Sky Conditions

The compared results of the two SDDs for clear-sky and cloudy-sky conditions are presented in Table 5. There are apparent differences in the r values between the clear-sky and cloudy-sky conditions in all of all cases. These ranged from 0.95–0.97 for clear-skies, whereas for the cloudy-skies, it ranged from 0.51–0.72 among the stations.

Similarly, the ranges of bias were much higher in the cloudy-skies than in the clear-skies. The lowest bias was a 2 W/m^2 (0.3%) overestimation at Maydan in clear-sky conditions whereas in the same station it reached 30 W/m^2 (24%) in cloudy-sky conditions. The highest bias for the clear-sky conditions was recorded at Batufa station, which was −32 W/m^2 (−4.7%). The same station had the highest bias for the cloudy-sky conditions, which was 89 W/m^2 (84%). These were recorded for HC3v5. The bias for CRSv3 was the same as HC3v3 for the low range in the clear-sky conditions. In others, the ranges start from −45 W/m^2 (−6.6%) underestimation at Batufa station to 7 W/m^2 (1%) overestimation at Kalar station, while for cloudy-skies, it ranged from 33 (29%) to 90 (85%) W/m^2 respectively. The variety of the two SDDs in term of bias is shown in Figure 7, which illustrates a moderate difference between cloudy-sky and clear-sky conditions from one station to the other. In addition, the range of bias was much lower in clear-skies than cloudy-skies and the bias in all-sky conditions was lower than that in clear-skies except for two stations for both SDDs where the bias was higher to some degree in the all-sky conditions than in the clear-skies (Figure 7).

Similarly, the RMSE was much higher in the cloudy-skies than in clear-skies for the SDDs in each station. For example, at Halsho station for CRSv3 it was 61 W/m^2 (9%) in clear-skies and increased sharply to 127 W/m^2 (120%) under cloudy conditions. Nearly the same situation can be seen for Batufa station for HC3v5. The RMSE for both SDDs was lower than 14% at all of the stations for the clear-skies while it was above 58% for the cloudy-skies (Figure 8, Table 5). The RMSE in clear-skies was lower than in the case of all-sky conditions in all of the study areas for both SDDs (Figure 8).

The smooth scatter density plot for Batufa, Kalar and Maydan of both SDDs separately shows that the density of observations were above the 1:1 line, and the direction of distribution was towards the high value of SDD which recorded high overestimation and high RMSE in cloudy-sky conditions (Figures 9–11). In contrast, in the clear-sky conditions for nearly all of the stations, the distribution of observations was near the 1:1 line. This leads to low RMSE in clear-skies compared to cloudy-sky and all-sky conditions (Figures 9–11). The observations under the 1:1 line illustrated a negative bias at Batufa station, whereas the opposite—i.e., positive bias—occurred at Kalar station, and a minimal bias was seen at Maydan station relative to the normal distribution.

Overall, the results of the validation varied from one station to another, and they are acceptable according to bias and RMSE. At some stations, the results were disappointing. The points around the stations had nearly the same ranges of bias and RMSE compared to the station location. In most of the cases, the bias and RMSE of HC3v5 were lower than CRSv3. The bias and RMSE were lower for the clear-sky and all-sky conditions than for the cloudy-skies.

4. Discussion

The validation results demonstrate good agreement between the ground data and SDDs in all-sky and clear-sky conditions (average r = 0.95, bias under 6% and RMSE under 21%), unlike the results for the cloudy-sky conditions (average r = 0.61, bias above 16% and RMSE above 61%). The results from the two neighbouring points at each station are close to the results at the station location with an average difference of 2%. Overall the performance of SDDs are in agreement with those from similar studies in other areas [39,41,42,48], also showing a better performance for HC3v5 over CRSv3 (Figures 7 and 8). This is mainly related to the inputs into the models for creating each dataset, whether it is H2 or H4 (see Section 2.3).

4.1. All-Sky Conditions

The high rates of positive bias and RMSE at Kalar compared to all of the other stations (Figures 7–11) might be due to the quality of the recorded data [46] and a partial shadow on the sensor at that station, because its mean is lower than that at the other stations by nearly 30 W/m^2 (Table 3), while those data have passed the quality check. However, a similar positive bias and RMSE are reported by other studies [41,42]. The low rates of bias (0–2%) were recorded for HC3v5 at the stations of Hojava, Halsho, Bazian and Maydan, and for CRSv3 for the same stations except for Hojava, whose bias rate reaches −3.8% for all-sky conditions (Table 3, Figure 7). These show that the GHI ground data are explained well by satellite data (Figures S1–S5), while the resolution of satellite imagery in that area is 5 km (Figure 2). The underestimations of bias (1–6%) and RMSE (12–20%)

were recorded at most stations (Table 3, Figures 7 and 8). Comparable percentages were recorded in similar studies in other areas and climate regions, namely Egypt [41], Brazil [39], and some BSRN stations [48]. The reasons are partly related to the local condition of the station, inputs to the Heliosat method—especially the atmospheric optical depth owing to its unavailability—, various cloud types, the resolution of satellite images and the aerosols effect [37,39,55]. This is because in some cases, the GHI ground data are well explained by SDDs (Figures S1–S5), but in other cases only some error rates were recorded (Table 3). Those rates are quite reasonable for hourly GHI [42]. Those rates of bias can be corrected or modified in some ways [14,18,56,57].

The low variabilities between the ground data and both SDDs are seen in Figures 5 and 6. This might be related to the geographical location and climatic condition, and another reason is that the data were aggregated, concealing some random errors between the two datasets. Hence, some error rates in winter months are related to the difficulty of the Heliosat methods to estimate GHI in cloudy conditions [31,51]. The performance of the clearness index (Table 4) is nearly the same for GHI in at all-sky conditions, which is related to the above-mentioned reasons.

The interesting side of this study is that the results of the validation for both SDDs with the two neighbouring points at each station separately are slightly closer to those at the station location. The differences range between 0–2% for bias and RMSE for each point at most of the locations (Table 3, Figures 7 and 8). The ±1–4% difference between the station location and the neighbouring points with the station ground data GHI are mainly related to the elevation above sea level for each location. Other factors might be related to local land surface types such as land and water, agriculture and bare soil. This indicates that the GHI from SDDs can be used for regional planning for various purposes, and the ground data GHI can be used for neighbouring areas when there is a limitation of ground data. This validation is also considered to add further weight to the assumption of near uniformity of solar radiation in a 25 km^2 area [23,53].

4.2. Clear-Sky and Cloudy-Sky Conditions

The validation results for both SDDs of GHI with the ground data for clear-sky conditions showed good agreement according to RMSE, which decreased at most stations (Figures 7–11). This is partly related to the inputs to the H2 method, especially in incorporating the visible images of MSG in cloud-free conditions into the model. Similarly, the increased performance of HC3v5 for clear-sky conditions has been reported [41] in Egypt. However, the remaining residuals of clear-sky conditions are caused by the factors that have been mentioned in the all-sky conditions above. However, the bias increased to some degree for both SDDs in most of the stations, which were recording underestimations for clear-sky conditions. This is partially related to the increase of the mean GHI ground data in clear-sky conditions. It has also been recorded by several studies [19,27,37], which show that the bias is underestimated for clear-skies.

The study investigated the performance of SDDs on cloudy-sky conditions, reflected in the very low performance of both HC3v5 and CRSv3 according to the high ratio of bias and RMSE (Figures 7 and 8). A close look at the samples of smooth scatter plots (Figures 9–11) shows how far the observations and their density are from the 1:1 line. This is related to difficulties in analysing cloudy pixels of MSG images [31]; the clouds prevented the ground being viewed from the sensor aboard the satellite [42], and as such it is hard to differentiate between cloud albedo and ground albedo [51]. These factors lead to an overestimation of GHI as shown in all of the stations (Table 5) for bias, and much higher RMSE (Figures 6 and 7). Indeed, in some of the cases, it is above the mean of the observations. Similar high residuals for cloudy conditions have been reported in the literature [19,27,37,41]. This indicates that the GHI ground data are well explained by the SDDs in clear-sky conditions, whereas they are not explained well in cloudy-sky conditions (Figures S6–S15). The results of high bias and RMSE indicate that further research is required to correct the errors under cloudy-sky conditions, whereas several studies have done bias corrections for all-sky conditions [14,18,56,57].

The limitations of this study are the different data timescales from one station to another and the limited information available for some parameters, such as the aerosols and local atmospheric properties. This might lead to a challenge to fully explain the reasons behind the results at each station.

However, the validation results vary from one station to another and are near the World Meteorological Organisation (WMO) standard, whereby the bias should be less than 3 W/m^2 and 95% of errors should not exceed 20 W/m^2 [50]. However, the validation results in a minority of stations are above the WMO standard. Therefore, it is probable that the SDDs can be used for modelling and mapping solar radiation with some modification and bias correction.

5. Conclusions

The study has validated hourly GHI from two SDDs, which are HC3v5 and CRSv3, with ground data from nine stations in northeast Iraq for all-sky, clear-sky and cloudy-sky conditions in the station pixels and with two other pixels around the station in all-sky conditions. The temporal changes of ground data GHI were well represented by both SDDs; r was above 0.94 for the all-sky and clear-sky conditions, and above 0.82 for the clearness index in most cases, while for cloudy-skies it was between 0.51–0.72. The bias was negative (underestimation) for most of the cases except for two HC3v5 and three CRSv3 cases, in which it was positive (overestimation); all of the bias ranges were smaller than 8% (W/m^2) of the mean GHI in all-sky and clear-sky conditions, whereas for cloudy-sky conditions, it was positive and varied from one station to another, by 17–85% (W/m^2) of the mean GHI. The same applies to RMSE. It ranged between 8–20% (W/m^2) in all of the stations for all-sky and clear-sky conditions. In contrast, the range was much higher in cloudy-sky conditions: above 56%. The differences between neighbouring pixels and at-station pixels in the SDDs compared to the ground data of GHI for each site are very small, varying by 2% in most cases. The overall performance of HC3v5 is better than that of CRSv3.

Despite the ratio of errors at some stations, the SDDs are closely related to the ground data at most of the stations. However, the resolution of MSG images is 5 km in the case study. The SDDs represent hourly GHI well, and this can be used to map solar resources and possibly for modelling GHI with ground data in areas with a low number of stations.

Further studies about the bias corrections in the SDDs are needed, especially for cloudy-sky conditions. Further research would also be useful for validating the SDDs in other climates. Some studies are also required to address the inputs to the Heliosat method, according to regional and local factors, for a better estimation of GHI from satellite images.

Supplementary Materials: The following are Available online at http://www.mdpi.com/2072-4292/10/10/1651/s1, Figures S1–S5: Cumulative frequency function of ground data compared to SDDs for all-sky conditions for the available period of data pairs. The closer red and green lines (SDDs) are to the black line (ground data) shows better performance of SDDs. A difference between lines shows the errors. Figures S6–S15 as in Figure S1 but for clear-sky and cloudy-sky conditions respectively.

Author Contributions: B.A. conducted this research manuscript as part of his Ph.D. studies at the University of Leicester, supervised by H.B. and C.J., E.W., C.T. and M.M. are contributed to the study by writing—review & editing. The manuscript was prepared by B.A. with suggestions and corrections from all co-authors. All authors have approved the final draft.

Funding: The funders had no role in the design of the study; in the collection, analyses, or interpretation of data; in the writing of the manuscript, and in the decision to publish the results.

Acknowledgments: The Higher Committee for Education Development in Iraq (HCED) funded this study as a scholarship, the Center for Landscape and Climate Research (CLCR) and the National Centre for Earth Observation (NERC) supported it. The authors are extremely grateful for the assistance of the Directorate of Meteorology—Sulaymaniyah and KRG Ministry of electricity for providing meteorological data. The authors are grateful to Soda Service for allowing access and free use of GHI SDD of CRSv3 data and for a subscription to use the HC3v5 data. A special thanks to Prof. Lucien Wald, MINES ParisTech-France for his oral advice towards the study.

Conflicts of Interest: The authors declare no conflict of interest.

References

1. Despotovic, M.; Nedic, V.; Despotovic, D.; Cvetanovic, S. Evaluation of empirical models for predicting monthly mean horizontal diffuse solar radiation. *Renew. Sustain. Energy Rev.* **2016**, *56*, 246–260. [CrossRef]
2. Kanters, J.; Wall, M. A planning process map for solar buildings in urban environments. *Renew. Sustain. Energy Rev.* **2016**, *57*, 173–185. [CrossRef]
3. Mohanty, S.; Patra, P.K.; Sahoo, S.S. Prediction and application of solar radiation with soft computing over traditional and conventional approach—A comprehensive review. *Renew. Sustain. Energy Rev.* **2016**, *56*, 778–796. [CrossRef]
4. Paulescu, E.; Blaga, R. Regression models for hourly diffuse solar radiation. *Sol. Energy* **2016**, *125*, 111–124. [CrossRef]
5. Polo, J.; Zarzalejo, L.; Cony, M.; Navarro, A.; Marchante, R.; Martin, L.; Romero, M. Solar radiation estimations over India using meteosat satellite images. *Sol. Energy* **2011**, *85*, 2395–2406. [CrossRef]
6. Urraca, R.; Martinez-de-Pison, E.; Sanz-Garcia, A.; Antonanzas, J.; Antonanzas-Torres, F. Estimation methods for global solar radiation: Case study evaluation of five different approaches in central Spain. *Renew. Sustain. Energy Rev.* **2017**, *77*, 1098–1113. [CrossRef]
7. Gherboudj, I.; Ghedira, H. Assessment of solar energy potential over the united arab emirates using remote sensing and weather forecast data. *Renew. Sustain. Energy Rev.* **2016**, *55*, 1210–12224. [CrossRef]
8. Qin, J.; Chen, Z.; Yang, K.; Liang, S.; Tang, W. Estimation of monthly-mean daily global solar radiation based on modis and trmm products. *Appl. Energy* **2011**, *88*, 2480–2489. [CrossRef]
9. Xie, Y.; Yu, T.; Gu, X.; Zhao, L.; Zhang, L.; Wang, G. The estimation of surface daily reflected solar radiation using landsat-7 etm+ and empirical models. In Proceedings of the 2012 5th International Congress on Image and Signal Processing (CISP), Chongqing, China, 16–18 October 2012; IEEE: Piscataway, NJ, USA, 2012; pp. 1109–1113.
10. Cano, D.; Monget, J.-M.; Albuisson, M.; Guillard, H.; Regas, N.; Wald, L. A method for the determination of the global solar radiation from meteorological satellite data. *Sol. Energy* **1986**, *37*, 31–39. [CrossRef]
11. Rigollier, C.; Lefèvre, M.; Wald, L. The method heliosat-2 for deriving shortwave solar radiation from satellite images. *Sol. Energy* **2004**, *77*, 159–169. [CrossRef]
12. Beyer, H.G.; Costanzo, C.; Heinemann, D. Modifications of the heliosat procedure for irradiance estimates from satellite images. *Sol. Energy* **1996**, *56*, 207–212. [CrossRef]
13. Eissa, Y.; Chiesa, M.; Ghedira, H. Assessment and recalibration of the heliosat-2 method in global horizontal irradiance modeling over the desert environment of the UAE. *Sol. Energy* **2012**, *86*, 1816–1825. [CrossRef]
14. Qu, Z.; Gschwind, B.; Lefevre, M.; Wald, L. Improving helioclim-3 estimates of surface solar irradiance using the mcclear clear-sky model and recent advances in atmosphere composition. *Atmos. Meas. Tech.* **2014**, *7*, 3927–3933. [CrossRef]
15. Janjai, S.; Pankaew, P.; Laksanaboonsong, J. A model for calculating hourly global solar radiation from satellite data in the tropics. *Appl. Energy* **2009**, *86*, 1450–1457. [CrossRef]
16. Janjai, S. A method for estimating direct normal solar irradiation from satellite data for a tropical environment. *Sol. Energy* **2010**, *84*, 1685–1695. [CrossRef]
17. Janjai, S.; Pankaew, P.; Laksanaboonsong, J.; Kitichantaropas, P. Estimation of solar radiation over cambodia from long-term satellite data. *Renew. Energy* **2011**, *36*, 1214–1220. [CrossRef]
18. Polo, J.; Wilbert, S.; Ruiz-Arias, J.A.; Meyer, R.; Gueymard, C.; Suri, M.; Martín, L.; Mieslinger, T.; Blanc, P.; Grant, I. Preliminary survey on site-adaptation techniques for satellite-derived and reanalysis solar radiation datasets. *Sol. Energy* **2016**, *132*, 25–37. [CrossRef]
19. Zhang, X.; Liang, S.; Wang, G.; Yao, Y.; Jiang, B.; Cheng, J. Evaluation of the reanalysis surface incident shortwave radiation products from NCEP, ECMWF, GSFC, and JMA using satellite and surface observations. *Remote Sens.* **2016**, *8*, 225. [CrossRef]
20. Polo, J. Solar global horizontal and direct normal irradiation maps in Spain derived from geostationary satellites. *J. Atmos. Sol.-Terr. Phys.* **2015**, *130–131*, 81–88. [CrossRef]
21. Journée, M.; Bertrand, C. Geostatistical merging of ground-based and satellite-derived data of surface solar radiation. *Adv. Sci. Res.* **2011**, *6*, 1–5. [CrossRef]
22. Journée, M.; Müller, R.; Bertrand, C. Solar resource assessment in the Benelux by merging meteosat-derived climate data and ground measurements. *Sol. Energy* **2012**, *86*, 3561–3574. [CrossRef]

23. Bojanowski, J.S.; Vrieling, A.; Skidmore, A.K. Calibration of solar radiation models for Europe using meteosat second generation and weather station data. *Agric. For. Meteorol.* **2013**, *176*, 1–9. [CrossRef]

24. Roerink, G.J.; Bojanowski, J.S.; de Wit, A.J.W.; Eerens, H.; Supit, I.; Leo, O.; Boogaard, H.L. Evaluation of Msg-derived global radiation estimates for application in a regional crop model. *Agric. For. Meteorol.* **2012**, *160*, 36–47. [CrossRef]

25. Sanchez-Lorenzo, A.; Wild, M.; Brunetti, M.; Guijarro, J.A.; Hakuba, M.Z.; Calbó, J.; Mystakidis, S.; Bartok, B. Reassessment and update of long-term trends in downward surface shortwave radiation over Europe (1939–2012). *J. Geophys. Res. Atmos.* **2015**, *120*, 9555–9569. [CrossRef]

26. Sanchez-Lorenzo, A.; Wild, M.; Trentmann, J. Validation and stability assessment of the monthly mean cm SAF surface solar radiation dataset over Europe against a homogenized surface dataset (1983–2005). *Remote Sens. Environ.* **2013**, *134*, 355–366. [CrossRef]

27. Amillo, A.; Huld, T.; Müller, R. A new database of global and direct solar radiation using the Eastern Meteosat Satellite, models and validation. *Remote Sens.* **2014**, *6*, 8165–8189. [CrossRef]

28. Riihelä, A.; Carlund, T.; Trentmann, J.; Müller, R.; Lindfors, A. Validation of cm SAF surface solar radiation datasets over Finland and Sweden. *Remote Sens.* **2015**, *7*, 6663–6682. [CrossRef]

29. Ineichen, P.; Barroso, C.S.; Geiger, B.; Hollmann, R.; Marsouin, A.; Mueller, R. Satellite application facilities irradiance products: Hourly time step comparison and validation over Europe. *Int. J. Remote Sens.* **2009**, *30*, 5549–5571. [CrossRef]

30. Moradi, I.; Mueller, R.; Alijani, B.; Kamali, G.A. Evaluation of the heliosat-II method using daily irradiation data for four stations in Iran. *Sol. Energy* **2009**, *83*, 150–156. [CrossRef]

31. Schillings, C.; Meyer, R.; Mannstein, H. Validation of a method for deriving high resolution direct normal irradiance from satellite data and application for the arabian peninsula. *Sol. Energy* **2004**, *76*, 485–497. [CrossRef]

32. AL-Jumaily, K.J.; AL-Salihi, A.M.; Al-Tai, O.T. Evaluation of meteosat-8 measurements using daily global solar radiation for two stations in iraq. *Energy Environ.* **2010**, *1*, 635–642.

33. Riihelä, A.; Kallio, V.; Devraj, S.; Sharma, A.; Lindfors, A. Validation of the Sarah-e satellite-based surface solar radiation estimates over India. *Remote Sens.* **2018**, *10*, 392. [CrossRef]

34. Gueymard, C.A.; Wilcox, S.M. Assessment of spatial and temporal variability in the US solar resource from radiometric measurements and predictions from models using ground-based or satellite data. *Sol. Energy* **2011**, *85*, 1068–1084. [CrossRef]

35. Habte, A.; Sengupta, M.; Wilcox, S. Comparing measured and satellite-derived surface irradiance. In Proceedings of the ASME 2012 6th International Conference on Energy Sustainability collocated with the ASME 2012 10th International Conference on Fuel Cell Science, Engineering and Technology, San Diego, CA, USA, 23–26 July 2012; American Society of Mechanical Engineers: New York, NY, USA, 2012; pp. 561–566.

36. Lave, M.; Weekley, A. Comparison of high-frequency solar irradiance: Ground measured vs. In Satellite-derived. In Proceedings of the 2016 IEEE 43rd Photovoltaic Specialists Conference (PVSC), Portland, OR, USA, 5–10 June 2016; IEEE: Piscataway, NJ, USA, 2016; pp. 1101–1106.

37. Xia, S.; Mestas-Nuñez, A.; Xie, H.; Vega, R. An evaluation of satellite estimates of solar surface irradiance using ground observations in San Antonio, Texas, USA. *Remote Sens.* **2017**, *9*, 1268. [CrossRef]

38. SoDa. Solar Radiation Data. Available online: http://www.soda-pro.com/ (accessed on 20 March 2017).

39. Thomas, C.; Wey, E.; Blanc, P.; Wald, L. Validation of three satellite-derived databases of surface solar radiation using measurements performed at 42 stations in Brazil. *Adv. Sci. Res.* **2016**, *13*, 81–86. [CrossRef]

40. Thomas, C.; Saboret, L.; Wey, E.; Blanc, P.; Wald, L. Validation of the new helioclim-3 version 4 real-time and short-term forecast service using 14 BSRN stations. *Adv. Sci. Res.* **2016**, *13*, 129–136. [CrossRef]

41. Eissa, Y.; Korany, M.; Aoun, Y.; Boraiy, M.; Abdel Wahab, M.; Alfaro, S.; Blanc, P.; El-Metwally, M.; Ghedira, H.; Hungershoefer, K.; et al. Validation of the surface downwelling solar irradiance estimates of the helioclim-3 database in egypt. *Remote Sens.* **2015**, *7*, 9269–9291. [CrossRef]

42. Marchand, M.; Al-Azri, N.; Ombe-Ndeffotsing, A.; Wey, E.; Wald, L. Evaluating meso-scale change in performance of several databases of hourly surface irradiation in south-eastern Arabic Pensinsula. *Adv. Sci. Res.* **2017**, *14*, 7–15. [CrossRef]

43. Kottek, M.; Grieser, J.; Beck, C.; Rudolf, B.; Rubel, F. World map of the köppen-geiger climate classification updated. *Meteorol. Z.* **2006**, *15*, 259–263. [CrossRef]

44. GDMS. Ministry of Transport and Communications 2016. General Directorate of Meteorology & Seismology. Available online: http://gdms-krg.org/ku/ (accessed on 14 May 2017).

45. Long, C.N.; Dutton, E.G. *BSRN Global Network Recommended QC Tests*; v2.0; PANGAEA: Bremerhaven, Germany, 2002.

46. Ameen, B.; Balzter, H.; Jarvis, C. Quality control of global horizontal irradiance estimates through bsrn, toacs and air temperature/sunshine duration test procedures. *Climate* **2018**, *6*, 69. [CrossRef]

47. Geiger, M.; Diabaté, L.; Ménard, L.; Wald, L. A web service for controlling the quality of measurements of global solar irradiation. *Sol. Energy* **2002**, *73*, 475–480. [CrossRef]

48. Thomas, C.; Wey, E.; Blanc, P.; Wald, L.; Lefèvre, M. Validation of helioclim-3 version 4, helioclim-3 version 5 and macc-rad using 14 bsrn stations. *Energy Procedia* **2016**, *91*, 1059–1069. [CrossRef]

49. Eissa, Y.; Munawwar, S.; Oumbe, A.; Blanc, P.; Ghedira, H.; Wald, L.; Bru, H.; Goffe, D. Validating surface downwelling solar irradiances estimated by the mcclear model under cloud-free skies in the United Arab Emirates. *Sol. Energy* **2015**, *114*, 17–31. [CrossRef]

50. Lefèvre, M.; Wald, L. Validation of the mcclear clear-sky model in desert conditions with three stations in Israel. *Adv. Sci. Res.* **2016**, *13*, 21–26. [CrossRef]

51. Blanc, P.; Gschwind, B.; Lefèvre, M.; Wald, L. The helioclim project: Surface solar irradiance data for climate applications. *Remote Sens.* **2011**, *3*, 343–361. [CrossRef]

52. Mueller, R.; Trentmann, J.; Träger-Chatterjee, C.; Posselt, R.; Stöckli, R. The role of the effective cloud albedo for climate monitoring and analysis. *Remote Sens.* **2011**, *3*, 2305–2320. [CrossRef]

53. Bouchouicha, K.; Razagui, A.; Bachari, N.E.I.; Aoun, N. Estimation of hourly global solar radiation using msg-hrv images. *Int. J. Appl. Environ. Sci.* **2016**, *11*, 351–368.

54. Quej, V.H.; Almorox, J.; Arnaldo, J.A.; Saito, L. Anfis, svm and ann soft-computing techniques to estimate daily global solar radiation in a warm sub-humid environment. *J. Atmos. Sol.-Terr. Phys.* **2017**, *155*, 62–70. [CrossRef]

55. Kaskaoutis, D.G.; Kambezidis, H.D.; Dumka, U.C.; Psiloglou, B.E. Dependence of the spectral diffuse-direct irradiance ratio on aerosol spectral distribution and single scattering albedo. *Atmos. Res.* **2016**, *178–179*, 84–94. [CrossRef]

56. Frank, C.W.; Wahl, S.; Keller, J.D.; Pospichal, B.; Hense, A.; Crewell, S. Bias correction of a novel european reanalysis data set for solar energy applications. *Sol. Energy* **2018**, *164*, 12–24. [CrossRef]

57. Polo, J.; Martín, L.; Vindel, J.M. Correcting satellite derived dni with systematic and seasonal deviations: Application to India. *Renew. Energy* **2015**, *80*, 238–243. [CrossRef]

remote sensing

MDPI

Article

Impact of Insolation Data Source on Remote Sensing Retrievals of Evapotranspiration over the California Delta

Martha Anderson [1],*, George Diak [2], Feng Gao [1], Kyle Knipper [1], Christopher Hain [3], Elke Eichelmann [4], Kyle S. Hemes [4], Dennis Baldocchi [4], William Kustas [1] and Yun Yang [1]

[1] Hydrology and Remote Sensing Laboratory, USDA-ARS, Beltsville, MD 20705, USA; feng.gao@ars.usda.gov (F.G.); kyle.knipper@ars.usda.gov (K.K.); bill.kustas@ars.usda.gov (W.K.); yun.yang@ars.usda.gov (Y.Y.)
[2] Space Sciences and Engineering Center, University of Wisconsin-Madison, Madison, WI 53706, USA; george.diak@ssec.wisc.edu
[3] NASA Marshall Space Flight Center, Huntsville, AL 35805, USA; christopher.hain@nasa.gov
[4] Department of Environmental Science, Policy and Management, University of California, Berkeley, CA 94720, USA; eeichelm@berkeley.edu (E.E.); khemes@berkeley.edu (K.S.H.); baldocchi@berkeley.edu (D.B.)
* Correspondence: martha.anderson@ars.usda.gov; Tel.: +1-301-504-6616

Received: 19 November 2018; Accepted: 4 January 2019; Published: 22 January 2019

Abstract: The energy delivered to the land surface via insolation is a primary driver of evapotranspiration (ET)—the exchange of water vapor between the land and atmosphere. Spatially distributed ET products are in great demand in the water resource management community for real-time operations and sustainable water use planning. The accuracy and deliverability of these products are determined in part by the characteristics and quality of the insolation data sources used as input to the ET models. This paper investigates the practical utility of three different insolation datasets within the context of a satellite-based remote sensing framework for mapping ET at high spatiotemporal resolution, in an application over the Sacramento–San Joaquin Delta region in California. The datasets tested included one reanalysis product: The Climate System Forecast Reanalysis (CFSR) at 0.25° spatial resolution, and two remote sensing insolation products generated with geostationary satellite imagery: a product for the continental United States at 0.2°, developed by the University of Wisconsin Space Sciences and Engineering Center (SSEC) and a coarser resolution (1°) global Clouds and the Earth's Radiant Energy System (CERES) product. The three insolation data sources were compared to pyranometer data collected at flux towers within the Delta region to establish relative accuracy. The satellite products significantly outperformed CFSR, with root-mean square errors (RMSE) of 2.7, 1.5, and 1.4 $MJ \cdot m^{-2} \cdot d^{-1}$ for CFSR, CERES, and SSEC, respectively, at daily timesteps. The satellite-based products provided more accurate estimates of cloud occurrence and radiation transmission, while the reanalysis tended to underestimate solar radiation under cloudy-sky conditions. However, this difference in insolation performance did not translate into comparable improvement in the ET retrieval accuracy, where the RMSE in daily ET was 0.98 and 0.94 mm d^{-1} using the CFSR and SSEC insolation data sources, respectively, for all the flux sites combined. The lack of a notable impact on the aggregate ET performance may be due in part to the predominantly clear-sky conditions prevalent in central California, under which the reanalysis and satellite-based insolation data sources have comparable accuracy. While satellite-based insolation data could improve ET retrieval in more humid regions with greater cloud-cover frequency, over the California Delta and climatologically similar regions in the western U.S., the CFSR data may suffice for real-time ET modeling efforts.

Keywords: evapotranspiration; insolation; surface energy balance; data fusion; water resource management

1. Introduction

Evapotranspiration (ET) describes the rate of exchange of water vapor between the land surface and the atmosphere, including water directly evaporated from the soil, open water, and other surfaces (E), as well as water consumed and transpired by plants in the process of biophysical development (T). The ability to accurately estimate ET spatially and temporally over large areas is critical to a broad range of applications, for example, in managing water resources, in assessing agricultural water use, and in monitoring ecosystem health and the impacts of drought [1]. Demands for real-time access to daily ET information at spatial scales from field to globe are ever increasing in support of food and water security applications [2]. To address these data needs will require timely and accurate methods for mapping ET based on modeling and/or remote sensing, and accurate estimates of forcing variables to serve as the model input.

The primary factor governing ET in many cases is the solar radiation load, which largely determines the energy available at the land surface to evaporate water [3,4]. Solar energy drives the surface energy balance: the partitioning of net (incoming minus outgoing) longwave and shortwave radiation primarily between sensible heat, latent heat (equivalent to ET, but in energy units), and soil heat conduction flux. Without an accurate estimate of the incoming shortwave radiation, the ability of any ET model to predict evaporative fluxes is severely limited. In addition to insolation, ET is modulated by soil moisture availability; vegetation amount, structure, and health; and meteorological conditions such as air temperature, vapor pressure deficit, and wind speed. An ET model with broad utility will consider each of these factors.

Given the critical importance of the solar radiation input, it is informative to assess the impact of insolation data sources on the accuracy and utility of operational ET methods. In this study, we investigate this in the context of a multi-scale surface energy balance algorithm based on remote sensing measurements of land-surface temperature (LST). The Atmosphere-Land Exchange Inverse (ALEXI) model and associated flux disaggregation algorithm (DisALEXI) use LST data derived from satellite-based thermal infrared (TIR) imaging systems to map ET and other surface energy fluxes at resolutions of 30 m at a landscape scale to 5 km at continental to global scales. ALEXI/DisALEXI datasets are being used for applications in drought monitoring, irrigation management, yield prediction, and in investigating changes in water-use accompanying landcover and land use change [5–9]. The choice of insolation product used in these applications will have ramifications for both model accuracy and operational feasibility.

In this study, we investigated the impact of different sources of solar global irradiance (insolation) data in an application of ALEXI/DisALEXI over central California, focusing on the Sacramento–San Joaquin River Delta region. This area was the focus of an ET model intercomparison study assessing the utility of remotely sensed consumptive use estimates in informing the response to California's Sustainable Groundwater Management Act (SGMA) [10]. ALEXI/DisALEXI is also being used to develop irrigation management strategies for viticulture as part of the Grape Remote sensing and Atmospheric Profile and Evapotranspiration eXperiment (GRAPEX [11]). Both applications have specific requirements regarding data latency, spatial and temporal resolution, period of record, accuracy, and reliability of data delivery. Continental to global scale applications will have additional requirements regarding the data domain coverage.

Here, we build on the baseline ET results presented by Anderson et al. [12] driven by modeled insolation from the Climate Forecast System Reanalysis (CFSR) global dataset generated at hourly timesteps and at a 0.25° spatial resolution. We compare the CFSR insolation to hourly satellite-based insolation datasets developed over the continental United States (CONUS) at 0.2°, using data from the Geostationary Operational Environmental Satellites (GOES) and the coarser resolution (1°) global Clouds and the Earth's Radiant Energy System (CERES) satellite-based insolation product. The impact of the insolation data source on ET retrievals at daily to yearly timescales is also assessed, with the purpose of identifying the optimal inputs for different water resource applications.

2. Materials and Methods

2.1. Study Domain

The study domain is shown in Figure 1 and is consistent with that used in the study presented in Anderson et al. [12]. This region includes the California Delta region at the confluence of the Sacramento and San Joaquin Rivers, which serves as a major hub for water supply within the state of California. Medellín-Azuara et al. [10] describe an ET model intercomparison study conducted over the CA Delta, commissioned by the Delta Water Master. Timeseries of daily ET data were produced with an ensemble of ET models for the water years 2015–2016 and intercompared to assess the level of agreement between the models, as well as the variability in model behavior for different landcover types. ALEXI/DisALEXI timeseries generated using CFSR insolation inputs for the ET model intercomparison study were further assessed by Anderson et al. [12] in comparison with observations collected at multiple flux towers within the modeling domain, sampling various representative landcovers, including vineyards, alfalfa, rice, and wetlands. The general locations of these towers are indicated in Figure 1. These include towers deployed on Sherman and Twitchell Islands to monitor water and carbon fluxes from a chronosequence of restored wetlands and converted rice fields in comparison with dryland agriculture [13]. Additionally, towers associated with the GRAPEX [11] study site in the Borden Ranch viticultural area outside of Lodi, CA were used in the assessment.

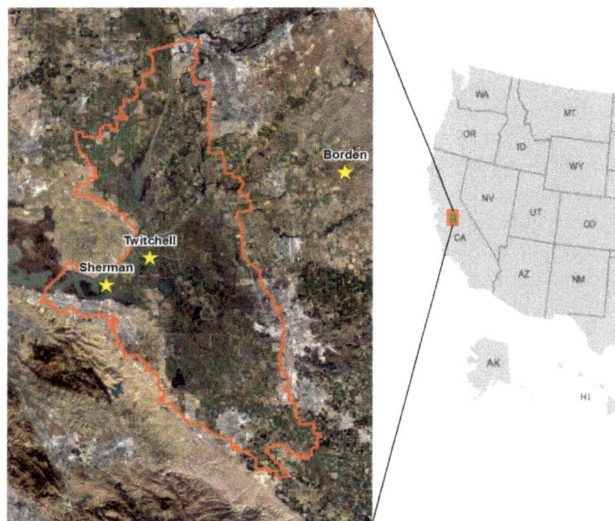

Figure 1. Study domain covering the Sacramento–San Joaquin Delta in California's Central Valley. The Legal Delta Area is delineated in red, with flux tower sites on Twitchell and Sherman Islands and in vineyards in the Borden Ranch area indicated with yellow stars.

The study period covered the water years of 2015 and 2016 (WY15-16), with a water year running from October 1 to September 30. This period came near the end of the extended severe drought (2012–2017) in California, which precipitated conversions in cropping and irrigation practices within the Central Valley. In this paper, the results of Anderson et al. [12] were revisited to assess the dependence of ET retrieval accuracy on the accuracy of the insolation data used to drive the energy balance model.

2.2. ET Remote Sensing Framework

2.2.1. ALEXI/DisALEXI

Surface energy fluxes were computed over the study domain using the Atmosphere-Land Exchange Inverse (ALEXI) surface energy balance model and the associated flux disaggregation technique, DisALEXI. This modeling system is described in Anderson et al. [12] and prior studies (e.g., [14–16]) and is only briefly summarized here. Land-surface exchanges in ALEXI/DisALEXI are governed by the Two-Source Energy Balance (TSEB) parameterization described originally by Norman et al. [17] with improvements in Kustas et al. [18,19]. In the TSEB, the soil and canopy energy budgets (designated with subscripts 's' and 'c', respectively) are solved separately:

$$RN_C = H_C + \lambda E_C$$
$$RN_S = H_S + \lambda E_S + G \qquad (1)$$
$$RN = H + \lambda E + G$$

where RN is net radiation, H is sensible heat, λE is latent heat, and G is the flux of heat into the substrate below the canopy. Sensible heat flux from the vegetation canopy (H_C) and substrate (H_S-typically soil) combine in series to form the net flux H, as constrained by the component temperature estimates T_C and T_S and the above-canopy air temperature T_A. The component temperatures are extracted via the system of model equations from radiances inferred by the bulk directional surface radiometric temperature:

$$T_{RAD}(\theta)^4 = f(\theta)T_C{}^4 + [1 - f(\theta)]T_S{}^4 \qquad (2)$$

where $f(\theta)$ and $T_{RAD}(\theta)$ are the directional vegetation cover fraction and radiometric temperature at the view angle of the thermal sensor, θ. Net radiation is computed using an analytical canopy light interception formulation described in Campbell and Norman [20] and Anderson et al. [21]. The treatment of solar irradiance follows the methods described in Weiss and Norman [22] for partitioning observations of incoming shortwave radiation into visible and near infrared direct beam and diffuse components.

For spatially distributed implementation of the TSEB over a landscape, a time-differential approach can be used to replace boundary conditions in near-surface air temperature T_A, which are difficult to prescribe with adequate accuracy and minimal bias with respect to the remotely sensed T_{RAD} data. The regional ALEXI modeling framework applies the TSEB at two times during the morning hours (approximately an hour after sunrise and an hour before local noon), with energy closure over this interval provided by a simple atmospheric boundary layer (ABL) growth model [23]. As early as 1993, Diak and Whipple [24] demonstrated the strong linkage between ABL growth and land-surface heating. In that study, an ABL model was used to diagnose values of sensible heating, but only at rawinsonde locations. ALEXI works well for the same reason, based on its ability to diagnose the time-change of the boundary layer height and its relation to sensible heat. However, the parameterization and the use of forecast model initial conditions allows ALEXI to be run at any location where time-changes of land-surface temperature, solar energy values, and initial atmospheric conditions are available, not just at rawinsonde locations as in Diak and Whipple [24].

Model surface temperature inputs at the two times are typically acquired via geostationary satellites, and sensitivity tests show that the model is sensitive to the morning change in T_{RAD}, but is minimally sensitive to time-invariant absolute errors in T_{RAD} retrieval [23]. In this approach, T_A is diagnosed at the boundary between the surface and ABL models rather than being prescribed as a model input. Daily (24-h) integrated latent heat flux is computed by scaling the λE derived at the second T_{RAD} observation time (pre-noon) using the local solar radiation curve (see Section 2.2.2). The daily ET (*ETd*; mm d^{-1}) can be obtained from the daily latent heat flux (λEd; MJ·m^{-2}·d^{-1}) using the latent heat of vaporization (λ, approximately 2.45 MJ·m^{-2} to evaporate 1 mm of water at 20 °C).

The spatial resolution of ALEXI flux estimates is constrained by the resolution of the high temporal frequency LST observation source (here, geostationary satellites)—generally on the order of several km. For generating higher resolution ET maps, capable of resolving individual crop fields or land management units, an ALEXI flux disaggregation approach (DisALEXI) was developed as described in References [25,26]. In the DisALEXI step, the TSEB is executed over a gridded model domain using higher resolution LST retrievals from Landsat (30 m, spatially sharpened as described in Section 2.3.1) or MODIS (1 km native or 500 m sharpened) with an initial T_A boundary derived from meteorological analyses. This nominal boundary condition is then iteratively adjusted at the ALEXI pixel scale until the DisALEXI *ETd* field reaggregates to the ALEXI baseline flux [14]. At the final step in the iteration, the modified T_A field is spatially smoothed to remove boxy ALEXI-scale artifacts in the output flux maps.

2.2.2. Upscaling to Daily ET Using Insolation

In the TIR-based version of ALEXI and DisALEXI, direct retrievals of ET are obtainable only on days when the skies are clear at the acquisition times of the satellite images used (for ALEXI, the full time period between morning acquisitions is required to be clear). In the current approach, the instantaneous ET at the overpass time was upscaled to a 24-h flux by conserving the ratio of ET to insolation flux, following recommendations by Cammalleri et al. [27]. That study compared several potential flux metrics for upscaling, including ratios with available energy, net radiation, and reference ET, and found that the best performance over a range of flux sites was obtained using insolation. Ryu et al. [28] also found insolation to be a robust scaling factor using data from the global flux tower datasets sampling multiple biomes. Here, the ratio (f_{SUN}) of instantaneous ET (*ETi*) to insolation (*Rsi*) flux was computed at the satellite overpass time:

$$f_{SUN} = ETi/Rsi \tag{3}$$

and then the daily ET flux (*ETd*) was estimated from the 24-h integrated insolation (*Rsd*) for that day:

$$ETd = f_{SUN} \times Rsd \tag{4}$$

In ALEXI, which is typically run using surface temperature data acquired by geostationary satellites with high temporal frequency (15-min), a Savitsky–Golay filter is first applied to the f_{SUN} timeseries at the pixel level to minimize day-to-day variability due to undetected cloud contamination, and then the smoothed timeseries is gap-filled using linear or spline interpolation. ALEXI *ETd* is then reconstructed using Equation (4) applied to the daily insolation maps. An all-sky retrieval system is under development using LST information derived from microwave Ka-band data (several km resolution), which can see through clouds, circumventing the need for gap filling at the ALEXI scale [29]. To obtain optimal correspondence with the ALEXI timeseries, MODIS retrievals (near daily overpass), a similar temporal smoothing and gap-filling procedure is applied to the ratio of MODIS to ALEXI *ETd*.

2.2.3. Data Fusion

Landsat overpasses are generally too infrequent (8–16 days or longer, depending on cloud cover and the number of concurrently operating Landsat platforms) to justify the kind of temporal gap-filling described in Section 2.2.2. Instead, a data fusion methodology has been employed to fuse Landsat (high spatial/low temporal resolution) and MODIS (low spatial/high temporal resolution) ET timeseries into a single *ETd* "datacube", with daily timesteps and 30-m pixels. The Spatial and Temporal Adaptive Reflectance Fusion Model (STARFM [30]) compares MODIS and Landsat image pairs on days when both are available, computes spatial weighting statistics, and then applies these weights to downscale MODIS images between clear-sky Landsat overpasses. STARFM was originally developed to fuse surface reflectance imagery, but it has also been applied successfully to ET datasets developed over a variety of agricultural and forested landscapes in the U.S. [6,7,15,16,31,32] and in Spain [33].

In this data fusion strategy, the direct Landsat ET retrievals on clear Landsat overpass dates serve as tie points in the temporal reconstruction. Time behavior between these clear-date retrievals is governed by changes in the MODIS ET at a coarser scale. Therefore, the accuracy of the reconstructed daily datacube is largely sensitive to the magnitude of the solar radiation estimates on these clear Landsat days, and secondarily to variations in the daily insolation between overpasses.

2.3. Insolation Datasets

Model inputs to the ET data fusion system are described in detail by Anderson et al. [12]. Here, we focus on examining three insolation datasets that could be used as input to ALEXI/DisALEXI and other ET mapping approaches, representing major classes of reanalysis and remotely sensed solar radiation products available today.

2.3.1. CFSR

Meteorological inputs of air temperature (used as the initial T_A boundary condition in the DisALEXI iteration), vapor pressure, atmospheric pressure, and wind speed were obtained from the near-real-time CFSR at a 0.25° resolution [34,35] and at 3-h intervals with global coverage, maintained by the National Centers for Environmental Prediction (NCEP) (College Park, MD, USA). CFSR also generates hourly gridded insolation data at a 0.25° resolution, which were used in the ET timeseries retrievals for the CA Delta described by Anderson et al. [12]. In CFSR, shortwave radiation transfer through the atmosphere is modeled using methods described by Chou et al. [36] and Hou et al. [37]. Near-real-time CFSR data are available from the NCEP with a latency of around six hours (http://nomads.ncep.noaa.gov/pub/data/nccf/com/cfs/prod/cfs). Retrospective CFSR data are available from NOAA's National Centers for Environmental Information (NCEI) from 1979 to present (https://www.ncdc.noaa.gov/data-access/model-data/model-datasets/climate-forecast-system-version2-cfsv2).

2.3.2. SSEC

To contrast with the coarse-scale CFSR modeled insolation, we also tested an hourly satellite-based insolation product generated at a 0.2° resolution by the University of Wisconsin Space Science and Engineering Center (SSEC) (Madison, WI, USA) using visible data from the GOES-East and GOES-West platforms [38]. The insolation model runs in the computational environment of the Man-computer Interactive Data Access System (McIDAS [39]), developed and distributed by the SSEC. Near-real-time insolation fields from this system have been used for regional- and continental-scale land surface carbon and water flux assessments [5,26,40–42], solar energy spatial and temporal variability analyses [43], subsurface hydrologic modeling efforts, and currently for agricultural forecasting products from the University of Wisconsin-Madison Extension (Madison, WI, USA) [44].

The radiative transfer (RT) models used to estimate insolation from the GOES data are simple, physically-based, and depend upon calibration (digital counts to energy) of the GOES-visible sensors (Figure 2). The basic equation set has been presented in Diak et al. [45], but was modified from the absorption coefficient form to the calculation of transmittances for the various atmospheric processes similar to Reference [46]. With knowledge of the solar constant, date, and time, the model calculates how much solar energy is entering the earth-atmosphere system for any latitude–longitude location. The role of the satellite measurement is to estimate how much of this incident energy escapes back to space. Subsequently, the atmospheric RT model, based on principles of energy conservation, is used to partition the energy remaining in the earth-atmosphere system into its various components, importantly surface insolation. Separate models for the clear and cloudy atmosphere are employed. A detailed description of the insolation model is provided in Diak [38].

GOES Insolation Model

$B \leq B_0 + \delta$
Use Clear-Sky Model

$B > B_0 + \delta$
Use Cloudy-Sky Model

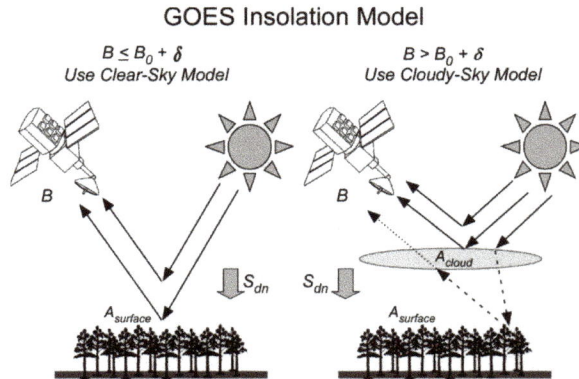

Figure 2. Graphical depiction of the physical model employed for (left-hand side) clear-sky conditions and (right-hand side) cloudy-sky conditions; B refers to the brightness observed by the satellite, while B_0 is the clear-air brightness threshold; Sdn refers to the downward shortwave radiation flux; and $A_{surface}$ and A_{cloud} refer to the surface and cloud albedos, respectively.

Despite its relative simplicity, recent studies have found the quality of results from this simple model to be similar to that from more complex formulations. Paech et al. [47] used the simple model as a basis for evaluating evapotranspiration in Florida and found the model insolation statistics to be similar to results from the NOAA satellite-based insolation model. The insolation results presented in this study were also similar in quality to the results for detailed for the NOAA model by Wonsick et al. [48]. In Diak [38], seasonal results of this insolation model were evaluated against in-situ pyranometer measurements at 45 sites of the United States Climate Research Network (USCRN) (Washington, DC, USA). Accuracy for hourly insolation estimates was 15% to 20%, while the accuracy range for the daily integrated insolation total ranged from 5% to 10%. Calibration of the visible channel has been an issue for the GOES satellites; their sensitivity degrades significantly with time and up until GOES-16 (brought on-line on 1 January 2018) there was no on-board calibration—it was done vicariously. The insolation model for the period described here used calibration coefficients provided by NOAA, and a Real Earth™ display of the prior day's insolation can be viewed at https://www.ssec.wisc.edu/insolation/.

Half-hourly and daily-integrated insolation estimates for the previous day in ASCII text format are available on the SSEC anonymous ftp site (ftp://prodserv1.ssec.wisc.edu) in the subdirectory insolation_high_res for no charge. The half-hourly data from the GOES-East and -West were downloaded and consolidated onto a single 0.2° grid covering the CONUS. These datasets have periodic gaps in coverage due to satellite outages or issues with model development, which were filled using CFSR to generate fully filled data grids.

2.3.3. CERES

For comparison with the CONUS GOES-based SSEC product, we also tested a global 1° insolation product developed by NASA's Langley Research Center using data from the CERES satellite instruments. The CERES Synoptic Radiative Fluxes and Clouds (SYN) product at a spatial resolution of 1° uses 3-h geostationary and MODIS radiances, cloud properties, MODIS aerosol observations, and atmospheric profiles provided by the NASA Global Modeling and Assimilation Office (GMAO) model to more accurately model the diurnal variability between CERES observations [49,50]. One disadvantage with the CERES SYN product is that it is not available in near-real-time and it usually has a latency in availability on the order of 3 to 6 months. Retrospective products are available at https://ceres.larc.nasa.gov/products.php?product=SYN1deg.

2.4. Flux Datasets

Micrometeorological data from eight eddy covariance (EC) systems operating within the study domain were used by Anderson et al. [12] to evaluate the fused CFSR-based WY15-16 ET timeseries over different landcover types (Table 1). Two flux towers were located in the Borden Ranch viticultural area north of Lodi, CA in adjacent vineyards with Pinot noir vines of age ~12 years (Lodi1) and 9 years (Lodi2) during the experiment timeframe. These are operated as part of the GRAPEX experiment. Also operating during the study period were four AmeriFlux EC towers installed on Twitchell Island, at the core of the Delta region. These towers include two sites in water-intensive crops: rice (US-Twt; doi:10.17190/AMF/1246140) and alfalfa (US-Tw3; doi:10.17190/AMF/1246149). Two additional towers are sited in restored wetlands: the East End wetland (US-Tw4; doi:10.17190/AMF/1246151), restored in 2014, is largely vegetated with some open water channels, while the West Pond site (US-Tw1; doi:10.17190/AMF/1246147) dates back to 1997, and the vegetation is now fully closed. Two AmeriFlux tower sites on nearby Sherman Island were also used, sampling the mature Mayberry wetland (US-Myb; doi:10.17190/AMF/1246139) established in 2010, and the sparsely vegetated US-Sne wetland (doi:10.17190/AMF/1418684), converted from pasture and newly flooded in December of 2016. Details regarding the Twitchell and Sherman Island measurements are provided by Eichelmann [13].

Table 1. Flux towers used in the analysis.

Site	Tower	Name	Cover	Latitude	Longitude
Borden Ranch	Lodi1		vineyard	38.2894	−121.1178
Borden Ranch	Lodi2		vineyard	38.2805	−121.1176
Twitchell Island	US-Tw1	West Pond	old wetland	38.1073	−121.6468
Twitchell Island	US-Tw4	East End	young wetland	38.1028	−121.6413
Twitchell Island	US-Twt		rice	38.1087	−121.6530
Twitchell Island	US-Tw3		alfalfa	38.1151	−121.6468
Sherman Island	US-Myb	Mayberry	intermediate wetland	38.0498	−121.7650
Sherman Island	US-Sne	Sherman	new wetland	38.0369	−121.7547

All the flux sites, except US-Myb, were instrumented with pyranometers to measure the incoming shortwave radiation (solar radiation) at 30-min intervals. For the Lodi towers, insolation measurements were obtained with the incoming shortwave component of a Kipp & Zonen CNR-1 four-component net radiometer. The two towers were located less than 1 km apart and daily totals agreed to within 2%, with an R^2 of 0.99. Twitchell and Sherman Island shortwave radiation was measured using the pyranometer component of Hukseflux NR01 four-way net radiometers, except at the Twitchell rice site, where a Kipp & Zonen CM11 pyranometer was employed. In a detailed comparison experiment conducted in Fall 2016, the radiation sensors for all Twitchell and Sherman Island sites were deployed at the same location next to the Twitchell rice measurement tower for three weeks. All instruments measuring incoming shortwave radiation, regardless of the time deployed in the field or make and model, showed very close agreement with measurements within 5% of each other. When compared to a recently factory calibrated 'golden standard', all sensors had R^2 values above 0.986 with regression slopes between 0.972 and 1.002.

The EC technique is known to yield turbulent flux estimates that do not fully close the observed energy budget, yielding closure errors given by $1 - (H + \lambda E)/(RN - G)$ typically on the order of 10–20% or higher in some cases [51,52]. For comparison with model estimates, which assume closure, observed latent heat fluxes over non-wetland surfaces have been closed using the residual of the energy balance. For the wetland sites, closure has not been attempted due to uncertainties in measuring the heat storage, G, within the water substrate given the dynamic bathymetry and fluctuating water tables. For those sites, observed λE was increased by a nominal 10% to account for flow distortion effects associated with non-orthogonal sonic anemometer configurations in the EC systems used here as in References [53–56].

3. Results

3.1. Insolation Product Evaluation at the Flux Tower Sites

Representative maps of daily insolation over CONUS for day of year (DOY) 200 (19 July) in 2017 from the CFSR, CERES, and SSEC insolation datasets are shown in Figure 3. While general cloud features are consistent between the datasets, the maps differ significantly in terms of spatial resolution and fidelity of structure. CFSR clearly reflect a more artificial, model-based treatment of cloud processes, while the satellite inputs to SSEC and CERES generate more realistic structures albeit at very different spatial resolutions.

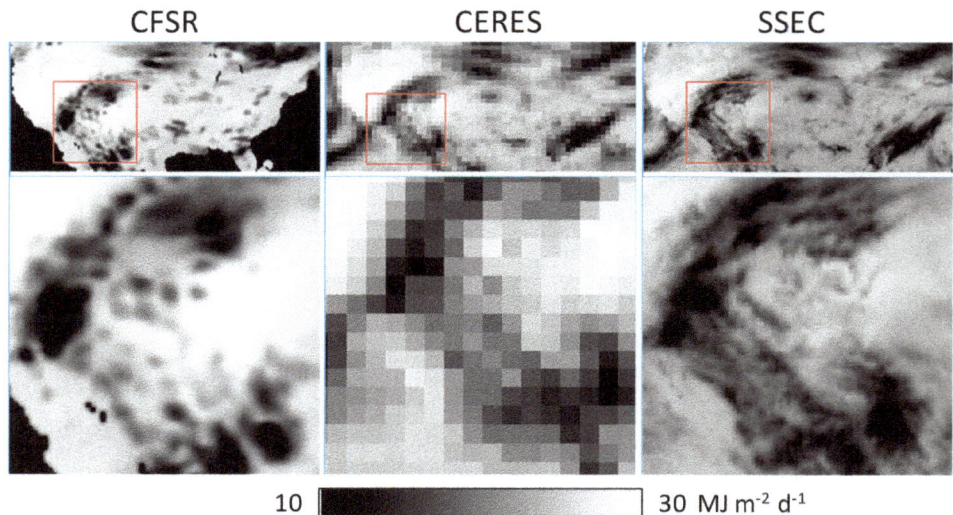

Figure 3. Comparison of daily insolation maps for 2017 DOY 200 from the CFSR, CERES, and SSEC products. A zoom in over the southwestern U.S. is included to highlight differences in structure and resolution between the products.

Table 2 provides statistical metrics comparing insolation estimates from the three products with fluxes observed at the flux sites equipped with pyranometers, as well as for all the sites combined. These metrics included the mean observed flux (<O>), mean bias error (MBE) in model minus observations, root mean square error (RMSE), Nash–Sutcliff coefficient of efficiency (NSE [57]), coefficient of determination (R^2), mean absolute error (MAE), and the relative error (RE = MAE/<O>). Graphical representations of MAE, RE, and MBE at daily to annual timesteps are shown in Figure 4, while Figure 5 shows scatter plots versus observations at daily timesteps for each site collecting pyranometer data.

At daily timesteps, the SSEC and CERES satellite-based products had significantly lower errors in comparison with the CFSR, with the RE reducing from approximately 0.09 to 0.05. However, the MBE was lower with CFSR at daily to annual timesteps. This may have resulted from better calibration of the clear-sky upper envelope in the CFSR over this part of the CONUS in comparison with the satellite products, which must additionally account for annual degradation in the imaging systems. Of the two satellite products, SSEC generally had a lower RE but marginally higher bias over this timeframe in comparison with CERES. (Note that some of the tower sites did not have a full annual record of insolation measurements. These missing site-years resulted in a sign flip in the MBE in CFSR insolation at the yearly timestep.)

Table 2. Statistical metrics of insolation dataset performance at flux sites on clear Landsat overpass dates (LS day) and at daily, weekly, monthly, and yearly timesteps, reported in units of $MJ \cdot m^{-2} \cdot d^{-1}$.

Timescale	Tower	N	<O>	CFSR								CERES							SSEC				
				MBE	RMSE	NSE	R2	MAE	RE	MBE	RMSE	NSE	R2	MAE	RE	MBE	RMSE	NSE	R2	MAD	RE		
DAILY	Lodi1	761	19.05	0.34	2.92	0.87	0.89	1.84	0.097	0.37	1.86	0.95	0.95	1.13	0.059	0.16	1.33	0.97	0.97	0.76	0.040		
	Lodi2	740	19.54	0.09	2.99	0.86	0.88	1.86	0.095	-0.10	1.35	0.97	0.97	0.79	0.040	-0.10	1.35	0.97	0.97	0.79	0.040		
	US-Sne	162	24.72	-0.08	1.78	0.94	0.95	1.07	0.043	-1.18	1.60	0.95	0.99	1.37	0.056	-1.18	1.60	0.95	0.99	1.37	0.056		
	US-Tw1	687	19.49	0.40	2.57	0.91	0.92	1.68	0.086	0.03	1.41	0.97	0.98	1.13	0.058	0.03	1.41	0.97	0.98	1.13	0.058		
	US-Tw3	797	18.99	-0.21	2.68	0.91	0.92	1.60	0.084	-0.46	1.48	0.97	0.98	1.07	0.056	-0.46	1.48	0.97	0.98	1.07	0.056		
	US-Tw4	762	20.24	-0.53	2.74	0.90	0.91	1.63	0.080	-0.82	1.63	0.96	0.98	1.23	0.061	-0.82	1.63	0.96	0.98	1.23	0.061		
	US-Twt	792	19.26	0.12	2.66	0.90	0.91	1.65	0.086	-0.18	1.29	0.98	0.98	0.83	0.043	-0.18	1.29	0.98	0.98	0.83	0.043		
LS DAY	ALL	365	23.39	0.38	1.98	0.90	0.91	1.17	0.050	0.04	1.40	0.95	0.95	0.90	0.038	-0.31	1.15	0.97	0.97	0.90	0.038		
DAILY	ALL	4701	19.61	0.02	2.74	0.90	0.91	1.69	0.086	-0.23	1.52	0.97	0.97	1.04	0.053	-0.27	1.43	0.97	0.98	0.98	0.050		
WEEKLY	ALL	661	19.76	0.03	1.41	0.97	0.97	1.04	0.052	-0.26	1.08	0.98	0.99	0.82	0.042	-0.29	1.03	0.98	0.99	0.78	0.039		
MONTHLY	ALL	151	20.03	0.08	0.94	0.99	0.99	0.77	0.038	-0.27	0.97	0.98	0.99	0.75	0.038	-0.31	0.95	0.98	0.99	0.72	0.036		
YEARLY	ALL	11	19.26	-0.62	0.89	-0.58	0.20	0.73	0.038	-0.77	1.00	-1.01	0.16	0.83	0.043	-0.81	1.05	-1.18	0.13	0.87	0.045		

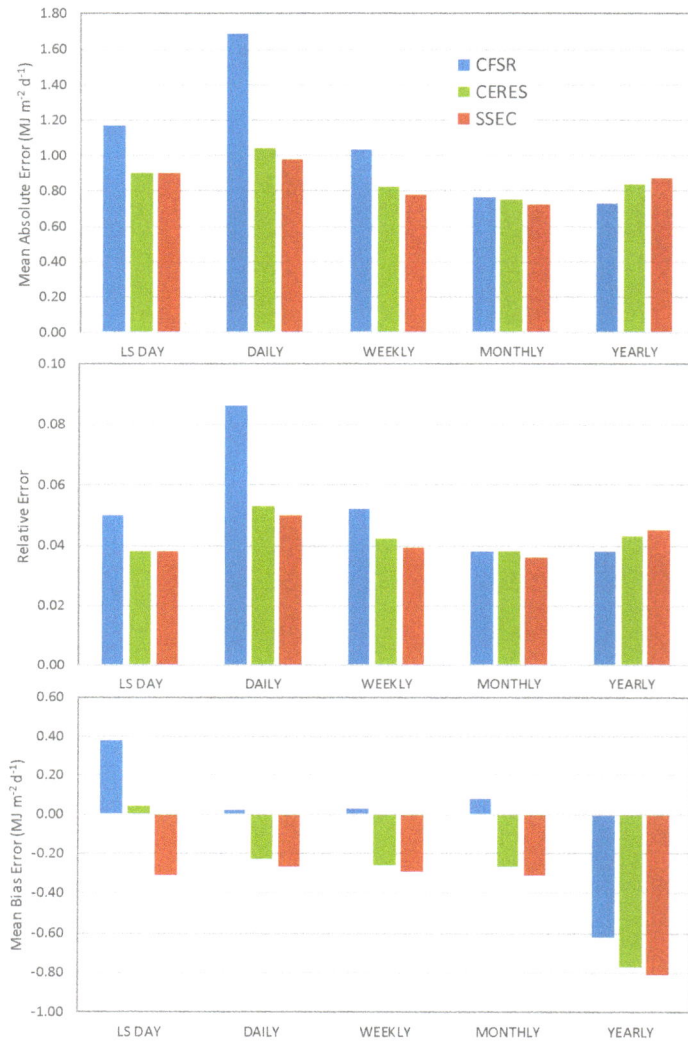

Figure 4. Mean absolute error (MAE), relative error (RE), and mean bias error (MBE) in insolation products on clear Landsat overpass dates (LS DAY) and at daily, weekly, monthly, and yearly timesteps over the study period.

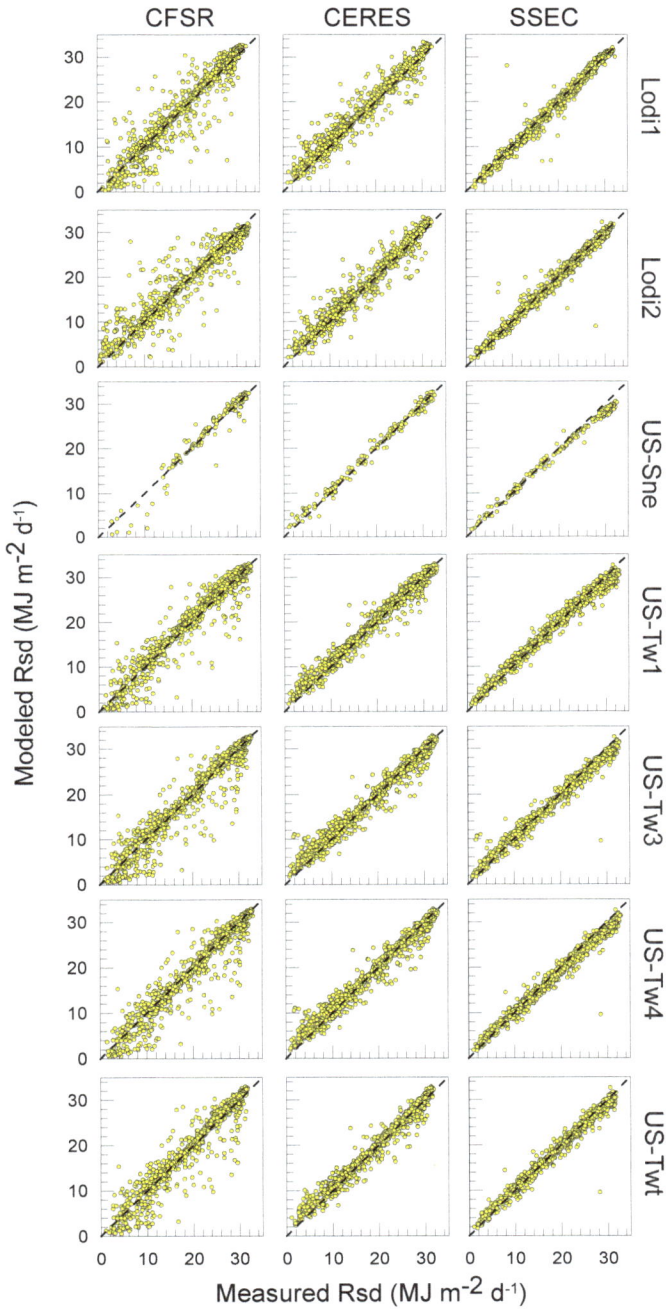

Figure 5. Comparison of the daily insolation fluxes from the CFSR, SSEC, and CERES datasets with pyranometer observations at the flux sites within the study region.

Timeseries comparisons at two sites showing significant improvement in the satellite versus modeled insolation (Lodi2 and US-Twt) further illuminated the error characteristics of the three insolation datasets (Figure 6). All three captured the predominantly cloudless conditions past mid-2016, a virtually rain-free period which served to exacerbate drought conditions in the Central Valley. CFSR tended to overpredict cloud effects (also see scatter plots in Figure 5), while SSEC best captured the strength and timing of cloud impacts on incident solar radiation at the scale of observation.

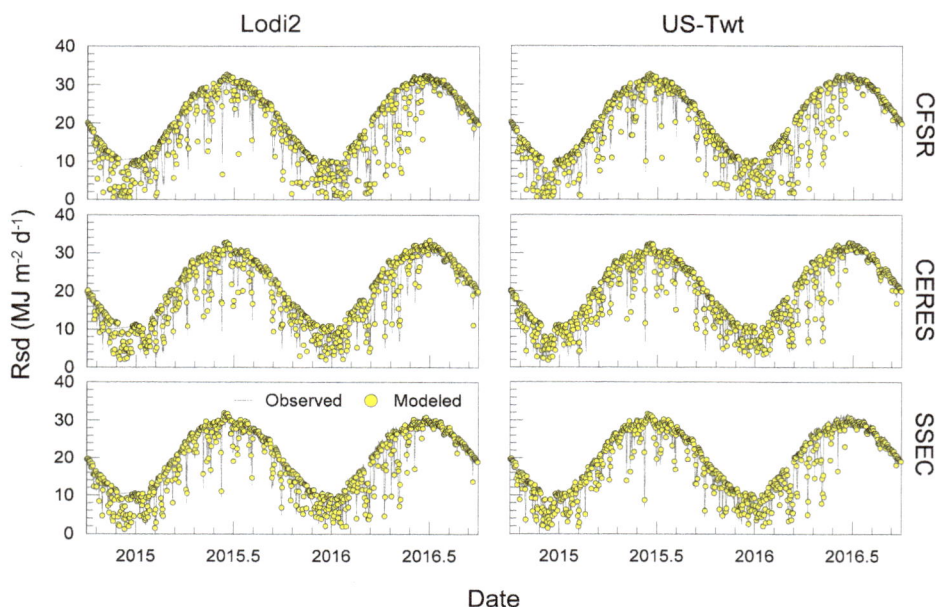

Figure 6. Daily modeled and observed insolation timeseries at the Lodi2 and US-Twt flux sites.

3.2. Energy Balance on Landsat Dates

In comparison with the daily results, insolation product performance, in terms of the MAE and RE, was more similar between the CFSR and the satellite insolation datasets on the Landsat overpass days when conditions were clear at the time of overpass and a direct retrieval of ET could be obtained (Figure 4 and Table 2). CFSR tended to overestimate daily insolation on these days, while SSEC underestimated by a similar amount on average over the two water years. CERES had minimal bias on clear Landsat overpass days. As noted in Section 2.2.3, these days are critical to the full daily reconstruction as these direct Landsat ET retrievals define key tie points in the data fusion process.

Scatter plots of modeled versus measured insolation from the CFSR and SSEC datasets on clear Landsat overpass dates are shown in Figure 7, along with the derived net radiation and partitioning between sensible, latent, and soil heating diagnosed with DisALEXI. These two insolation products were selected as a demonstration to bracket the range of expected DisALEXI performance. Agreement with observations for all flux components was marginally improved using the SSEC insolation product, primarily due to the reduction of outliers. While MAE in insolation was reduced from 1.2 to 0.9 $MJ \cdot m^{-2} \cdot d^{-1}$ moving from CFSR to SSEC, respectively, the impact on net radiation was less apparent, with a reduction from 1.6 to 1.5 $MJ \cdot m^{-2} \cdot d^{-1}$. Latent heat (and ET, in units of mass) performance on Landsat dates was similar for each of the insolation datasets, with a RE of 0.19 in both cases and a MAE of 1.8 $MJ \cdot m^{-2} \cdot d^{-1}$ (0.75 $mm \cdot d^{-1}$). The lack of improvement in the ET retrievals on Landsat dates using SSEC insolation was consistent with the reasonable capability of the model reanalysis to capture insolation fluxes on predominantly clear days.

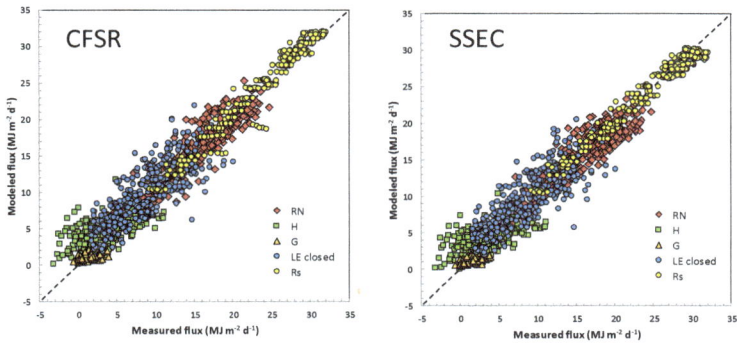

Figure 7. Energy flux partitioning generated with DisALEXI on clear Landsat dates using CFSR and SSEC insolation data as input compared to the tower flux observations.

3.3. Daily ET from Data Fusion

Statistics describing the performance of the full daily ET timeseries generated via the Landsat-MODIS data fusion process are provided in Table 3 at daily timesteps and aggregated to weekly, monthly, and annual averages, and visually summarized in Figure 8. These metrics elucidate the impact of the insolation data source on the accuracy of daily ET retrievals from the data fusion system. While there was some improvement in ET at the daily timestep using the SSEC insolation, with the MAE decreasing from 0.72 with CFSR to 0.69 mm d^{-1} (RE of 0.23 to 0.22), the impact was not large—particularly given the large improvement in accuracy in the daily insolation drivers (RE of 0.09 to 0.05). As demonstrated in Figure 9, showing scatter plot comparisons of the measured and modeled ET for each tower site, the effect of the satellite insolation was primarily to reduce the magnitude of outliers, and the degree of improvement varied from site to site. The value of the satellite-based dataset diminished for applications at weekly and longer timescales, where the performance of the two insolation datasets was more similar (Figure 8).

The largest improvements in ET reconstruction using the SSEC insolation occurred at the Lodi2 (vineyard) site where the MAE decreased from 0.52 to 0.43 mm d^{-1}, and at US-Twt (rice) with a MAE reduction from 0.91 to 0.83 mm d^{-1} (Table 3). Assessment of the ET timeseries at US-Twt demonstrated the nature of the improvement in performance—primarily a more realistic portrayal of day-to-day ET variability with the SSEC-based retrievals, particularly on cloudy days (Figure 10). Given the quality of the SSEC product, residual ET errors must be due to errors in other inputs or in the modeling assumptions.

Table 3. Statistical metrics of ET retrieval performance (mm d^{-1}) at flux sites on clear Landsat overpass dates (LS Day) and at daily, weekly, monthly, and yearly timesteps using insolation inputs from the CFSR and SSEC datasets.

Timescale	Tower	N	<O>	CFSR MBE	CFSR RMSE	CFSR NSE	CFSR R2	CFSR MAE	CFSR RE	SSEC MBE	SSEC RMSE	SSEC NSE	SSEC R2	SSEC MAE	SSEC RE
DAILY	Lodi1	450	3.44	0.11	0.80	0.76	0.77	0.63	0.184	0.10	0.76	0.78	0.79	0.60	0.175
	Lodi2	438	3.35	0.10	0.67	0.75	0.77	0.52	0.156	0.07	0.60	0.80	0.80	0.45	0.135
	US-Myb	818	3.11	0.10	0.81	0.84	0.85	0.63	0.203	0.07	0.75	0.86	0.86	0.58	0.188
	US-Sne	161	3.70	−0.31	0.84	0.53	0.74	0.70	0.190	−0.40	0.82	0.56	0.71	0.65	0.176
	US-Tw1	818	3.03	−0.11	0.99	0.79	0.79	0.80	0.265	−0.13	0.98	0.79	0.80	0.78	0.258
	US-Tw3	817	2.77	−0.39	1.09	0.56	0.64	0.81	0.291	−0.42	1.08	0.57	0.65	0.78	0.281
	US-Tw4	818	3.57	0.16	0.77	0.90	0.90	0.60	0.169	0.12	0.77	0.90	0.90	0.60	0.167
	US-Twt	817	2.96	0.48	1.24	0.74	0.80	0.91	0.306	0.44	1.14	0.78	0.82	0.83	0.281
LS DAY	ALL	330	3.90	0.04	0.96	0.74	0.75	0.75	0.192	−0.07	0.94	0.75	0.75	0.75	0.191
DAILY	ALL	5241	3.10	0.08	0.95	0.78	0.79	0.72	0.233	0.05	0.91	0.80	0.80	0.69	0.221
WEEKLY	ALL	720	3.18	0.09	0.76	0.84	0.85	0.57	0.179	0.05	0.75	0.85	0.85	0.56	0.174
MONTHLY	ALL	169	3.19	0.08	0.58	0.90	0.91	0.44	0.137	0.04	0.57	0.90	0.91	0.42	0.132
YEARLY	ALL	10	3.02	0.07	0.28	0.02	0.64	0.24	0.078	0.04	0.28	0.02	0.63	0.23	0.076

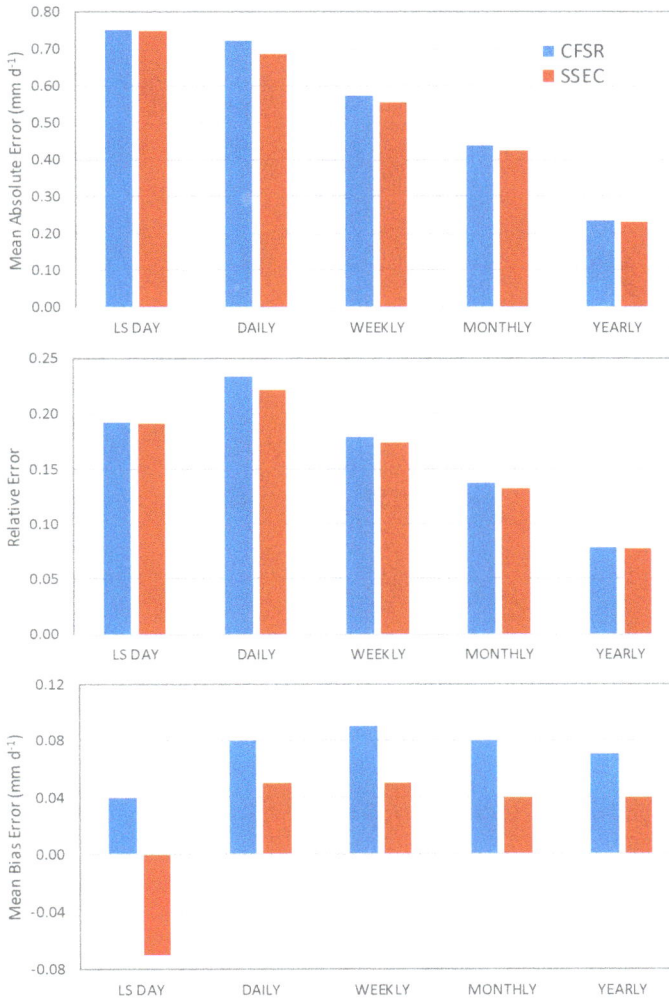

Figure 8. MAE, RE, and MBE in ET retrievals using the CFSR and SSEC insolation inputs on clear Landsat overpass dates (LS DAY) and at daily, weekly, monthly, and yearly timesteps over the study period.

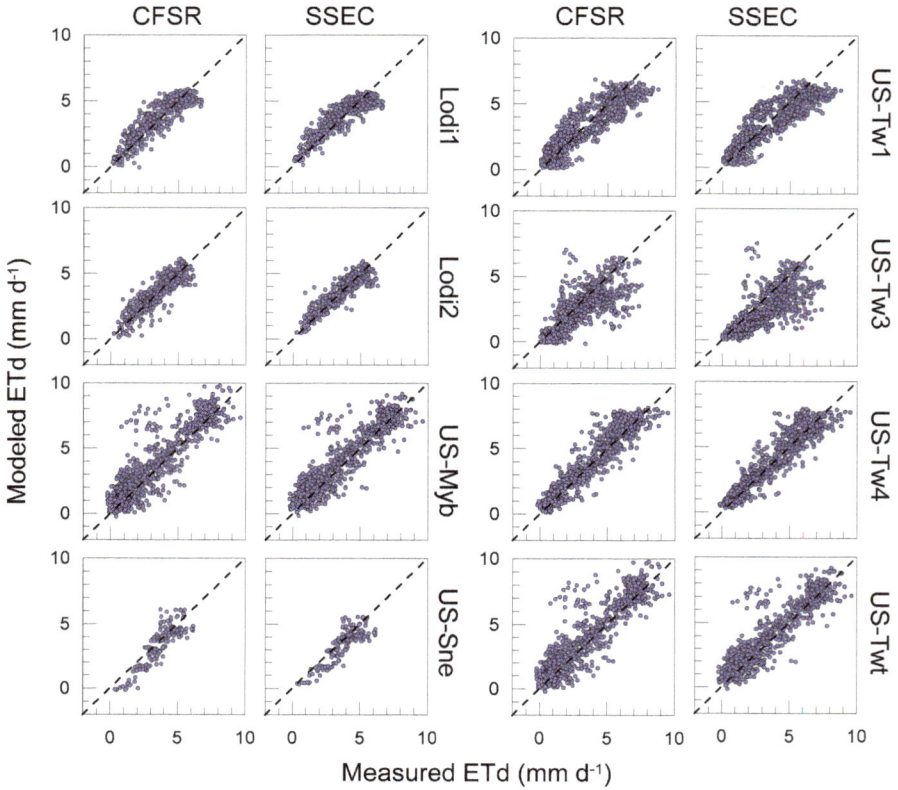

Figure 9. Comparison of daily observed and modeled ET fluxes generated using the CFSR and SSEC insolation datasets.

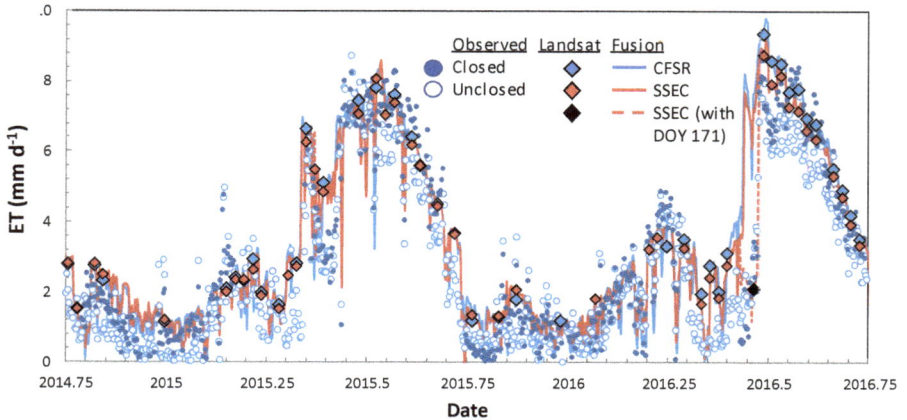

Figure 10. Timeseries of daily ET observations (unclosed and closed circles) at US-Twt and retrievals using the CFSR (light blue) and SSEC (red) insolation inputs (diamonds: on Landsat dates, solid lines: fused timeseries). The fused timeseries generated using the Landsat retrieval on DOY 171 (black diamond) is also shown (red dotted line).

Remote Sens. **2019**, *11*, 216

In general, the tower sites with the largest ET retrieval errors (US-Twt, US-Tw3, and US-Tw1) tended to be in landcovers where ET was more decoupled from the solar radiation load as evidenced in the flux tower observations (Figure 11). In the densely vegetated West Pond wetland (US-Tw1), increases in spring ET were delayed relative to the solar radiation curve due to a dense mat of litter material that inhibited new vegetative growth [13]. In the case of US-Twt (rice) and US-Tw3 (alfalfa), this decoupling was due to management activities, such as regulated flooding/drainage (rice) and periodic cuttings (alfalfa) [12,13]. Some portion of this high temporal resolution structure in water use phenology was missed by the ET data fusion system, leading to higher errors at these sites that were not resolved with the improved insolation inputs. For the rice site, a critical Landsat scene on DOY 171 in 2016 was omitted from the fusion process due to clouds and contrails in the southern part of the domain that were not captured by the Landsat cloud mask. Including this scene, which occurred at an inflection in moisture dynamics just prior to reflooding of the rice paddy, reduced the RMSE in ETd at US-Tw3 from 1.14 to 0.97 mm d^{-1} at the daily timescale using the SSEC insolation inputs (Figure 11), and from 0.91 to 0.88 mm d^{-1} for all the sites combined. In this case, the excess error was in part due to inadequate temporal sampling by Landsat during times of rapid change occurring at spatial scales too small to be well-captured in the MODIS timeseries. Landsat temporal sampling also poses a challenge in reproducing observed fluxes over alfalfa, where monthly cuttings may not coincide with a Landsat overpass and subsequent vegetation regrowth is rapid.

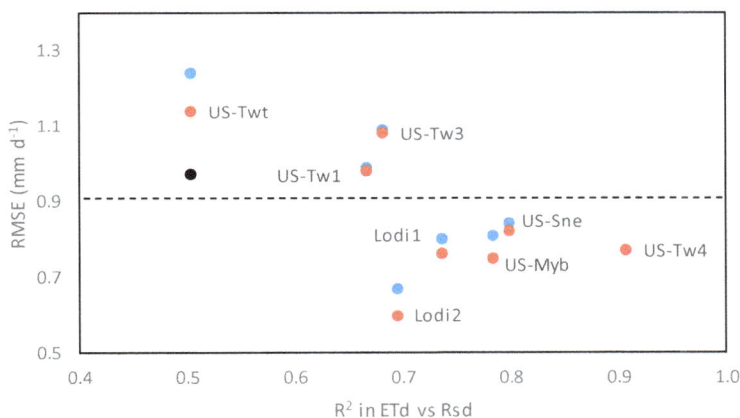

Figure 11. Root mean square error (RMSE) in daily ET retrievals using the CFSR (blue dots) and SSEC (red dots) inputs vs. the R^2 value computed between daily solar radiation and ET (unclosed) observed at the tower sites. The dotted line indicates the RMSE obtained with all sites combined. Also shown is the RMSE at US-Twt from a fusion experiment including the Landsat ET retrieval on DOY 171 in 2016 (black dot), omitted from the standard run due to extensive cloud cover in the southern part of the modeling domain.

Heterogeneous conditions with the tower footprint scale impacted model-measurement agreement at the US-Tw3 and US-Tw1 sites. The US-Tw3 site was highly variable in terms of stand health, and eddy covariance footprint analyses show off-field contributions may influence the tower fluxes [13]. The restored wetland sampled by the US-Tw1 tower is quite narrow (about 90 m wide near the tower—a few Landsat pixels). In both cases, off-site contributions to the tower fluxes were likely and a rigorous daily footprint analysis would be required to optimize the selection of model pixels for comparison with tower fluxes (see also Figure 4 in [12]).

4. Discussion

4.1. How Representative Are These Results?

The results presented here may be specific to the study site in Central California, which is dominated in the summer months by clear conditions under which the CFSR performance is optimal and clear-sky Landsat retrievals are frequent, facilitating quick correction in the fused ET timeseries if an individual retrieval is poor. A more significant improvement in ET accuracy will likely be obtained in regions with higher climatological frequency of cloud cover, such as in the northwest and eastern U.S. In these regions, fewer direct ET retrievals will be available, serving as tie points in the fusion reconstruction. Enhanced errors in the CFSR representation of insolation under clouds will further degrade performance between clear Landsat overpasses. A global intercomparison of multiple reanalysis insolation products, including CFSR, with ground-based observations and with the CERES satellite product revealed significant spatiotemporal patterns in bias and the RMSE due in part to biases in the cloud fraction and model treatment of aerosols [58].

Work is underway to intercompare CFSR and SSEC-based ET timeseries at other flux sites distributed across the U.S. to draw generalized conclusions regarding the impact of insolation dataset on ET retrieval accuracy.

4.2. Which Is the Optimal Insolation Datasource?

Selection of an optimal insolation data source for ET mapping depends on many factors and may vary by application and region of interest. The three datasets investigated here have distinct advantages and disadvantages.

CFSR has obvious advantages for operational use, including low data latency (available next day), hourly timesteps, complete and global coverage, and a long period of records generated with a consistent processing system. The latter is particularly important for climatological and trend analyses, which can be very sensitive to discontinuous changes in processing of inputs [59]. Disadvantages include lower accuracy at daily timesteps, artificial spatial structures, significant bias in some regions, and much lower spatial resolution in comparison with the ET model pixel scale.

In contrast, the spatial resolution for the SSEC GOES product is comparable to the ALEXI pixel scale. The precision obtained over the CA Delta sites was excellent, and the product is freely available with low data latency (available next day) and at half-hourly timesteps. Importantly, in ALEXI the solar radiation and thermal inputs are optimally consistent, both spatially and temporally, when both are derived from imagery from the same geostationary platform. Disadvantages are limitations in spatial domain (currently CONUS only), frequent gaps that require filling, and moderate bias and shifts in calibration, e.g., due to new satellites and sensor degradation. There has been no reprocessing for a consistent long-term archive.

CERES, despite its relatively coarse spatial resolution, showed comparable RMSE to the SSEC product over the flux sites in central CA, and lower bias errors. At hourly timesteps and with complete global coverage, this dataset holds significant promise for retrospective ET mapping over a range of spatial scales. Unfortunately, at present, the latency in data delivery (6-month lag) makes CERES insolation unsuitable for real-time applications.

5. Conclusions

This study examined the impacts of insolation data source on a multi-sensor TIR remote sensing data fusion approach for mapping actual evapotranspiration at field scale and daily timesteps. Pyranometer observations collected in central California for water years 2015–2016 were compared to data from three gridded insolation datasets, including the Climate Forecast System Reanalysis (CFSR), and geostationary satellite-based products over the CONUS (SSEC) and the globe (CERES). Daily relative errors in the two satellite datasets (RE = 0.05) were approximately half that from the reanalysis data (RE = 0.09), resulting from the improved capacity to capture cloud occurrence and impact on surface insolation.

The resulting improvement in ET retrievals at these tower sites was less notable using the satellite-based insolation inputs, with a RE of 0.23 and 0.22 for CFSR and SSEC at daily timesteps, respectively, and a RMSE of 0.95 and 0.91 mm d^{-1}. The lack of significant improvement in ET performance suggests that other modeling inputs or errors dominate over the set of sites examined here. This may be due in part to the relatively clear-sky conditions prevalent in central California during the peak growing season. CFSR errors are smaller under the clear-sky conditions when direct satellite retrievals can be conducted. Residual model errors in runs using the high quality SSEC insolation inputs are largest at tower sites where the observed ET is less coupled to solar radiation forcings, such as in highly managed agricultural fields (rice and alfalfa) and in a wetland system where storage fluxes and dense vegetation residue delay spring ET increases. These sites also tended to be spatially complex, with significant off-site contributions to tower flux measurements necessitating rigorous daily footprint analyses to better clarify the spatial model performance. A greater improvement in ET retrievals using satellite-based insolation may be expected in regions where evaporative fluxes are energy limited over a greater portion of the growing season.

The optimal insolation data source depends on the data latency requirements and spatial extent of the application in question. Both CFSR and SSEC datasets are available in near-real-time (less than one day latency); however, SSEC coverage currently includes only the continental U.S. While the global CERES insolation product has error statistics similar to that of SSEC, the latency is currently close to 6 months, precluding real-time use. Over the California Delta and climatologically similar regions in the western U.S., the CFSR data may suffice for real-time ET modeling efforts.

Author Contributions: Conceptualization: M.A., G.D., F.G.; Data Fusion system: F.G., C.H., Y.Y., M.A., Image Processing: G.D., K.K., M.A., Data collection: D.B., E.E., K.H., W.K.; Formal analysis: M.A., G.D., D.B., E.E., K.H., K.K. Writing—original draft: M.A., G.D.; Writing—review and editing: D.B., E.E., K.H., W.K., K.K.

Funding: This work was funded in part by the NASA MEASURES and Land Cover Land Use Change programs and the NASA ECOSTRESS Earth Ventures Instrument project [80NSSC18K0483].

Acknowledgments: The authors would like to acknowledge the contribution of UC Berkeley Biomet lab technician Daphne Szutu in conducting an extensive radiometer comparison experiment at the Twitchell Rice site.

Conflicts of Interest: The authors declare no conflict of interest.

References

1. Fisher, J.B.; Melton, F.S.; Middleton, E.M.; Hain, C.R.; Anderson, M.C.; Allen, R.G.; McCabe, M.F.; Hook, S.; Baldocchi, D.D.; Townsend, P.A.; et al. The future of evapotranspiration: Global requirements for ecosystem functioning, carbon and climate feedbacks, agricultural management, and water resources. *Water Resour. Res.* **2017**, *53*, 2618–2626. [CrossRef]

2. Garcia Garcia, L.E.; Rodriguez, D.J.; Wijnen, M.M.P.; Pakulski, I.; Serrat Capdevila, A.; Garcia Ramirez, D.A.; Tayebi, N.; Guerschman, J.P.; Donohue, R.J.; Niel, T.G.V.; et al. *Earth Observation for Water Resources Management: Current Use and Future Opportunities for the Water Sector*; The World Bank: Washington, DC, USA, 2016; pp. 33–38.

3. Monteith, J.L. Evaporation and Environment. *Symp. Soc. Exp. Biol.* **1965**, *19*, 205–234.

4. Priestley, C.H.B.; Taylor, R.J. On the assessment of surface heat flux and evaporation using large-scale parameters. *Mon. Weather Rev.* **1972**, *100*, 81–92. [CrossRef]

5. Anderson, M.C.; Kustas, W.P.; Norman, J.M.; Hain, C.R.; Mecikalski, J.R.; Schultz, L.; Gonzalez-Dugo, M.P.; Cammalleri, C.; d'Urso, G.; Pimstein, A.; et al. Mapping daily evapotranspiration at field to continental scales using geostationary and polar orbiting satellite imagery. *Hydrol. Earth Syst. Sci.* **2011**, *15*, 223–239. [CrossRef]

6. Yang, Y.; Anderson, M.C.; Gao, F.; Hain, C.; Kustas, W.P.; Meyers, T.; Crow, W.; Finocchiaro, R.G.; Otkin, J.A.; Sun, L.; et al. Impact of tile drainage on evapotranspiration (ET) in South Dakota, USA based on high spatiotemporal resolution et timeseries from a multi-satellite data fusion system. *J. Sel. Top. Appl. Earth Obs. Remote Sens.* **2017**, *10*, 2550–2564. [CrossRef]

7. Yang, Y.; Anderson, M.C.; Gao, F.; Hain, C.R.; Semmens, K.A.; Kustas, W.P.; Normeets, A.; Wynne, R.H.; Thomas, V.A.; Sun, G. Daily Landsat-scale evapotranspiration estimation over a managed pine plantation in North Carolina, USA using multi-satellite data fusion. *Hydrol. Earth Syst. Sci.* **2017**, *21*, 1017–1037. [CrossRef]

8. Yang, Y.; Anderson, M.C.; Gao, F.; Wardlow, B.; Hain, C.R.; Otkin, J.A.; Yang, Y.; Sun, L.; Dulaney, W. Field-scale mapping of evaporative stress indicators of crop yield: An application over Mead, NE. *Remote Sens. Environ.* **2018**, *210*, 387–402. [CrossRef]

9. Knipper, K.R.; Kustas, W.P.; Anderson, M.C.; Alfieri, J.G.; Prueger, J.H.; Hain, C.R.; Gao, F.; Yang, Y.; McKee, L.G.; Nieto, H.; et al. Evapotranspiration estimates derived using thermal-based satellite remote sensing and data fusion for irrigation management in California vineyards. *Irrig. Sci.* **2018**, in press. [CrossRef]

10. Medellín-Azuara, J.; Paw U, K.T.; Jin, Y.; Kent, E.; Clay, J.; Wong, A.; Bell, A.; Anderson, M.; Howes, D.; Melton, F.S.; et al. A comparative study for estimating crop evapotranspiration in the Sacramento-San Joaquin Delta. 2018. Available online: https://watershed.ucdavis.edu/project/delta-et (accessed on 21 December 2018).

11. Kustas, W.P.; Anderson, M.C.; Alfieri, J.G.; Knipper, K.; Torres-Rua, A.; Parry, C.K.; Nieto, H.; Agam, N.; White, A.; Gao, F.; et al. The Grape Remote sensing Atmospheric Profile and Evapotranspiration eXperiment. *Bull. Am. Meteorol. Soc.* **2018**, *99*, 1791–1812. [CrossRef]

12. Anderson, M.C.; Gao, F.; Knipper, K.; Hain, C.; Dulaney, W.; Baldocchi, D.D.; Eichelmann, E.; Hemes, K.S.; Yang, Y.; Medellín-Azuara, J.; et al. Field-scale assessment of land and water use change over the California Delta using remote sensing. *Remote Sens.* **2018**, *10*, 889. [CrossRef]

13. Eichelmann, E.; Hemes, K.S.; Knox, S.H.; Oikawa, P.; Chamberlain, S.D.; Sturtevant, C.; Verfaillie, J.; Baldocchi, D.D. The effect of land cover type and structure on evapotranspiration from agricultural and wetland sites in the Sacramento/San Joaquin River Delta, California. *Agric. For. Meteorol.* **2018**, *256–257*, 179–195. [CrossRef]

14. Anderson, M.C.; Kustas, W.P.; Alfieri, J.G.; Hain, C.R.; Prueger, J.H.; Evett, S.R.; Colaizzi, P.D.; Howell, T.A.; Chavez, J.L. Mapping daily evapotranspiration at landsat spatial scales during the BEAREX'08 field campaign. *Adv. Water Resour.* **2012**, *50*, 162–177. [CrossRef]

15. Cammalleri, C.; Anderson, M.C.; Gao, F.; Hain, C.R.; Kustas, W.P. A data fusion approach for mapping daily evapotranspiration at field scale. *Water Resour. Res.* **2013**, *49*, 1–15. [CrossRef]

16. Cammalleri, C.; Anderson, M.C.; Gao, F.; Hain, C.R.; Kustas, W.P. Mapping daily evapotranspiration at field scales over rainfed and irrigated agricultural areas using remote sensing data fusion. *Agric. For. Meteorol.* **2014**, *186*, 1–11. [CrossRef]

17. Norman, J.M.; Kustas, W.P.; Humes, K.S. A two-source approach for estimating soil and vegetation energy fluxes from observations of directional radiometric surface temperature. *Agric. For. Meteorol.* **1995**, *77*, 263–293. [CrossRef]

18. Kustas, W.P.; Norman, J.M. Use of remote sensing for evapotranspiration monitoring over land surfaces. *Hydrol. Sci. J.* **1996**, *41*, 495–516. [CrossRef]

19. Kustas, W.P.; Norman, J.M. Evaluation of soil and vegetation heat flux predictions using a simple two-source model with radiometric temperatures for partial canopy cover. *Agric. For. Meteorol.* **1999**, *94*, 13–29. [CrossRef]

20. Campbell, G.S.; Norman, J.M. *An Introduction to Environmental Biophysics*; Springer: New York, NY, USA, 1998.

21. Anderson, M.C.; Norman, J.M.; Meyers, T.P.; Diak, G.R. An analytical model for estimating canopy transpiration and carbon assimilation fluxes based on canopy light-use efficiency. *Agric. For. Meteorol.* **2000**, *101*, 265–289. [CrossRef]

22. Weiss, A.; Norman, J.M. Partitioning solar radiation into direct and diffuse, visible and near-infrared components. *Agric. For. Meteorol.* **1985**, *34*, 205–213. [CrossRef]

23. Anderson, M.C.; Norman, J.M.; Diak, G.R.; Kustas, W.P.; Mecikalski, J.R. A two-source time-integrated model for estimating surface fluxes using thermal infrared remote sensing. *Remote Sens. Environ.* **1997**, *60*, 195–216. [CrossRef]

24. Diak, G.R.; Whipple, M.S. Improvements to models and methods for evaluating the land-surface energy balance and 'effective' roughness using radiosonde reports and satellite-measured skin temperature data. *Agric. For. Meteor.* **1993**, *63*, 189–218. [CrossRef]

25. Norman, J.M.; Anderson, M.C.; Kustas, W.P.; French, A.N.; Mecikalski, J.R.; Torn, R.D.; Diak, G.R.; Schmugge, T.J.; Tanner, B.C.W. Remote sensing of surface energy fluxes at 10^1-m pixel resolutions. *Water Resour. Res.* **2003**, *39*. [CrossRef]

26. Anderson, M.C.; Norman, J.M.; Mecikalski, J.R.; Torn, R.D.; Kustas, W.P.; Basara, J.B. A multi-scale remote sensing model for disaggregating regional fluxes to micrometeorological scales. *J. Hydrometeor.* **2004**, *5*, 343–363. [CrossRef]

27. Cammalleri, C.; Anderson, M.C.; Kustas, W.P. Upscaling of evapotranspiration fluxes from instantaneous to daytime scales for thermal remote sensing applications. *Hydrol. Earth Syst. Sci.* **2014**, *18*, 1885–1894. [CrossRef]

28. Ryu, Y.; Baldocchi, D.D.; Black, T.A.; Detto, M.; Law, B.E.; Leuning, R.; Miyata, A.; Reichstein, M.; Vargas, R.; Amman, C.; et al. On the temporal upscaling of evapotranspiration from instantaneous remote sensing measurements to 8-day mean daily-sums. *Agric. For. Meteorol.* **2012**, *152*, 212–222. [CrossRef]

29. Holmes, T.; Hain, C.; Crow, W.; Anderson, M.C.; Kustas, W.P. Microwave implementation of two-source energy balance approach for estimating evapotranspiration. *Hydrol. Earth Syst. Sci.* **2018**, *22*, 1351–1369. [CrossRef]

30. Gao, F.; Masek, J.; Schwaller, M.; Hall, F.G. On the blending of the Landsat and MODIS surface reflectance: Predicting daily Landsat surface reflectance. *IEE Trans. Geosci. Remote. Sens.* **2006**, *44*, 2207–2218.

31. Semmens, K.A.; Anderson, M.C.; Kustas, W.P.; Gao, F.; Alfieri, J.G.; McKee, L.; Prueger, J.H.; Hain, C.R.; Cammalleri, C.; Yang, Y.; et al. Monitoring daily evapotranspiration over two California vineyards using Landsat 8 in a multi-sensor data fusion approach. *Remote Sens. Environ.* **2015**. [CrossRef]

32. Sun, L.; Anderson, M.C.; Gao, F.; Hain, C.R.; Alfieri, J.G.; Sharifi, A.; McCarty, G.; Yang, Y.; Yang, Y. Investigating water use over the Choptank River watershed using a multi-satellite data fusion approach. *Water Resour. Res.* **2017**, *53*, 5298–5319. [CrossRef]

33. Carpintero, E.; González-Dugo, M.P.; Hain, C.; Gao, F.; Andreu, A.; Kustas, W.P.; Anderson, M.C. Continuous evapotranspiration monitoring and water stress at watershed scale in a mediterranean oak savanna. *Proc. SPIE* **2016**, *9998*, 99980N. [CrossRef]

34. Saha, S.; Moorthi, S.; Pan, H.-L.; Wu, X.; Coauthors. The NCEP Climate Forecast System Reanalysis. *Bull. Am. Meteorol. Soc.* **2010**, *91*, 1015–1057. [CrossRef]

35. Dee, D.P.; Balmaseda, M.; Engelen, R.; Simmons, A.J.; Thépaut, J.M. Toward a consistent reanalysis of the climate system. *Bull. Am. Meteorol. Soc.* **2013**, *95*, 1235–1248. [CrossRef]

36. Chou, M.D.; Suarez, M.J.; Ho, C.H.; Yan, M.M.H.; Lee, K.T. Parameterizations for cloud overlapping and shortwave single-scattering properties for use in general circulation and cloud ensemble models. *J. Clim.* **1998**, *11*, 202–214. [CrossRef]

37. Hou, Y.; Moorthi, S.; Campana, K. *Parameterization of Solar Radiation Transfer in the NCEP Models*; NCEP Office Note 441; NCEP: Camp Spring, MD, USA, 2002; 46p.

38. Diak, G.R. Investigations of improvements to an operational GOES-satellite-data-based insolation system using pyranometer data from the U.S. Climate Reference Network (USCRN). *Remote Sens. Environ.* **2017**, *195*, 79–95. [CrossRef]

39. Lazzara, M.A.; Benson, J.M.; Fox, R.J.; Laitsch, D.J.; Rueden, J.P.; Santek, D.A.; Wade, D.M.; Whittaker, T.M.; Young, J.T. The man computer interactive data access system: 25 years of interactive processing. *Bull. Am. Meteorol. Soc.* **1999**, *80*, 271–284. [CrossRef]

40. Mecikalski, J.M.; Diak, G.R.; Anderson, M.C.; Norman, J.M. Estimating fluxes on continental scales using remotely-sensed data in an atmosphere-land exchange model. *J. Appl. Meteorol.* **1999**, *38*, 1352–1369. [CrossRef]

41. Mecikalski, J.R.; Sumner, D.M.; Jacobs, J.M.; Pathak, C.S.; Paech, S.; Douglas, E.M. Use of visible geostationary operational meteorological satellite imagery in mapping reference and potential evapotranspiration over Florida. In *Evapotranspiration*; Labedzki, L., Ed.; IntechOpen: London, UK, 2011; pp. 229–254.

42. Anderson, M.C.; Norman, J.M.; Mecikalski, J.R.; Otkin, J.A.; Kustas, W.P. A climatological study of evapotranspiration and moisture stress across the continental U.S. based on thermal remote sensing: I. Model formulation. *J. Geophys. Res.* **2007**, *112*, D10117. [CrossRef]

43. Teegavarapu, R.S.V.; Pathak, C.S.; Mecikalski, J.R.; Srikishen, J. Optimal solar radiation sensor network design using spatial and geostatistical analyses. *J. Spat. Sci.* **2016**, *61*, 69–97. [CrossRef]

44. Diak, G.R.; Anderson, M.C.; Bland, W.L.; Norman, J.M.; Mecikalski, J.M.; Aune, R.M. Agricultural management decision aids driven by real-time satellite data. *Bull. Am. Meteorol. Soc.* **1998**, *79*, 1345–1355. [CrossRef]

45. Diak, G.R.; Gautier, C. Improvements to a simple physical model for estimating insolation from GOES data. *J. Clim. Appl. Meteorol.* **1983**, *22*, 505–508. [CrossRef]

46. Gautier, C.; Landsfeld, M. Surface solar radiation flux and cloud radiative forcing for the Atmospheric Radiation Measurement (ARM) Southern Great Plains (SGP): A satellite, surface observations, and radiative transfer model study. *J. Atmos. Sci.* **1997**, *54*, 1289–1307. [CrossRef]

47. Paech, S.J.; Mecikalski, J.R.; Sumner, D.M.; Pathak, C.S.; Wu, Q.; Islam, S.; Sangoyomi, T. A calibrated, high-resolution GOES satellite solar insolation product for a climatology of Florida evapotranspiration. *J. Am. Water Resour. Assoc.* **2009**, *45*, 1328–1342. [CrossRef]

48. Wonsick, M.M.; Pinker, R.T.; Meng, W.; Nguyen, L. Evaluation of surface shortwave flux estimates from GOES: Sensitivity to sensor calibration. *J. Atmos. Ocean. Technol.* **2006**, *23*, 927–935. [CrossRef]

49. Young, D.F.; Minnis, P.; Doelling, D.R.; Gibson, G.G.; Wong, T. Temporal interpolation methods for the Clouds and the Earth's Radiant Energy System (CERES) experiment. *J. Appl. Meteorol.* **1998**, *37*, 572–590. [CrossRef]

50. Smith, G.L.; Priestley, K.J.; Loeb, N.G.; Wielicki, B.A.; Charlock, T.P.; Minnis, P.; Doelling, D.R.; Rutan, D.A. Clouds and Earth Radiant Energy System (CERES), a review: Past, present and future. *Adv. Space Res.* **2011**, *48*, 254–263. [CrossRef]

51. Twine, T.E.; Kustas, W.P.; Norman, J.M.; Cook, D.R.; Houser, P.R.; Meyers, T.P.; Prueger, J.H.; Starks, P.J.; Wesely, M.L. Correcting eddy-covariance flux underestimates over a grassland. *Agric. For. Meteorol.* **2000**, *103*, 279–300. [CrossRef]

52. Wilson, K.; Goldstein, A.; Falge, E.; Aubinet, M.; Baldocchi, D.; Berbigier, P.; Bernhofer, C.; Ceulemans, R.; Dolman, H.; Field, C.; et al. Energy balance closure at Fluxnet sites. *Agric. For. Meteorol.* **2002**, *113*, 223–243. [CrossRef]

53. Kochendorfer, J.; Meyers, T.P.; Frank, J.M.; Massman, W.J.; Heuer, M.W. How well can we measure the vertical wind speed? Implications for fluxes of energy and mass. *Bound.-Layer Meteor.* **2012**, *145*, 383–398. [CrossRef]

54. Frank, J.M.; Massman, W.J.; Ewers, B.E. Underestimates of sensible heat flux due to vertical velocity measurement errors in non-orthogonal sonic anemometers. *Agric. For. Meteorol.* **2013**, *171–172*, 72–81. [CrossRef]

55. Horst, T.W.; Semmer, S.R.; Maclean, G. Correction of a non-orthogonal, three-component sonic anemometer for flow distortion by transducer shadowing. *Bound.-Layer Meteor.* **2015**, *155*, 371–395. [CrossRef]

56. Frank, J.M.; Massman, W.J.; Ewers, B.E. A bayesian model to correct underestimated 3-D wind speeds from sonic anemometers increases turbulent components of the surface energy balance. *Atmos. Meas. Tech.* **2016**, *9*, 5933–5953. [CrossRef]

57. Nash, L.E.; Sutcliffe, J.V. River flow forecasting through conceptual models—Part 1: A discussion of principles. *J. Hydrol.* **1970**, *10*, 282–290. [CrossRef]

58. Zhang, X.; Liang, S.; Wang, G.; Yao, Y.; Jiang, B.; Cheng, J. Evaluation of the reanalysis surface incident shortwave radiation products from NCEP, ECMWF, GSFC, and JMA using satellite and surface observations. *Remote Sens.* **2016**, *8*, 225. [CrossRef]

59. Sheffield, J.; Goteti, G.; Wood, E.F. Development of a 50-year high-resolution global dataset of meteorological forcings for land surface modeling. *J. Clim.* **2006**, *19*, 3088–3111. [CrossRef]

remote sensing

MDPI

Article

Improvement in Surface Solar Irradiance Estimation Using HRV/MSG Data

Filomena Romano [1,*], Domenico Cimini [1,2], Angela Cersosimo [1], Francesco Di Paola [1], Donatello Gallucci [1], Sabrina Gentile [1,2], Edoardo Geraldi [1,3], Salvatore Larosa [1], Saverio T. Nilo [1], Elisabetta Ricciardelli [1], Ermann Ripepi [1] and Mariassunta Viggiano [1]

[1] Institute of Methodologies for Environmental Analysis, National Research Council (IMAA/CNR), 85100 Potenza, Italy; domenico.cimini@imaa.cnr.it (D.C.); angela.cersosimo@imaa.cnr.it (A.C.); francesco.dipaola@imaa.cnr.it (F.D.P.); donatello.gallucci@imaa.cnr.it (D.G.); sabrina.gentile@imaa.cnr.it (S.G.); edoardo.geraldi@cnr.it (E.G.); salvatore.larosa@imaa.cnr.it (S.L.); saverio.nilo@imaa.cnr.it (S.T.N.); elisabetta.ricciardelli@imaa.cnr.it (E.R.); ermann.ripepi@imaa.cnr.it (E.R.); mariassunta.viggiano@imaa.cnr.it (M.V.)

[2] Center of Excellence Telesensing of Environment and Model Prediction of Severe events (CETEMPS), University of L'Aquila, 67100 L'Aquila, Italy

[3] Institute for Archaeological and Monumental Heritage, National Research Council (IBAM/CNR), 85100 Potenza, Italy

* Correspondence: filomena.romano@imaa.cnr.it; Tel.: +39-0971-427266

Received: 10 July 2018; Accepted: 13 August 2018; Published: 15 August 2018

Abstract: The Advanced Model for the Estimation of Surface Solar Irradiance (AMESIS) was developed at the Institute of Methodologies for Environmental Analysis of the National Research Council of Italy (IMAA-CNR) to derive surface solar irradiance from SEVIRI radiometer on board the MSG geostationary satellite. The operational version of AMESIS has been running continuously at IMAA-CNR over all of Italy since 2017 in support to the monitoring of photovoltaic plants. The AMESIS operative model provides two different estimations of the surface solar irradiance: one is obtained considering only the low-resolution channels (SSI_VIS), while the other also takes into account the high-resolution HRV channel (SSI_HRV). This paper shows the difference between these two products against simultaneous ground-based observations from a network of 63 pyranometers for different sky conditions (clear, overcast and partially cloudy). Comparable statistical scores have been obtained for both AMESIS products in clear and cloud situation. In terms of bias and correlation coefficient over partially cloudy sky, better performances are found for SSI_HRV (0.34 W/m^2 and 0.995, respectively) than SSI_VIS (-33.69 W/m^2 and 0.862) at the expense of the greater run-time necessary to process HRV data channel.

Keywords: solar irradiance; MSG; SEVIRI; HRV; AMESIS

1. Introduction

The deployment and optimization of photovoltaic system plants require accurate knowledge of the temporal and spatial variability of surface solar radiation, which is directly linked to solar energy production. The characterization of the temporal and spatial distribution of surface solar irradiance is very challenging due to cloud presence and, in complex orography, to shadowing and altitude/slope effects [1,2]. Clouds play a major role in the Earth–atmosphere system, affecting the incoming solar and outgoing thermal energy with interactions with the other atmospheric components. Clouds absorb and even more scatter radiation in the shortwave part of the spectrum, they absorb radiation in the longwave part (which is emitted by Earth's surface and lower troposphere) and they reemit it downwards and into space [3]. It is very difficult to estimate the radiative effects of clouds accurately, due to their high temporal and spatial variability. Ideally, in situ measurements provide the best method for obtaining solar irradiance estimation at surface [4], but the network of these ground instruments is not sufficiently dense [5,6]; therefore, methods involving extrapolation statistics based on surface measurements or satellite observations are commonly exploited to complement ground-based measurements and cover areas with no ground instruments installed [7–9]. The geostationary orbit covers the Earth's surface with a high time resolution, but its weakness is the increase in the pixel apparent size with latitude and longitude. By contrast, polar satellites, rotating at a much lower altitude, have a higher spatial resolution but a restricted temporal coverage. Therefore, only geostationary data allow the diurnal cycle of the solar irradiance at the Earth's surface to be captured. However, the availability of satellite-derived solar radiation measurements has often proven to be spatially and temporally inadequate for many applications [10]; in particular, broken clouds are difficult to detect, especially small clouds that can make the solar surface irradiance estimation less accurate [11]. At present, many different methods are exploited for solar radiation mapping, including the interpolation of surface measurements and the orographic downscaling of satellite data [10,12–14]. The accuracy and the suitability of interpolation and orographic downscaling approaches for complex-orography regions are still under study. In recent years, several parametric, look-up table (LUT), and statistical methods have been developed to retrieve the surface solar irradiance from satellite observations. Parametric methods estimate the surface solar irradiance relying on the parameterization of surface and atmospheric variables [15–17]. One weakness of these methods resides in the spatial and temporal variability of some variables, especially cloud, aerosol, and water vapor distribution. Look-up table methods are based on a pre-established radiative-transfer database, and have been shown to give satisfactory results [18–22]. In addition, hybrid methods have been extensively used. Hybrid methods combine the robustness of a physical approach, based on radiative transfer models, with a simplified cloud/aerosol model using a cloud index. Thus, the clear-sky solar irradiance is firstly computed by means of a LUT or a radiative transfer model, then it is used to compute the cloudy solar irradiance depending upon a cloud index derived from satellite observations [23–25]. Also, artificial neural networks (ANN) have been employed to estimate solar irradiation using different sets of inputs [26–28]. The study presented in [29] developed an ANN to retrieve solar radiation with hourly cloud parameters derived by combining low-Earth-orbit observations (from MODIS) with geostationary imagery (from Multifunctional Transport Satellite, MTSAT).

Cloudy and cloud-free satellite pixels must be distinguished before the estimation of solar irradiance surface. The main difficulties in cloud detection are due to the reduced cloud-surface contrast [30]. To improve cloud detection from satellite observations, many efforts have been made and different methods have been proposed. Most of the techniques based on satellite image algorithms use the threshold method [31–36] or statistical procedures [37,38]. Artificial intelligence techniques are increasingly being used in areas of prediction and classification, these algorithms are robust and very flexible [39,40]. In the last few years, the Spinning Enhanced Visible and InfraRed Imager (SEVIRI) High-Resolution Visible (HRV) channel on board the Meteosat Second Generation (MSG) has been used for solar irradiance estimation [41,42] and short-term forecasting [43].

This work focuses on the improvement of the spatial resolution of surface solar irradiance by using higher-resolution imaging; this allows us to avoid downscaling techniques to achieve ~1 km and, at the same time, to apply downscaling techniques starting from ~1 km when higher spatial resolution is required. In this work, we use observations from the SEVIRI. Solar radiance varies rapidly in time and space and therefore requires instruments with high spatial and temporal resolution. Among the operational instruments in Europe, the most suitable temporal and spatial resolution trade-off is from SEVIRI on MSG-4 (0°N, 0°E). SEVIRI images are also available at higher temporal resolution (5 min) through the Rapid Scanning Service (RSS) from MSG-3 (0°N, 9.5°E). The Advanced Model for the Estimation of Surface Solar Irradiance (AMESIS) [19] operative model (described in the following sections) that we have developed is able to produce surface solar irradiance using both the IR-VIS and IR-VIS + HRV SEVIRI channel combinations. The implementation of AMESIS on RSS images is currently under development. However, this would not improve the spatial resolution, which is the main focus of this work. Similarly, benefits are expected from the ever-increasing spatial resolution of new and future satellite instruments, e.g., the Flexible Combined Imager (FCI) on the Meteosat Third Generation series planned from 2021 onwards. The SEVIRI High-Resolution Visible (HRV) channel is able to resolve sub-pixel clouds not detectable by the other SEVIRI VIS channels. The main purpose of this paper is to understand whether the use of the HRV channel at a higher resolution can improve the accuracy of the solar irradiance estimation in particular over partially cloudy sky. The paper shows a comparison between these two products against ground-based in situ observations.

2. Data and Methods

The SEVIRI instrument is designed to support numerical weather forecasting and nowcasting over Europe and Africa [44,45]. The SEVIRI radiometer is the main payload on board the MSG geostationary satellite series, operated by the European Organization for the Exploitation of Meteorological Satellites (EUMETSAT). The MSG satellites—namely, Meteosat-8, -9, -10 and -11—operate over Europe, Africa and the Indian Ocean in geostationary orbit 36,000 km above the equator. SEVIRI, a 50-cm-diameter aperture line-by-line scanning radiometer, has twelve spectral channels, eleven at low and one at high spatial resolution. The low spatial resolution channels are in infrared and visible bands, with a spatial sampling of 3 km and an actual instantaneous field of view leading to a spatial resolution of about 4.8 km at the sub-satellite point (SSP). The high spatial resolution channel, the HRV, is a broadband (0.3–1.1 µm) channel with a spatial sampling of 1 km and an actual instantaneous field of view leading to spatial resolution of about 1.67 km at the SSP. The HRV images are acquired on a reduced Earth area.

A ground-based pyranometer network is deployed and maintained by the regional agency for environmental protection (ARPA) of Lombardy region in Italy (http://www.arpalombardia.it/). ARPA Lombardia performs data acquisition, processing and quality control at either 10 or 60 min temporal intervals. The network consists of 63 pyranometers manufactured by Kippen and Zonen. The instrument specification states ~3% measurement uncertainty. Overall instrument maintenance is performed every ~2 years. Following WMO guidelines, automatic data quality tests have been implemented by ARPA Lombardia [46,47]. The pyranometer stations reported in Table 1 are located at different altitudes (see Figure 1).

Table 1. Identity code (ID), location, and altitude of ground observation stations.

ID_Station	Latitude	Longitude	Altitude (m)	ID_Station	Latitude	Longitude	Altitude (m)
132	45.659851	19.659089	211	1347	45.950001	9.4600000	1713
146	46.013573	9.6624260	1824	110	44.963810	10.767801	22
595	45.633198	9.5564108	190	139	45.156864	10.797858	19
586	45.826481	10.097358	192	214	45.187836	10.887401	15
847	46.040707	9.7975283	1954	148	45.547844	8.8476763	182
1360	45.885708	9.9469538	564	166	45.157330	10.824207	25
119	45.881668	9.6473770	700	695	45.412109	10.683630	113
129	46.025890	10.342786	362	671	45.151348	10.860067	22
596	45.620644	9.6118517	182	100	45.496109	9.2578459	120
1077	45.880001	9.7700005	1138	102	45.186337	9.4865770	140
1325	45.810440	10.375551	1068	140	45.281292	8.9889202	100
1365	45.673401	10.340300	911	147	45.541992	9.2059374	142
1367	45.877102	10.447216	775	502	45.472557	9.2226477	122
134	45.433056	10.039325	93	620	45.470985	9.1894445	122
145	45.891499	10.188790	222	106	44.823254	9.1953688	500
846	46.174320	10.471110	2108	114	45.320095	9.2646246	88
1075	46.169319	10.341660	659	125	45.232506	8.6830683	106
1366	46.038010	9.1413231	291	512	45.671303	9.2342949	250
1378	45.750465	10.736627	291	642	45.194016	9.1649857	77
136	45.162968	10.059123	44	672	45.039421	8.9145050	74
141	45.717659	9.0858936	310	133	46.105179	9.5694208	800
150	45.121067	10.195482	39	143	46.493744	10.207782	2320

Table 1. Cont.

ID_Station	Latitude	Longitude	Altitude (m)	ID_Station	Latitude	Longitude	Altitude (m)
629	45.365639	9.7042618	79	848	46.477097	10.205837	2660
677	45.141880	10.044145	43	836	46.147926	10.159731	1950
1202	46.181168	9.3250074	980	1343	46.366001	9.8999996	3032
1303	45.109543	10.069242	36	1346	46.417400	9.3619003	1880
109	45.260002	9.3799944	60	107	46.164913	9.8488007	307
111	45.930000	9.4799995	1234	108	46.235424	9.4269791	206
123	45.269161	9.5620594	67	835	46.460434	10.343612	43
127	45.701279	9.3094740	360	1273	46.240761	9.6324825	1191
706	45.233459	9.4002743	272	1342	46.144493	9.9752855	2440
1266	45.233459	9.6662951	65				

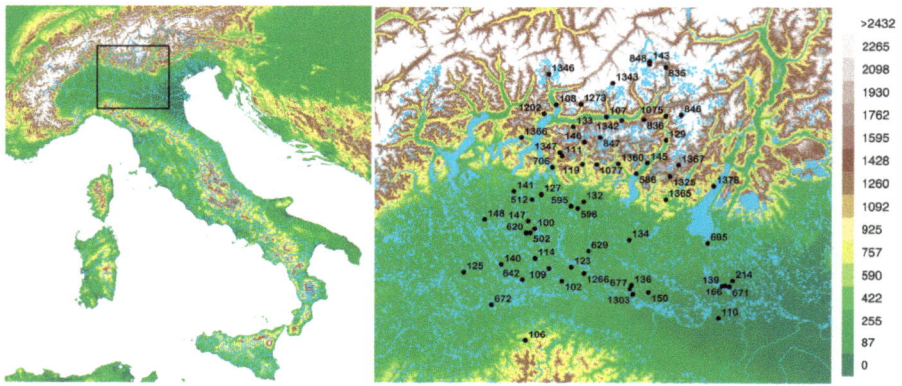

Figure 1. Geographical locations of the used meteorological ground observation stations.

AMESIS is the software package developed at IMAA-CNR that implements and operates the Advanced Model for the Estimation of Surface Solar Irradiance (AMESIS). AMESIS has been running at IMAA-CNR since 2017 in support to photovoltaic plant site monitoring. The operational model is an upgrade of the AMESIS version described in [19]; the new features consist mainly in providing two output products, one obtained using the HRV and the other using the lower resolution channels only. This feature was developed specifically for quantifying the improvements obtainable with the use of the HRV channel.

In this Section, we provide a brief overview of AMESIS, while for further detail we refer to [19]. AMESIS exploits MSG-SEVIRI spectral information (at 15 min resolution) by ingesting the values of radiance and brightness temperature from all SEVIRI channels to estimate surface solar irradiance. The first step of AMESIS consists of classifying the pixels as clear, cloudy, partially cloudy, or affected by aerosol presence [19,48]. A cloud phase retrieval is then applied to pixels identified as cloudy or partially cloudy. To this end, the cloud Classification Mask Coupling of Statistical and Physical methods (C_MACSP) algorithm is used [31,49]. C_MACSP, i.e., the cloud mask developed by us, has been validated widely with other operating products and also with 2B-GeoProf (CLOUDSAT Geometrical Profiling Product) [50]. GeoProf is one of the standard products produced by CLOUDSAT, a satellite mission designed to measure the vertical structure of clouds from space. In all the cases, except for those detected as clear after the first step, the model retrieves the microphysical optical parameters for clouds or aerosol. The cloud microphysical parameter retrieval uses radiances at visible (0.6–0.8 μm), near-infrared (1.6 μm), and infrared (10.8–12.0 μm) wavelengths as a first step, whereas as a second step, the retrieval also uses the other channels to improve the accuracy of cloud parameters, because the other channels (for instance, the channels at 3.9 μm) can contain useful information on particular types of clouds. Retrieving aerosol properties from satellite observations over land can be difficult because of the surface reflection, complex aerosol composition and aerosol absorption. Nonetheless, since aerosols are highly variable in time and space, we prefer to use information retrieved at the same time and location of solar irradiance estimation, rather than from spatially distant ground measurements or temporally distant satellite retrievals derived from other platforms. Aerosol information retrieved by AMESIS has been validated through comparisons with MODIS and AERONET. Validation of cloud and aerosol microphysical properties has been reported in [19]. Subsequently, the model retrieves the surface solar irradiance on the basis of the correspondent look-up tables. The look-up tables are periodically updated in order to take into account albedo/emissivity variations. The update is typically performed once every 8 days. The model, therefore, incorporates the effects due to aerosol, overcast and partially cloudy coverage; the retrieval of surface temperature, total integrated water vapor, cloud and aerosol microphysical parameters is achieved by using the low spatial resolution VIS and IR SEVIRI channels, whereas surface solar irradiance is retrieved through the high spatial resolution HRV channel. If the HRV pixel is classified as clear, AMESIS model retrieves the correct coefficient from the correspondent look-up table according to albedo, integrated water vapor, solar zenith, azimuth solar-satellite angles, and the elevation above sea level. If the HRV and the IR/VIS pixels are classified as cloudy, the procedure starts to retrieve cloudy microphysical parameters. Depending on the cloud microphysical parameter information obtained from this first step, the surface irradiance is estimated by using the correspondent look-up table. If the HRV and the IR/VIS pixels are classified, respectively, as cloudy and partially cloudy, the procedure starts to retrieve cloudy microphysical parameters for partially cloudy cases according to the cloud fraction index. In the second step, the surface irradiance is estimated by using the correspondent look-up table for the cloud microphysical parameters. To evaluate the performances (in terms of run-time and accuracy), the AMESIS operational version gives two different estimated solar surface irradiance products, one using IR-VIS channels only (called SSI_VIS hereafter), the other also using the HRV channel (called SSI_HRV hereafter).

3. Methodology

The AMESIS SSI_VIS and SSI_HRV products are validated against the surface solar irradiance product from the ground pyranometer network (called SSI_Ground hereafter). Data from these three sources were treated in order to check data quality, to find space/time colocation, and finally to compute statistical scores.

The AMESIS surface solar irradiance products are co-located with the ground-based products in order to associate each satellite pixel with the corresponding surface solar irradiance values measured by ground-based pyranometers. The temporal co-location is obtained as follows.

The SEVIRI imaging phase takes twelve minutes for acquisition, and three minutes for calibration, retrace and stabilization. The knowledge of the pixel scan time relative to the initial time of the full disk scan cycle is necessary for co-locating satellite-derived products with ground-based measurements. The scan time (minutes) is calculated for each MSG SEVIRI scan line (the scan line is completed in less than a second)

$$\Delta t(\text{lin}) = \frac{\text{lin} + \frac{(\text{lin}_{\text{sev}} - \text{lin}_{\text{tot}})}{2} - 1}{\text{lin}_{\text{sev}}} \cdot t_{\text{full}}$$

where lin_{sev} and lin_{tot} indicate the number of SEVIRI and total lines, and $t_{\text{full}} = \text{lin}_{\text{sev}}/(3\cdot100)$ is the full disk scan time in minutes, since 3 VIS/IR lines are scanned per MSG revolution and the MSG satellite rotates at 100 rotations per minute. Please note that the sensor has more lines than reported in the Level 1.5 image data: $\text{lin}_{\text{sev}} = 3750$ (11,250 for HRV) and $\text{lin}_{\text{tot}} = 3712$ (11,136 for HRV) [44].

An accurate georeferencing of satellite imagery is very important when computing the surface solar irradiance estimation for the monitoring of the incoming solar power to the PV power plants. To our knowledge, only few studies regarding the SEVIRI geometry accuracy have been reported in the open literature. An East-West image shift immediately after eclipse of up to 3 pixels with respect to the multispectral channels of Meteosat-8 with 3 km nominal resolution has been reported [51]. A shift of up to 8 pixels to the North on the HRV images over Alpine areas using lakes as landmarks has been reported in [52]. Such accuracy for HRV navigation does not fully satisfy the accuracy requirements for photovoltaic monitoring. Thus, a weekly automatic routine was developed to correct the navigation of IR/VIS and HRV channels. The navigation is based on a coastline-matching algorithm using the SEVIRI solar channels (HRV, 0.6 and 0.8 μm). The navigation procedure is, thus, restricted to daytime operation and assumes that the misalignment is linear in both north/south and west/east directions. For navigation correction, clear images have been used in order to have 70% or more coastline edge pixels visible.

Each AMESIS product is associated with the scan time of the satellite overpass. Pyranometer surface solar irradiances measured every 10 min have been interpolated at scan time of the satellite overpass.

The validation of the AMESIS products against the ground-based reference products is performed through the assessment of statistical scores. In addition to the correlation coefficient, four different statistical scores, from among the most commonly used, are calculated. These are the mean bias error (MBE), root mean square error (RMSE), mean absolute error (MAE) and mean absolute percentage error (MAPE) computed as follows:

$$corr = \frac{cov(p,s)}{\delta_p \delta_s}$$

where p denotes ground measured data, s denotes satellite products, cov is the covariance, δ_p is the standard deviation of p and δ_s is the standard deviation of s;

$$\text{MBE} = \frac{1}{n} \sum_{i=1}^{n} (s_i - p_i)$$

$$\text{RMSE} = \left[\frac{1}{n} \sum_{i=1}^{n} (s_i - p_i)^2 \right]^{1/2}$$

$$\text{MAE} = \frac{1}{n} \sum_{i=1}^{n} |s_i - p_i|$$

$$\text{MAPE} = \frac{100\%}{n} \sum_{i=1}^{n} \left| \frac{(p_i - s_i)}{p_i} \right|$$

4. Results and Discussion

From the data listed in Table 2, three datasets were derived (clear, cloudy and partially cloudy) by using cloud detection AMESIS modules for low (VIS) [31,49] and high spatial (HRV) [53] resolution. In the validation phase, we did not find features that could be related to geographic/altitude characteristics of ground-based stations. Conversely, the quality of the estimates seems to be mostly related to meteorological and albedo conditions, particularly the cloud/aerosol and surface emissivity values used. The pre-established LUTs have been calculated for different altitudes (0 to 3500 m). The min/max statistical scores obtained for each station in clear sky are 0.993/0.996 correlation coefficient, 2.33/3.95 W/m^2 MBE, 18.52/27.03 W/m^2 RMSE, and 2.36/9.28 W/m^2 MAE.

Table 2. Periods considered in the AMESIS validation.

Year	Month	Days
2017	June	10 to 30
2017	October	1 to 20
2018	March	10 to 30

The datasets were compared against ground observation data.

Figures 2–4 show the surface solar irradiance estimated by AMESIS and measured by ground pyranometers for different days under different meteorological situations. For cloudy (Figure 3) and clear pixels (Figure 2) the AMESIS SSI_VIS and SSI_HRV products are comparable, while for partially covered sky (Figure 4), we can see how the SSI_HRV product is closer to the ground-measured data with respect to the SSI_VIS product. This is likely due to the fact that HRV, thanks to its resolution, better captures the weather conditions at the pyranometer site, especially in cloud-broken sky. At the same time, the HRV channel also exhibits a difference in mountainous areas, better capturing the sharp changes of albedo. The resolution of the other channels is low compared to the scale of peculiarities of the surface and broken clouds. Other authors have investigated the potential to increase the fraction of cloud-free observations by increasing sensor resolution. Particularly, some authors have also demonstrated that the relative gain in cloud-free observations as a function of sensor resolution depends on cloud coverage regions and seasons [54].

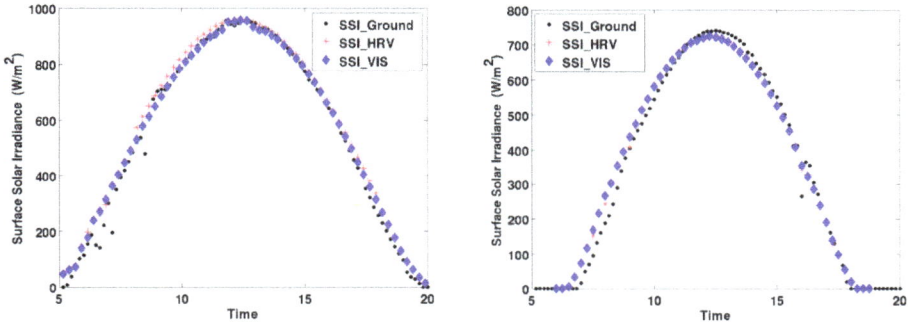

Figure 2. Surface solar irradiance estimated in clear sky by AMESIS, both SSI_HRV and SSI_VIS, and measurements by ground-based pyranometers. (**Left column**): 10 June 2017 (Lat = 45.269261, Long = 9.5620594). (**Right column**): 14 March 2018 (Lat = 45.121067, Long = 10.195482).

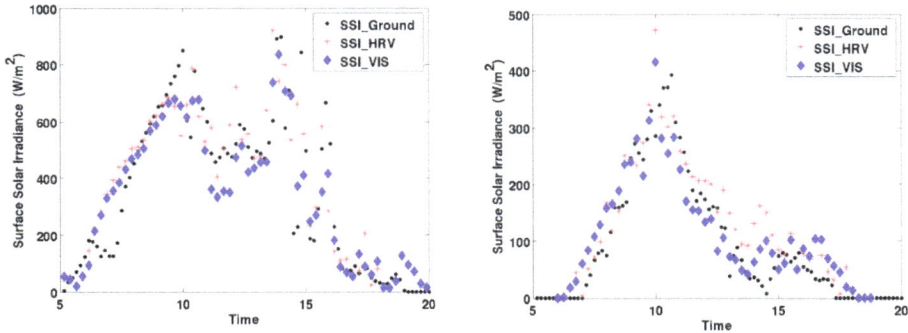

Figure 3. As in Figure 2, but for cloudy sky. (**Left column**): 15 June 2017 (Lat = 45.891499, Long = 10.188790). (**Right column**): 18 March 2018 (Lat = 45.187836, Long = 10.887401).

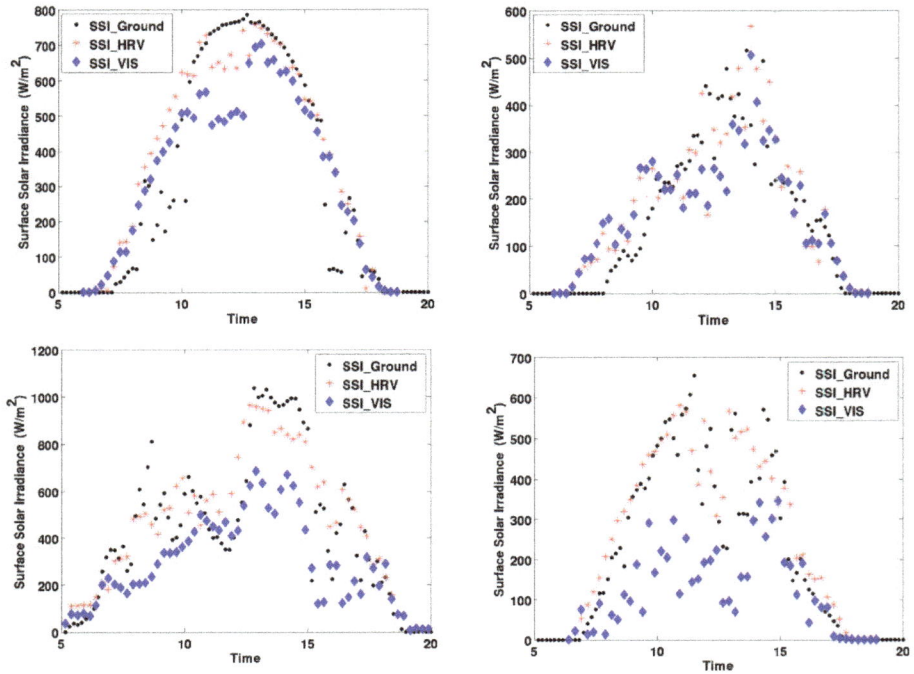

Figure 4. As in Figure 2, but for partially cloudy sky. First row; (**Left column**): 16 March 2018 (Lat = 46.105179, Long = 9.5694208). (**Right column**): 12 March 2018 (Lat = 46.013573, Long = 9.6624260). Second row; (**Left column**): 15 June 2017 (Lat = 45.633198, Long = 9.5564108). (**Right column**): 4 October 2017 (Lat = 45.715240, Long = 9.6892786).

Tables 3 and 4 show the statistical analysis by means of the RMSE, MBE, MAE, MAPE and correlation coefficient. Figures 5–7 show the scatterplot of the irradiance derived from AMESIS compared against the ground-based measurements for clear, overcast and partially cloudy sky.

Table 3. Results of statistical assessment for the SSI_VIS product with respect to SSI_Ground. Units of the statistical scores are specified in the bracket. N indicates the number of used pixels.

	N	CORR	MBE (W/m^2)	RMSE (W/m^2)	MAE (W/m^2)	MAPE
Clear	653	0.996	1.37	23.02	16.12	10.46
Cloudy	4564	0.969	−2.92	49.95	36.47	24.31
Partially cloudy	4028	0.862	−33.69	117.84	76.84	112.13

Table 4. As in Table 2, but for the SSI_HRV product with respect to SSI_Ground.

	N	CORR	MBE (W/m^2)	RMSE (W/m^2)	MAE (W/m^2)	MAPE
Clear	474	0.997	0.76	17.25	11.79	4.93
Cloudy	4558	0.968	0.86	53.88	41.91	33.07
Partially cloudy	4143	0.995	0.34	23.04	15.44	12.37

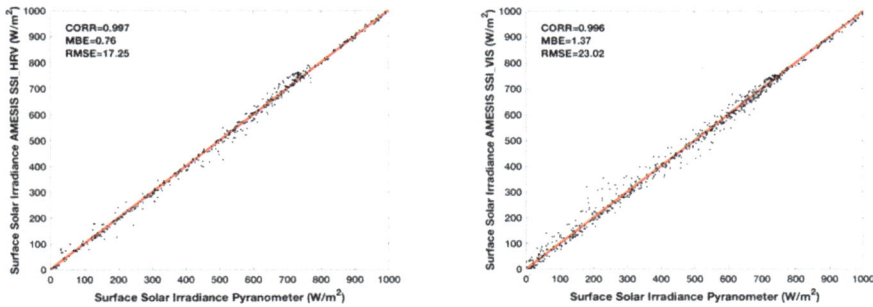

Figure 5. Scatterplot of the irradiance outputs derived from AMESIS compared against the ground-based irradiance for clear sky. (**Left**): SSI_HRV product, (**right**): SSI_VIS product.

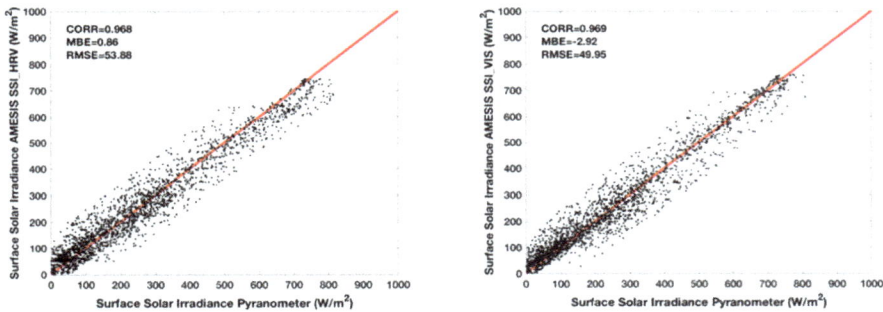

Figure 6. Same as Figure 5, but for cloudy sky. (**Left**): SSI_HRV product, (**right**): SSI_VIS product.

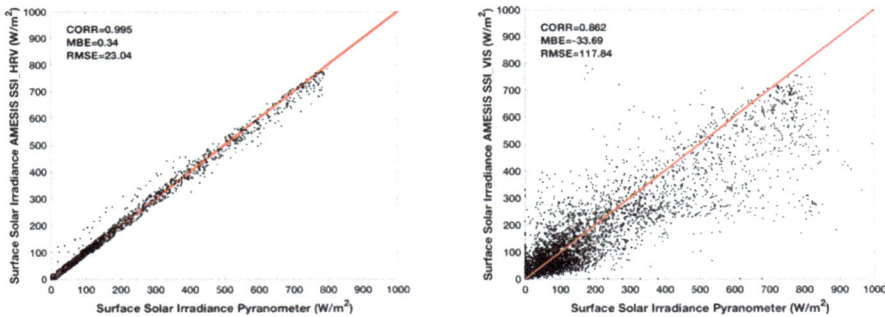

Figure 7. Same as Figure 5, but for partially cloudy sky. (**Left**): SSI_HRV product, (**right**): SSI_VIS product.

Small positive biases, of 1.30 and 0.76 W/m^2, respectively, for SSI_VIS and SSI_HRV, and correlation above 0.99 for both the products are found. This indicates, together with the other statistical scores, a very good agreement between AMESIS model and measurements for clear sky. For overcast cloudy sky, we found biases of −2.92 and 0.86 W/m^2, respectively, for SSI_VIS and SSI_HRV, and a correlation above 0.96 for both products. The RMSEs were 49.95 and 53.88 W/m^2, respectively, for SSI_VIS and SSI_HRV, indicating a good detection of clouds and an accurate estimation of their microphysical parameters, despite the limitations due to the SEVIRI low vertical resolution. The bias,

RMSE, and correlation coefficients from the comparison with ground-based measurements show that AMESIS provides cloudy sky products that are generally as accurate as other methods of similar a kind. A large negative bias of -33.69 W/m^2 and a correlation of 0.862 were found for SSI_VIS for partially cloudy sky. Dramatic improvements (bias 0.34 W/m^2 and 0.995 correlation) were found for SSI_HRV. This indicates that HRV is able to detect clear area contained within the largest pixel of low spatial resolution SEVIRI channels. Also, the RMSE is much larger for SSI_VIS (117.84 W/m^2) than for SSI_HRV (23.04 W/m^2). The RMSE in all cases is slightly higher than the MAE, except for VIS partially cloudy data; this means that there are values quite different from the measured ones. In fact, as seen from Figure 7, and also from the MBE value, the SSI_VIS often underestimates the measured values. The HRV channel is very useful for analyzing low-resolution pixels and estimating the fraction of the pixels covered with clouds. Additionally, MAPE values can be seen as acceptable, with the lowest value being found for the clear HRV dataset. The analysis of statistical scores suggests that the HRV channel, or more generally, a higher-resolution sensor, can improve the estimation of solar irradiance over partially cloudy sky. In the evaluation of the performances under different sky conditions, the statistical scores were also compared to those given by the validation of 4 different papers cited in the introduction. The first and the second [11,22] were selected because they report the statistics for different weather conditions and different satellite platforms; the third [42] was considered because it uses the HRV channel, whereas the fourth [2] is based on MSG/SEVIRI and reports on the validation against some alpine stations. The statistical scores reported in [11] for two different stations in clear sky are (min/max): 0.77/0.86 correlation coefficient, 47.80/-22.29 W/m^2 MBE and 191.81/140.77 W/m^2 RMSE. For cloudy sky, the statistical scores are 0.74/0.84 correlation coefficient, 80.66/40.31 W/m^2 MBE and 193.79/151.24 W/m^2 RMSE. The results of the comparison in [22] based on seven pyranometers showed that the estimated surface irradiance agreed with ground measurements with correlation coefficients of 0.94, 0.69, and 0.89, a bias of 26.4 W/m^2, -5.9 W/m^2, and 14.9 W/m^2 for clear-sky, cloudy sky, and all-sky conditions, respectively. The root mean square errors (RMSEs) of surface solar irradiance were 80.0 W/m^2 (16.8%), 127.6 W/m^2 (55.1%), and 99.5 W/m^2 (25.5%) for clear sky, cloudy sky (overcast sky and partly cloudy sky), and all sky (clear-sky and cloudy-sky) conditions, respectively. The min/max statistical scores reported in [42] for the different seasons are 0.94/0.98 correlation coefficient, -7.46/13.31 W/m^2 MBE, 64.25/97.23 W/m^2 RMSE. The last one [2], discussing the HelioMont model based on MSG/SEVIRI, reports the statistical scores for three alpine stations. The min/max statistical scores reported are 20/40 W/m^2 MAE, -5/17 MBE W/m^2 for clear sky, while 49/72 W/m^2 MAE, -17/-8 W/m^2 MBE are reported for cloudy sky. In the HelioMont model, correction methods are also implemented to account for the effects of topography, such as shadowing, reflection, local horizon elevation angle and sky view factor. All these statistical scores, compared with those obtained using AMESIS model (Table 4), suggest that our proposed method can successfully estimate surface solar irradiance and yields similar or sometimes better performances. Obviously, in order to properly evaluate the performances of the models it would be necessary to consider the same ground-data set. Concerning processing time, it may be useful to separate the process into two steps. During the first step, HRV calibration, navigation, co-location and cloud mask run simultaneously with VIS and IR channel processes (calibration, navigation, cloud detection and classification, aerosol and cloud microphysical parameter retrievals) and, therefore, HRV processing does not add significant delay to the low channel processing chain. During the second step, i.e., the final estimation of SSI, the HRV processing takes longer than the VIS processing by a factor of five. More precisely, the operative chain process 1,415,088 HRV pixels and 79,616 low-resolution pixels centered in Italy. The runtime is 3.48 min to estimate the SSI_VIS product, while it is 4.48 min for the SSI_HRV product. The processes were run on a server with 2 x Intel Xeon Gold 6132 2.6 GHz RAM 256 GB DDR4-2667MT/S.

5. Conclusions

The AMESIS model was developed at the Institute of Methodologies for Environmental Analysis of the National Research Council of Italy (IMAA-CNR) in order to infer surface solar irradiance from a SEVIRI radiometer on board the MSG geostationary satellite. The operational version of the AMESIS model has been running continuously at IMAA-CNR over all Italy for a year in order to estimate the accuracy of the continuous monitoring from space in comparison with real PV energy inputs and outputs within the framework of the SolarCloud project. This paper shows the difference between the two AMESIS products (SSI_VIS and SSI_HRV) against simultaneous ground-based observations from a network of 63 pyranometers for different sky conditions (clear, overcast and partially clear). Regarding the impact on PV systems, a dedicated field campaign is ongoing; the in-depth analysis will be the subject of future work. For clear conditions, both products are very accurate; indeed, a small positive bias of 1.37 and 0.76 W/m^2, respectively, for SSI_VIS and SSI_HRV, and a correlation above 0.99 for both products, was found. For overcast cloudy sky, we found biases of -2.92 and 0.86 W/m^2, respectively, for SSI_VIS and SSI_HRV, and a correlation above 0.96 for both products. A large negative bias, -33.69 W/m^2, and a correlation of 0.862 were found for SSI_VIS for partially cloudy sky, whereas a bias of 0.34 W/m^2 and a correlation value of 0.995 were found for SSI_HRV. This indicates that solar resource monitoring requires footprints that minimize the effect of clouds. The HRV channel is able to detect clear areas contained within the larger pixels of low spatial resolution SEVIRI channels, thus providing more accurate surface solar irradiance, while the SSI_VIS product underestimates the surface solar irradiance due to its lower resolution. During the first step, HRV processing does not add significant delay to the low channel processing chain. During the second step, i.e., the final estimation of SSI, the HRV processing takes longer than the VIS processing by a factor of five, due to the greater number of pixels to consider. To evaluate the AMESIS performance, our statistical scores were compared against those obtained in [2,11,22,42]; this comparison suggests that our proposed AMESIS model (SSI_HRV) can successfully estimate surface solar irradiance and yields similar or sometimes better performances. This confirms that the HRV channel, or more generally, a higher-resolution sensor, can improve the estimation of solar irradiance over partially cloudy sky.

Author Contributions: F.R., D.C., E.R. (Elisabetta Ricciardelli) and E.G., designed the research, wrote the paper and contributed to evaluation process. D.G., F.D.P., S.G., A.C., S.T.N., E.R. (Ermann Ripepi), S.L., and M.V. contributed to data processing. All the co-authors helped to revise the manuscript.

Funding: This work has been financed by the Italian Ministry of Economic Development (MISE) in the framework of the SolarCloud project, contract No. B01/0771/04/X24.

Conflicts of Interest: The authors declare no conflict of interest. The founding sponsors had no role in the design of the study; in the collection, analyses, or interpretation of data; in the writing of the manuscript, and in the decision to publish the results.

References

1. Roupioz, L.; Jia, L.; Nerry, F.; Menenti, M. Estimation of daily solar radiation budget at kilometer resolution over the Tibetan Plateau by integrating MODIS data products and a DEM. *Remote Sens.* **2016**, *8*, 504. [CrossRef]
2. Castelli, M.; Stöckli, R.; Zardi, D.; Tetzlaff, A.; Wagner, J.E.; Belluardo, G.; Zebisch, M.; Petitta, M. The HelioMont method for assessing solar irradiance over complex terrain: Validation and improvements. *Remote Sens. Environ.* **2014**, *152*, 603–613. [CrossRef]
3. BWielicki, B.; Cess, R.D.; King, M.D.; Randall, D.A.; Harrison, E.F. Mission to planet Earth: Role of clouds and radiation in climate. *Bull. Am. Meteor. Soc.* **1995**, *76*, 2125–2153. [CrossRef]
4. Iqbal, M. *An Introduction to Solar Radiation*; Academic Press: Cambridge, MA, USA, 1983.
5. Perez, R.; Seals, R.; Zelenka, A. Comparing satellite remote sensing and ground network measurements for the production of site/time specific irradiance data. *Sol. Energy* **1997**, *60*, 89–96. [CrossRef]
6. Zelenka, A.; Czeplak, G.; D'Agostino, V.; Josefson, W.; Maxwell, E.; Perez, R. *Techniques for Supplementing Solar Radiation Network Data*; International Energy Agency: Paris, France, 1992.

7. Mueller, R.; Behrendt, T.; Hammer, A.; Kemper, A. A New Algorithm for the Satellite-Based Retrieval of Solar Surface Irradiance in Spectral Bands. *Remote Sens.* **2012**, *4*, 622–647. [CrossRef]

8. Deneke, H.M.; Feijt, A.J.; Roebeling, R.A. Estimating surface solar irradiance from METEOSAT SEVIRI-derived cloud properties. *Remote Sens. Environ.* **2008**, *112*, 3131–3141. [CrossRef]

9. Qu, Z.; Gschwind, B.; Lefèvre, M.; Wald, L. Improving HelioClim-3 estimates of surface solar irradiance using the McClear clear-sky model and recent advances in atmosphere composition. *Atmos. Meas. Tech.* **2014**, *7*, 3927–3933. [CrossRef]

10. Journée, M.; Bertrand, C. Improving the spatio-temporal distribution of surface solar radiation data by merging ground and satellite measurement. *Remote Sens. Environ.* **2010**, *114*, 2692–2704. [CrossRef]

11. Xia, S.; Mestas-Nuñez, A.M.; Xie, H.; Vega, R. An Evaluation of Satellite Estimates of Solar Surface Irradiance Using Ground Observations in San Antonio, Texas, USA. *Remote Sens.* **2017**, *9*, 1268. [CrossRef]

12. Antonanzas-Torres, F.; Martínez de Pisón, F.J.; Antonanzas, J.; Perpiñán, O. Downscaling of global solar irradiation in complex areas in R. *J. Renew. Sustain. Energy* **2014**, *6*, 063105. [CrossRef]

13. Bessafi, M.; Oree, V.; Khoodaruth, A.; Jumaux, G.; Bonnardot, F.; Jeanty, P.; Delsaut, M.; Chabriat, J.-P.; Dauhoo, M.Z. Downscaling solar irradiance using DEM-based model in young volcanic islands with rugged topography. *Renew. Energy* **2018**, *126*, 584–593. [CrossRef]

14. Journée, M.; Bertrand, C. Geostatistical merging of ground-based and satellite-derived data of surface solar radiation. *Adv. Sci. Res.* **2011**, *6*, 1–5. [CrossRef]

15. Wang, H.M.; Pinker, R.T. Shortwave radiative fluxes from MODIS: Model development and implementation. *J. Geophys. Res.* **2009**, *114*. [CrossRef]

16. Huang, G.; Liu, S.; Liang, S. Estimation of net surface shortwave radiation from MODIS data. *Int. J. Remote Sens.* **2012**, *33*, 804–825. [CrossRef]

17. Sun, Z.; Liu, J.; Zeng, X.; Liang, H. Parameterization of instan-taneous global horizontal irradiance: Cloudy-sky component. *J. Geophys. Res.* **2012**, *117*. [CrossRef]

18. Mueller, R.; Matsoukas, C.; Gratzki, A.; Behr, H.; Hollmann, R. The CM–SAF operational scheme for the satellite based retrieval of solar surface irradiance–A LUT based eigenvector hybrid approach. *Remote Sens. Environ.* **2009**, *113*, 1012–1024. [CrossRef]

19. Geraldi, E.; Romano, F.; Ricciardelli, E. An Advanced Model for the Estimation of the Surface Solar Irradiance Under All Atmospheric Conditions Using MSG/SEVIRI Data. *IEEE Trans. Geosci. Remote Sens.* **2012**, *50*, 2934–2953. [CrossRef]

20. Lu, N.; Liu, R.; Liu, J.; Liang, S. An algorithm for estimating downward shortwave radiation from GMS 5 visible imagery and its evaluation over China. *J. Geophys. Res.* **2010**, *115*. [CrossRef]

21. Huang, G.; Ma, M.; Liang, S.; Liu, S.; Li, X. A LUT-based approach to estimate surface solar irradiance by combining MODIS and MTSAT data. *J. Geophys. Res.* **2011**, *116*. [CrossRef]

22. Zhang, H.; Huang, C.; Yu, S.; Li, L.; Xin, X.; Liu, Q. A Lookup-Table-Based Approach to Estimating Surface Solar Irradiance from Geostationary and Polar-Orbiting Satellite Data. *Remote Sens.* **2018**, *10*, 411. [CrossRef]

23. Hammer, A.; Heinemann, D.; Hoyer, C.; Lorenz, E.; Muller, R.; Beyer, H.G. Solar energy assessment using remote sensing technologies. *Remote Sens. Environ.* **2003**, *86*, 423–432. [CrossRef]

24. Rigollier, C.; Lefèvre, M.; Wald, L. The method Heliosat-2 for deriving shortwave solar radiation from satellite images. *Sol. Energy* **2004**, *77*, 159–169. [CrossRef]

25. Posselt, R.; Mueller, R.; Stöckli, R.; Trentmann, J. Remote sensing of solar surface radiation for climate monitoring—The CM-SAF retrieval in international comparison. *Remote Sens. Environ.* **2012**, *118*, 186–198. [CrossRef]

26. Linares-Rodriguez, A.; Ruiz-Arias, J.A.; Pozo-Vazquez, D.; Tovar-Pescador, J. An artificial neural network ensemble model for estimating global solar radiation from Meteosat satellite images. *Energy* **2013**, *61*, 636–645. [CrossRef]

27. Antonanzas-Torres, F.; Urraca, R.; Antonanzas, J.; Fernandez-Ceniceros, J.; Martinez-de-Pison, F.J. Generation of daily global solar irradiation with support vector machines for regression. *Energy Convers. Manag.* **2015**, *96*, 277–286. [CrossRef]

28. Zou, L.; Wang, L.; Lin, A.; Zhu, H.; Peng, Y.; Zhao, Z. Estimation of global solar radiation using an artificial neural network based on an interpolation technique in southeast China. *J. Atmos. Sol. Terr. Phys.* **2016**, *146*, 110–122. [CrossRef]

29. Tang, W.; Qin, J.; Yang, K.; Liu, S.; Lu, N.; Niu, X. Retrieving high-resolution surface solar radiation with cloud parameters derived by combining MODIS and MTSAT data. *Atmos. Chem. Phys.* **2016**, *16*, 2543–2557. [CrossRef]
30. Romano, F.; Cimini, D.; Nilo, S.T.; Di Paola, F.; Ricciardelli, E.; Ripepi, E.; Viggiano, M. The Role of Emissivity in the Detection of Arctic Night Clouds. *Remote Sens.* **2017**, *9*, 406. [CrossRef]
31. Ricciardelli, E.; Romano, F.; Cuomo, V. Physical and statistical approaches for cloud identification usingMeteosat Second Generation-Spinning Enhanced Visible and Infrared Imager Data. *Remote Sens. Environ.* **2008**, *112*, 2741–2760. [CrossRef]
32. Reuter, M.; Thomas, W.; Albert, P.; Lockhoff, M.; Weber, R.; Karlsson, K.G.; Fischer, J. The CM-SAF and FUB Cloud Detection Schemes for SEVIRI: Validation with Synoptic Data and Initial Comparison with MODIS and CALIPSO. *J. Appl. Meteorol. Climatol.* **2009**, *48*, 301–316. [CrossRef]
33. Derrien, M.; Le Gléau, H. MSG/SEVIRI cloud mask type from SAFNWC. *Int. J. Remote Sens.* **2005**, *26*, 4707–4732. [CrossRef]
34. Bley, S.; Deneke, H. A Threshold-based cloud mask for the high-resolution visible channel of Meteosat second generation SEVIRI. *Atmos. Meas. Tech.* **2013**, *6*, 2713–2723. [CrossRef]
35. Ackerman, S.A.; Strabala, K.I.; Menzel, W.P.; Frey, R.A.; Moeller, C.C.; Gumley, L.E. Discriminating clear sky from clouds with MODIS. *J. Geophys. Res.* **1998**, *103*, 132–141. [CrossRef]
36. Hocking, J.; Francis, P.N.; Saunders, R. Cloud detection in Meteosat Second Generation imagery at the Met Office. *Meteorol. Appl.* **2011**, *18*, 307–323. [CrossRef]
37. Amato, U.; Antoniadis, A.; Cuomo, V.; Cutillo, L.; Franzese, M.; Murino, L.; Serio, C. Statistical cloud detection from SEVIRI multispectral images. *Remote Sens. Environ.* **2008**, *112*, 750–766. [CrossRef]
38. Asmala, A.; Shaun, Q. Cloud masking for remotely sensed data using spectral and principal components analysis. *ETASR Eng. Technol. Appl. Sci. Res.* **2012**, *2*, 221–225.
39. Nair, M.S.; Lakshmanan, R.; Wilscy, M.; Tatavarti, R. Fuzzy logic-based automatic contrast enhancement of satellite images of ocean. *Signal Image Video Process.* **2011**, *5*, 69–80. [CrossRef]
40. Bose, A.; Mali, K. Fuzzy-based artificial bee colony optimization for gray image segmentation. *Signal Image Video Proc.* **2016**, *10*, 1089–1096. [CrossRef]
41. Boulifa, M.; Adane, A.; Rezagui, A.; Ameur, Z. Estimate of the Global Solar Radiation by Cloudy Sky Using HRV Images. *Energy Proc.* **2015**, *74*, 1079–1089. [CrossRef]
42. Bouchouicha, K.; Razagui, A.; Bachari, N.E.I.; Aoun, N. Estimation of Hourly Global Solar Radiation Using MSG-HRV images. *Int. J. Appl. Environ. Sci.* **2016**, *11*, 351–368, ISSN 0973-6077. Available online: https://www.ripublication.com/ijaes16/ijaesv11n2_01.pdf (accessed on 27 July 2018).
43. Hammer, A.; Kühnert, J.; Weinreich, K.; Lorenz, E. Short-Term Forecasting of Surface Solar Irradiance Based on Meteosat-SEVIRI Data Using a Nighttime Cloud Index. *Remote Sens.* **2015**, *7*, 9070–9090. [CrossRef]
44. Schmetz, J.; Pili, P.; Tjemkes, S.; Just, D.; Kerkmann, J.; Rota, S.; Ratier, A. An Introduction to Meteosat second generation (MSG). *Bull. Am. Meteorol. Soc.* **2002**, *83*, 977–992. [CrossRef]
45. Gallucci, D.; Romano, F.; Cersosimo, A.; Cimini, D.; Di Paola, F.; Gentile, S.; Geraldi, E.; Larosa, S.; Nilo, S.T.; Ricciardelli, E.; et al. Nowcasting Surface Solar Irradiance with AMESIS via Motion Vector Fields of MSG-SEVIRI Data. *Remote Sens.* **2018**, *10*, 845. [CrossRef]
46. Steinacker, R.; Haberli, C.; Pottschacher, W. A transparent method for the analysis quality evaluation of irregularly distributed noisy observational data. *J. Appl. Meteorol.* **2000**, *12*, 2303–2316. [CrossRef]
47. Lussana, C.; Uboldi, F.; Salvati, M.R. A spatial consistency test for surface observations from mesoscale meteorological networks. *Q. J. R. Meteorol. Soc.* **2010**, *136*, 1075–1088. [CrossRef]
48. Romano, F.; Ricciardelli, E.; Cimini, D.; Di Paola, F.; Viggiano, M. Dust Detection and Optical Depth Retrieval Using MSG-SEVIRI Data. *Atmosphere* **2013**, *4*, 35–47. [CrossRef]
49. Ricciardell, E.; Cimini, D.; Di Paola, F.; Romano, F.; Viggiano, M.A. A statistical approach for rain intensity differentiation using Meteosat Second Generation-Spinning enhanced visible and infrared imager observations. *Hidrol. Earth Syst. Sci.* **2014**, *18*, 2559–2576. [CrossRef]
50. Mace, G.G.; Zhang, Q. The CloudSat radar-lidar geometrical profile product (RL-GeoProf): Updates, improvements, and selected results. *J. Geophys. Res. Atmos.* **2014**, *119*. [CrossRef]
51. Hanson, C.; Mueller, J. Status of the SEVIRI Level 1.5 Data. In Proceedings of the Second MSG RAO Workshop (ESA SP-582, November 2004), Salzburg, Austria, 9–10 September 2004; Available online: http://earth.esa.int/workshops/msg_rao_2004/papers/4_hanson.pdf (accessed on 24 June 2018).

52. Dürr, B.; Zelenka, A. Deriving surface global irradiance over the Alpine region from Meteosat Second Generation by supplementing the HELIOSAT method. *Int. J. Remote Sens.* **2009**, *30*, 5821–5841. [CrossRef]

53. Nilo, S.T.; Romano, F.; Cermak, J.; Cimini, D.; Ricciardelli, E.; Cersosimo, A.; Di Paola, F.; Gallucci, D.; Gentile, S.; Geraldi, E.; et al. Fog Detection Based on Meteosat Second Generation-Spinning Enhanced Visible and InfraRed Imager High Resolution Visible Channel. *Remote Sens.* **2018**, *10*, 541. [CrossRef]

54. Krijger, J.M.; van Weele, M.; Aben, I.; Frey, R. Technical Note: The effect of sensor resolution on the number of cloud-free observations from space. *Atmos. Chem. Phys.* **2007**, *7*, 2881–2891. [CrossRef]

remote sensing

MDPI

Article

Validation of the SARAH-E Satellite-Based Surface Solar Radiation Estimates over India

Aku Riihelä [1,*], **Viivi Kallio** [1], **Sarvesh Devraj** [2], **Anu Sharma** [3] and **Anders V. Lindfors** [1]

1   Finnish Meteorological Institute, Erik Palménin aukio 1, P.O. Box 503, FI-00101 Helsinki, Finland;
    viivi.kallio@fmi.fi (V.K.); anders.lindfors@fmi.fi (A.V.L.)
2   The Energy and Resources Institute, Darbari Seth Block, IHC Complex, Lodhi Road,
    New Delhi 110003, India; sarvesh.devraj@teri.res.in or sarveshdevraj@gmail.com
3   TERI School of Advanced Studies, Vasant Kunj Institutional Area 10, New Delhi 110070, India;
    anu.sharma@teriuniversity.ac.in
*   Correspondence: aku.riihela@fmi.fi

Received: 15 December 2017; Accepted: 2 March 2018; Published: 3 March 2018

Abstract: We evaluate the accuracy of the satellite-based surface solar radiation dataset called Surface Solar Radiation Data Set-Heliosat (SARAH-E) against in situ measurements over a variety of sites in India between 1999 and 2014. We primarily evaluate the daily means of surface solar radiation. The results indicate that SARAH-E consistently overestimates surface solar radiation, with a mean bias of 21.9 W/m^2. The results are complicated by the fact that the estimation bias is stable between 1999 and 2009 with a mean of 19.6 W/m^2 but increases sharply thereafter as a result of rapidly decreasing (dimming) surface measurements of solar radiation. In addition, between 1999 and 2009, both in situ measurements and SARAH-E estimates described a statistically significant (at 95% confidence interval) trend of approximately -0.6 W/m^2/year, but diverged strongly afterward. We investigated the cause of decreasing solar radiation at one site (Pune) by simulating clear-sky irradiance with local measurements of water vapor and aerosols as input to a radiative transfer model. The relationship between simulated and measured irradiance appeared to change post-2009, indicating that measured changes in the clear-sky aerosol loading are not sufficient to explain the rapid dimming in measured total irradiance. Besides instrumentation biases, possible explanations in the diverging measurements and retrievals of solar radiation may be found in the aerosol climatology used for SARAH-E generation. However, at present, we have insufficient data to conclusively identify the cause of the increasing retrieval bias. Users of the datasets are advised to be aware of the increasing bias when using the post-2009 data.

Keywords: surface solar radiation; remote sensing; validation; India; solar radiation trends

---

1. Introduction

India has experienced an extended period of rapid economic growth driven largely by coal-based nonrenewable energy production [1]. Given the climatic and environmental issues inherent in reliance on nonrenewable energy sources, the Indian Government has recently launched an ambitious initiative to substantially increase solar energy production at a national level [2,3]. As a low-latitude country, India has a large solar resource at its disposal. However, for planning the locations and production capability of the solar energy plants, accurate mapping of the surface incoming solar radiation (SSR), or global radiation as it is also called, is required.

For spatiotemporally comprehensive mapping of SSR over the Indian subcontinent, satellite-based remote sensing offers the most cost effective solution. To answer this need, the Satellite Application Facility on Climate Monitoring (CM SAF), a project of the European Organization for the Exploitation of Weather Satellites (EUMETSAT), has developed and released a long-term dataset. This dataset, called Surface Solar Radiation Data Set—Heliosat, Meteosat-East (SARAH-E), Edition 1 [4,5], is a satellite-based climate data record of the solar radiative quantities over India and surrounding regions. The dataset is derived from observations of the reflected solar radiation from the geostationary Meteosat Indian Ocean Data Coverage (IODC) satellites. The dataset spans 15 years at a spatial resolution of 0.05°.

In addition to SARAH-E, there exists a growing number of satellite-derived datasets of SSR, calculated with a variety of algorithms for both regional [6,7] and global coverage [8–10]. A multitude of studies also exists on the validation of said datasets (e.g., [11–13]), but to our knowledge, satellite-based SSR estimates, including SARAH-E, have not been validated over India at the sub-continental scale. Therefore, to facilitate the use of this satellite-based dataset, we must be able to ascertain its accuracy in determining SSR, its seasonal cycle and any possible trends. The goal of this paper is to provide this information, specifically answering these research questions:

- What is the accuracy of the SARAH-E dataset against quality controlled in situ SSR measurements from sites all across India? Is the SSR estimation accuracy stable over time? Are there regional differences in SSR estimation bias?
- Are the source datasets in general robust and long-term enough for reliable trend determination? If yes, what can we say about trends in SSR over India over the past 15 years? Do the satellite-based estimates agree with in situ measurements over the magnitude and direction of the trends?

We use ~120 station-years of in situ SSR measurement data from 17 sites across India, maintained by the India Meteorological Department (IMD), as our reference. The sites cover a wide variety of environmental conditions, from rural mountainous areas to sites located in India's largest cities. The paper is organized as follows: We first summarize the primary parameters of the SARAH-E dataset, followed by a description of the in situ measurement dataset, its accuracy, and quality control measures. We then present the validation methods, metrics, and the study results themselves. The paper concludes with a discussion of the obtained results and the conclusions drawn.

Although the SARAH-E dataset also contains estimates for direct solar radiation, here we focus exclusively on investigation of the SSR estimate. In this study, when referring to a daily mean SSR, we mean the 24-h mean of insolation. Further, monthly mean SSR values are derived from averaging the daily means to enable comparability between the satellite estimates, provided at daily mean resolution, and the in situ data, provided on an hourly basis.

2. Materials and Methods

2.1. SARAH-E

The SARAH-E satellite dataset being evaluated covers 1999–2015, with a spatial resolution of 0.05 degrees, based on observations from the Meteosat Visible Infra-Red Imager (MVIRI) instruments on board the Meteosat First Generation (MFG) satellites 5 and 7 [4,5]. Here, we validate the daily means of SARAH-E. The year 2006 was not processed in the original release of SARAH-E; however, it was post-processed and provided for this validation study, to be later included in the publicly available data. In Figure 1, we illustrate the average SSR in June over the Indian subcontinent between 1999 and 2015, based on averaging the relevant daily means.

The retrieved SSR in SARAH-E is based on the following equation:

$$SSR = SSR_{CLR}(1 - CAL) \tag{1}$$

where SSR_{CLR} is the clear-sky irradiance and CAL is the effective cloud albedo, as observed by the satellite imager (*CAL* is also referred to as cloud index). The method, also known as the Heliosat method [14–16], is based on the concept that all solar radiation entering the Earth's atmosphere will either be reflected away by the cloud cover or will penetrate it to reach the surface as (attenuated) solar irradiance. Therefore, the absorbing and scattering impacts on SSR from atmospheric constituents such as aerosols or water vapor, or variable solar geometry conditions are accounted for in the calculation of the clear-sky irradiance SSR_{CLR}. Absorption and saturation effects in optically thick clouds (CAL > 0.8) are considered through adjustment of the cloud albedo. The clear-sky irradiance is obtained from a linear interpolation of a look-up-table (LUT) of precomputed irradiances from the libRadTran model (Mayer and Kylling, 2005) for a wide range of atmospheric states [5]. Full retrieval details are available in Gracia-Amillo et al. [4].

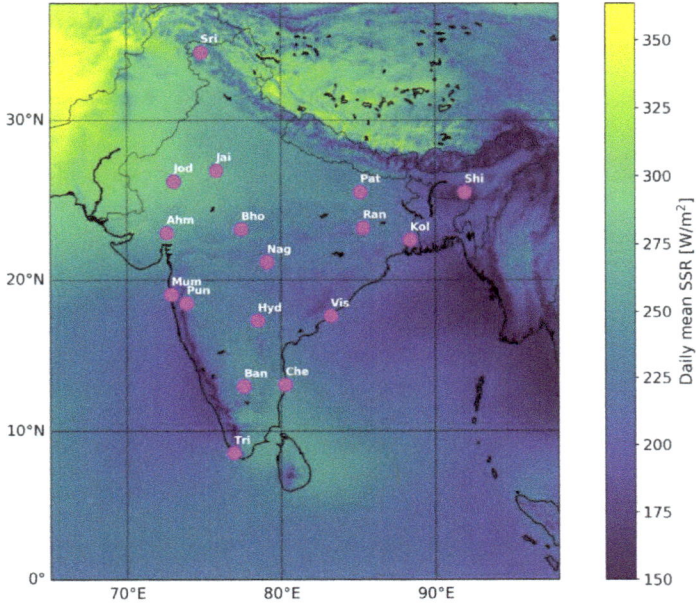

Figure 1. June mean SSR during 1999–2015 over the Indian subcontinent from SARAH-E data. Validation site locations shown with circular markers and partial site names. Decreased SSR along the western seaboard reflects the intense coastal cloudiness of the monsoon season.

The atmospheric state inputs considered in SARAH-E are the total atmospheric water vapor, aerosol optical depth (at 550 nm) and type, and ozone content. Water vapor data are from the monthly means of the ERA-Interim reanalysis [17]. Aerosols are considered as a climatology, calculated from the 2003–2010 monthly means of the Monitoring Atmospheric Composition and Climate (MACC) dataset [18,19]. Ozone content is considered through climatological values from standard profiles of the Max-Planck Institute of Air Chemistry [20].

Furthermore, the effective cloud albedo in SARAH-E is calculated from the following equation:

$$CAL = \frac{C_{VIS} - C_{CLS}}{C_{MAX} - C_{CLS}} \tag{2}$$

where all terms are based on observed digital counts of the MVIRI instrument, i.e., uncalibrated observed reflectivities. C_{VIS} is the observed count of the observation (pixel value) for which CAL is to be determined, C_{CLS} is an iteratively precomputed clear-sky count for the observed time slot and pixel (keeping in mind that the viewing geometry is fixed), and C_{MAX} is the typical (statistical) maximum reflectivity count for thick bright clouds, precomputed as the 95th percentile of observed counts in a region within 45.8°S–52.7°S and 40°E–56°E, known for being consistently very cloudy. Before usage in this equation, the MVIRI counts are corrected for the instrument dark current as well as variations in

Sun–Earth distance and Solar Zenith Angle. C$_{MAX}$ is also corrected for the effects of varying scattering angle between time slots. See Gracia-Amillo et al. [4] for further details on SARAH-E calculations.

2.2. In Situ SSR Measurements

The India Meteorological Department (IMD) operates a network of sites where SSR is observed on a regular basis. Time series of SSR observations from this network forms the primary in situ reference data source of this study. The locations of the sites are shown in Figure 1 and their primary attributes are listed in Table 1. The provided SSR observations were daytime hourly means only (05:00–20:00 Indian time), extended here into full 24-h days with nighttime hours set to zero irradiance.

Table 1. Characteristics of the India Meteorological Department (IMD) sites measuring Surface Solar Radiation (SSR).

Site	Latitude (°N)	Longitude (°E)	Data Coverage	Comments and Climate Regime
Bangalore	12.9716	77.5946	1999–2009	Urban, tropical
Jodhpur	26.2389	73.0243	1999–2014	Desert, arid
Bhopal	23.2599	77.4126	1999–2009	Humid, subtropical
Pune	18.5204	73.8567	1999–2014	Suburban
Sri Nagar	34.0837	74.7973	1999–2009	Valley, mountainous
Jaipur	26.9124	75.7873	1999–2009	Semi-arid, dry
Ranchi	23.3441	85.3096	1999–2010	Urban, hilly
Visakhapatnam	17.6868	83.2185	1999–2014	Coastal, urban
Chennai	13.0827	80.2707	1999–2014	Coastal, urban
Trivandrum	8.5241	76.9366	1999–2014	Coastal, tropical
Nagpur	21.1458	79.0882	1999–2010, 2013–2014	Urban, dry apart from monsoon
Shillong	25.5788	91.8933	1999–2014	Subtropical highland, low AOD
Ahmedabad	23.0225	72.5714	1999–2009, 2013–2014	Dry, semi-arid
Hyderabad	17.385	78.4867	1999–2014	Semi-arid
Kolkata	22.5726	88.3639	1999–2012	Sparse data, urban subtropical
Mumbai	19.076	72.8777	1999–2010, 2013–2014	Coastal, urban
Patna	25.5941	85.1376	1999–2009	Urban

The IMD irradiance measurement network is described in detail in Soni et al. [21]. The sites operate WMO Class 1 or 2 pyranometers, with periodic re-calibrations every two years against reference standards (cavity pyrheliometer as primary, PSP pyranometers as secondary) according to present knowledge. For most of the sites, a change of pyranometer instrumentation and data loggers took place in 2009 to equipment manufactured by Apogee Instruments and EKO Instruments. The pyranometers are cleaned and maintained daily at principal measurement sites (e.g., Pune). The provided hourly averages were subjected to further quality assurance procedures (see below).

The sites represent a wide variety of atmospheric and surface regimes, from the dry, arid conditions of Jodhpur and Jaipur to the yearlong tropical monsoon climate of Trivandrum in the south. The summer monsoon season (southwest monsoon; June–September) brings intensive cloudiness and rainfall to sites in the central and southern parts of the subcontinent. Sites in the north such as Sri Nagar are also affected by the monsoon, but to a much lesser degree [22]. Some sites, such as Chennai, also experience rains during the northeast monsoon during October–December, further altering their annual rainfall cycle.

In terms of aerosol properties, the sites again display a complex variety of conditions and annual cycles. During the monsoon, sea salts and marine dust form the dominant contribution to aerosol loading over coastal sites, also impacting inland sites in the central and southern parts of the subcontinent. On the other hand, rainfall scavenging during monsoon often lowers the overall aerosol loading of the atmosphere [23]. Anthropogenic industrial and automotive sources have a large impact on aerosol loading over the urban, densely industrialized sites such as Hyderabad or Bangalore [24,25]. Increased biomass burning during the cooler winter months leads to an increase in aerosol loading over several urban sites, such as Kolkata [26]. Dust storms are a major natural source of aerosol loading over the arid sites in Northern India (Jodhpur and Jaipur) [27]. Because of its location in the foothills of the Himalayas, Shillong is the only site in the study with a relatively clean atmosphere, with AOD generally in the range of 0.2–0.25 [28].

2.3. Quality Assurance of SSR Measurements

Quality control of in situ SSR measurements is a critical component to be undertaken prior to any analysis [29–31]. A series of quality control tests were made to ensure the robustness of the reference SSR dataset and to provide both liberal and conservative quality control limits for the dataset. The tests conducted were the BSRN (Baseline Surface Radiation Network) quality check made by Long and Dutton [32] for the use of irradiance data quality control in the BSRN program, a 0-limit for daytime values, and percentile based upper and lower limits. In addition to the tests, some basic quality control measures were taken. The data were found to have some measurement periods with a fixed SSR; these periods were all discarded as unphysical.

The BSRN quality check has a limit for physically possible and extremely rare values. The extremely rare-limit was used here, for which the lower limit is $-2 \, \text{W}/\text{m}^2$, and the upper limit is calculated with:

$$\text{BSRN}_{\text{upper}} = S_a \times 1.2 \times \mu_0^{1.2} + 50 \frac{\text{W}}{\text{m}^2} \tag{3}$$

where $S_a = S0/AU^2$, S0 is the solar constant at mean Earth–Sun distance, AU is the Earth–Sun distance in astronomical units, μ_0 equals the cosine of SZA, and SZA is the solar zenith angle [32]. In the calculation of the astronomical quantities and solar position, the python-pvlib package was used [33]. The 0-limit for daytime irradiance values was necessary as several days with $0 \, \text{W}/\text{m}^2$ measurements for daytime irradiance were found after applying the BSRN limits. All irradiances of $0 \, \text{W}/\text{m}^2$ for times when SZA < 90 degrees were considered unrealistic and thus removed. All irradiances exceeding the upper limit or below the lower limit in BSRN were removed. As only days with full coverage (24 available one-hour averages) are used in this analysis, this approach provides the highest coverage without affecting the data reliability.

To provide a more conservative quality control test, percentile based limits were calculated from the data to remove some of the physically possible, yet highly unlikely irradiances from both lower and higher ends of the data. This was made by dividing the data into bins by calendar month and time of day. The lower percentile limit was determined as the 1st percentile of each bin, and the upper limit as the 98th percentile with 20% of the 98th percentile added as follows:

$$\text{Percentile top limit} = \text{98th percentile} + \text{98th percentile} \times 0.2. \tag{4}$$

Note that the data used for the calculation of percentiles contain only the days for which full coverage remains after the BSRN and 0-limit corrections. Other percentiles were tested; however, the aforementioned limits were found to limit the erroneous values appropriately with little effect on the data distribution. The percentile limits were also combined to form the most conservative set of limits. As the validation metrics showed a dependence of less than $1 \, \text{W}/\text{m}^2$ on the chosen set of QA limits, we present only the in situ data that have passed the most rigorous QA testing in the subsequent analysis. The original dataset consisted of 74,719 days of observations (204.71 station-years). The various QA phases discarded a fraction of these observations as follows (cumulative and relative to the original dataset): BSRN testing, 17.34%; daytime 0-limit filtering, 31.27%; percentile top limit, 33.20%; lower percentile limit, 35.76%; and top and low percentile limit combined, 37.54%. Therefore, after the strictest set of QA measures, a total of 46,670 full days of observations (127.86 station-years) remained for the validation. A visualization of the limits employed is available as Figure S1.

As a further QA measure, we obtained level 1.5 (V3) measurement data on aerosols and water vapor from the Aerosol Robotic Network (AERONET) site at Pune, spanning 2004–2014 [34]. These cloud-screened AERONET measurements acted as inputs for the python-pvlib implementation of the Simplified Solis radiative transfer model [35], which provided an independent estimate of clear-sky SSR to compare against the in situ SSR measurements at Pune. The comparison is presented in Section 3. In addition to Pune, the AERONET site at Jaipur also possesses a relatively long-term measurement record. However, the Jaipur AERONET data begin in 2009 when the SSR measurements end at the IMD site, thus no comparison at this site was possible.

2.4. Validation Metrics and Methods

We have chosen to use validation methods and metrics applied in past studies (e.g., [11,13,36]) to facilitate easier comparison of results. Thus, for each site, we report the Mean Bias Difference, Mean Absolute Bias Difference and Root Mean Square Difference (MBD, MABD and RMSD, respectively):

$$MBD = \frac{1}{n} \sum_{i=1}^{n} (SIS_i - F_i) \tag{5}$$

$$MABD = \frac{1}{n} \sum_{i=1}^{n} |SIS_i - F_i| \tag{6}$$

$$RMSD = \sqrt{\frac{\sum_{i=1}^{n} (SIS_i - F_i)^2}{n}} \tag{7}$$

where SIS is the satellite-derived SSR, F is the site-measured SSR, and n is the number of available valid observations (in the period in question).

The in situ SSR measurement data were spatiotemporally collocated to the SARAH-E satellite-based estimates. The hourly in situ data were averaged into daily means, discarding any days with missing hourly measurements. The daily means were then spatially matched with SARAH-E daily means using a nearest-neighbor search. A nearest-neighbor collocation retains the basic impediment of a point-to-pixel comparison, however, the representativeness impact into SSR bias determination has been studied over Europe and generally found to be very small for a spatial resolution of 0.05 degrees, such as SARAH-E [37]. However, this result may not hold for sites in mountainous terrain with rapidly varying topography. In this study, the site at Sri Nagar is potentially such a site, but we have retained its measurements for completeness.

3. Results and Discussion

We begin with a general overview of SARAH-E performance. Figure 2 shows the temporal evolution of the SARAH-E SSR estimation bias against all in situ measurement sites with valid data. The result suggests very stable estimation precision and scatter between 1999 and 2009, but increasing and more variable estimation bias since then. The difference in annual mean SSR bias between 2013 and 2000 is 23.9 W/m^2, while the annual mean bias between 1999 and 2009 varied by less than 5 W/m^2.

We point out that the in situ data shown here have been filtered with the strictest set of tests, i.e., BSRN tests combined with discarding realistic but improbable daily means with SSR in the lowest 1%, or in the highest 98% + 0.2 × 98%. We point out that the mean bias or standard deviation of the full comparison varied by less than 1 W/m^2 between this set and the set filtered with the basic BSRN tests, making the choice of filter conditions of secondary important. Because IMD measurements were available only until 2014, the 2015 SARAH-E data are not included and do not affect the results.

Resolving the validation for each individual site shows that the upward trend in SSR estimation bias is not specific to any subset of the validation sites, but rather a common feature for many of the investigated sites (Figure 3). The year-to-year variation and level shifts in MBD over some sites such as Jodhpur or Ahmedabad are nearly always a result of variations in measured SSR; SARAH-E retrieved SSR generally varies much less from year to year. The validation metrics for the individual sites are also collected and shown in Table 2. For MBD, we also show value for both pre- and post-2009 periods to illustrate the scale of the change in bias. We further note that MBD and MABD are similar for many sites, suggesting that SARAH-E consistently overestimates the in situ SSR measurement, as shown in Figure 2.

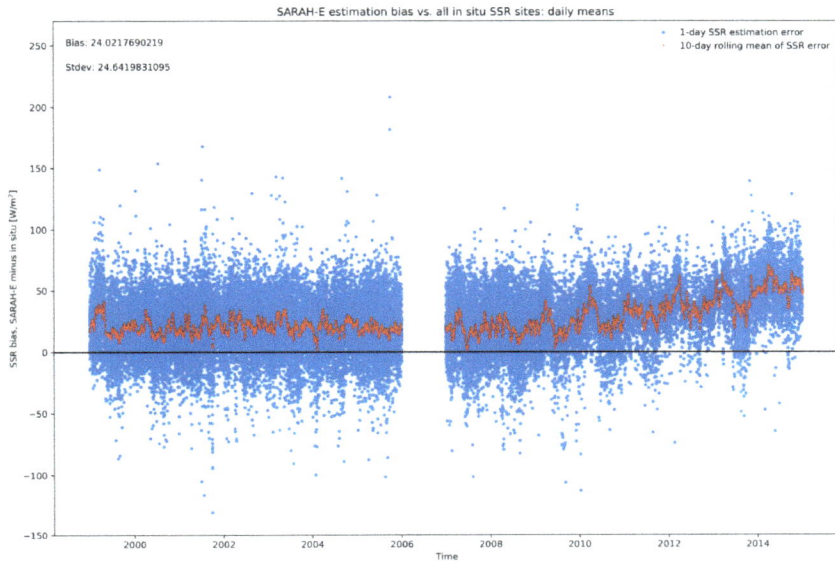

Figure 2. SARAH-E SSR estimation bias versus all in situ measurement sites. Blue markers are the daily mean SSR estimation errors (marker per site per day), while the red markers are the 10-day rolling mean of estimation errors.

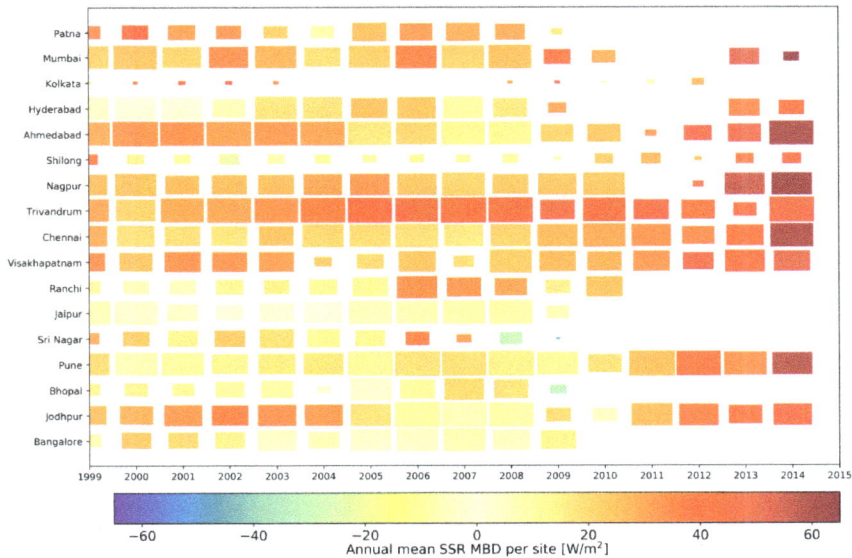

Figure 3. Annual mean SSR estimation bias per site (MBD, satellite minus in situ, W/m^2). Rectangle color indicates the magnitude of annual mean SSR bias, size is proportional to the amount of valid in situ measurement days in each year.

Table 2. Validation metrics for all sites (QA method: conservative).

Site	MBD (W/m^2)	MABD (W/m^2)	RMSD (W/m^2)	MBD (W/m^2) 1999–2009	MBD (W/m^2) 2010–2014
Bangalore	9.12	18.84	24.39	9.12	N/A
Jodhpur	24.60	26.64	31.42	21.18	37.71
Bhopal	10.45	18.57	24.09	10.45	N/A
Pune	20.92	25.60	31.46	14.69	42.84
Sri Nagar	17.04	26.71	35.12	17.04	N/A
Jaipur	5.30	13.83	18.04	5.30	N/A
Ranchi	20.92	23.92	29.60	20.26	32.67
Visakhapatnam	31.83	34.34	39.65	27.55	47.10
Chennai	26.33	30.19	37.44	20.43	47.12
Trivandrum	38.35	38.77	42.35	36.45	46.99
Nagpur	30.52	31.90	37.98	25.54	52.70
Shillong	7.85	30.15	37.61	-5.05	40.75
Ahmedabad	29.55	30.84	36.58	25.18	51.77
Hyderabad	14.62	22.16	28.22	12.35	41.65
Kolkata	27.89	35.35	43.07	38.22	37.34
Mumbai	27.85	31.25	36.29	25.99	53.35
Patna	28.93	32.98	38.92	28.93	N/A
All (mean)	**21.89**	**27.77**	**33.66**	**19.62**	**44.33**

The mean MBD and RMSD per site are geographically visualized in Figure 4. SSR retrievals over coastal validation sites have, in general, both a larger bias (MBD) and more scatter (RMSD) than retrievals further inland. Nagpur and Patna appear to be exceptions; at Nagpur, the metrics are affected by sharply decreasing measured SSR in 2013 and 2014, while at Patna, the larger MBD and RMSD are quite constant through time (Figure 3).

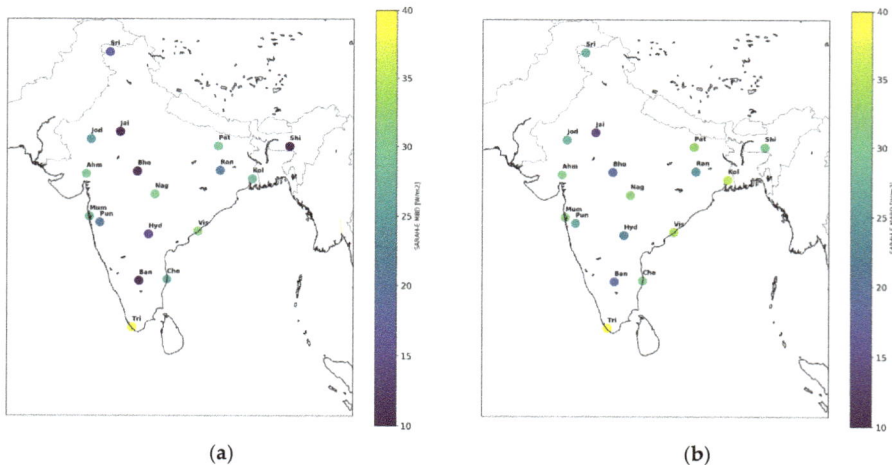

(a)   (b)

Figure 4. Geographical distribution across the validation sites of: (**a**) SARAH-E MBD; and (**b**) SARAH-E RMSD.

We also examine the SARAH-E estimation bias on monthly basis. Figure 5 illustrates these results, showing MBD (a) and relative MBD (b). The post-2009 increase in MBD is clearly seen to affect all months, though most prominently in the early and late part of the year, particularly when taking the seasonal SSR variation into account (Figure 5b). Interestingly, retrievals fare better during the monsoon season (April–June/July) contrary to expectations; a common finding in satellite-based SSR validation studies has been that biases grow during intensively cloudy periods, either because of incomplete cloud characterization from satellites, or small-scale variability in cloudiness which affects in situ measurements differently from the coarser satellite observations. The winter and autumn seasons are well-known to be less cloudy than summer over India.

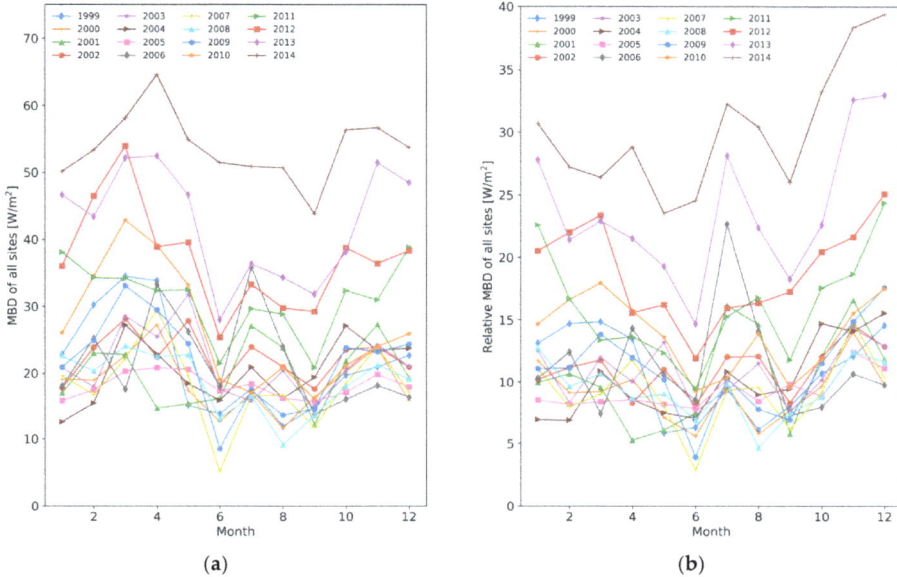

Figure 5. (**a**) Monthly mean MBD of SARAH-E vs. all validation sites resolved for 1999–2014; and (**b**) relative monthly mean MBD of SARAH-E vs. all validation sites resolved for 1999–2014.

All of the results shown here suggest the same conclusion: The mean SARAH-E SSR estimation accuracy has been 10–35 W/m^2 between 1999 and 2009 for any month of the year (with midsummer retrievals accurate to within 10–20 W/m^2), but has grown to 30–65 W/m^2 post-2009 for virtually all sites providing data for this period. These biases are on average larger than reported by Gracia Amillo et al. [4] when validating SARAH-E against Baseline Surface Radiation Network (BSRN) sites across Asia, although their analysis did not cover sites in India. In addition, their reported SSR estimation biases were in the range of −7–+14 W/m^2 for MBD, and 22–44 W/m^2 for MABD. This suggests that the biases seen here for SSR, particularly for 1999–2009, are relatively close to the expected performance envelope, although with a clear tendency for overestimation particularly after 2009. Next, we will investigate possible causes of this behavior using independent in situ measurements and stability analyses of both SARAH-E and the in situ SSR record.

Figure 6 shows a per-month stability analysis of both SARAH-E and in situ monthly mean SSR. We show the analysis as a violin plot to simultaneously compare the extents and distribution of the monthly mean SSR, averaged across all validation sites, across 1999–2014. Markers within the violin plots show the SSR values for each year. We immediately note that while the groups of SARAH-E monthly mean SSRs are relatively tightly packed and their distribution is usually fairly even, the in situ record distribution displays both wider scatter and a prominent lower tail for most months of the year. The tail is predominantly formed from values of the years 2010–2014.

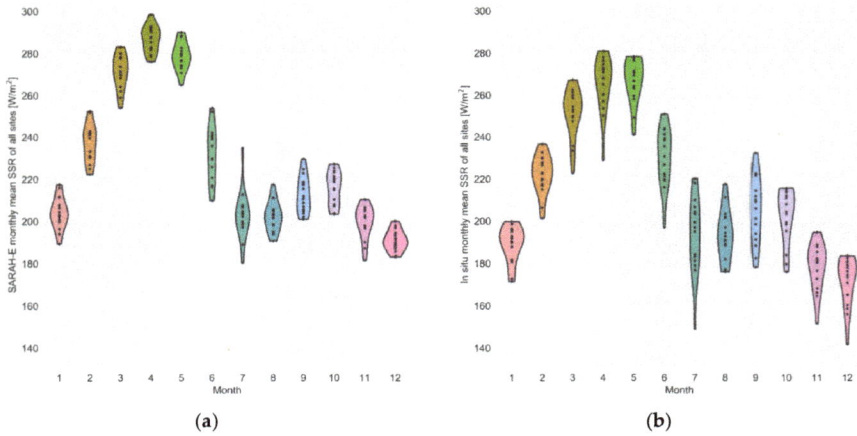

Figure 6. Monthly mean SSR over all validation sites between 1999 and 2014 from: (**a**) SARAH-E; and (**b**) in situ measurements. Violin plots indicate the kernel density estimates for each month, cropped to data extents. Markers within represent the SSR of the different years in the period.

While the differences post-2009 are clear, this comparison alone does not reveal if the increasing difference results from an undetected and uncompensated bias in the satellite retrievals or the in situ measurements. Soni et al. [21] showed a decreasing SSR trend from in situ measurements of a subset of the IMD sites used here, but for a longer analysis period which ended in 2009. They argued that the SSR measurement network is well-maintained and that a large factor in the decrease is the increasing in aerosol loading over India from anthropogenic emissions. This would imply that the atmospheric constituent records (aerosols, water vapor, etc.) used in SARAH-E generation do not adequately consider a substantial increase in aerosol loading over the subcontinent as a whole after 2010, since we observe an increasing bias for nearly all sites in that period.

To partially test this supposition, we extracted the AERONET level 1.5 measurement record at Pune, as described in Section 2. The record at Pune is the longest available for Indian AERONET sites, and the only one with a significant overlap with the SARAH data record. Level 1.5 is cloud-screened, although the final calibration of the data is not yet applied. Level 2 data are not yet available at Pune. Figure 7 shows the measured AOD at 440 and 675 nm at Pune between 2005 and 2015. The measurements show slight increasing trends in AOD at these wavelengths close to the one used in SARAH generation from the MACC dataset. However, the increases are modest in comparison to the magnitude of bias growth. Moreover, between 2010 and 2013, there is little evidence of an increasing aerosol loading, whereas the SARAH-E MBD was already seen to grow over several sites (Figure 3). Similarly, the single scattering albedo (SSA) of aerosols at 440 and 675 nm at Pune showed no evidence of a downward trend consistent with increased aerosol absorption (not shown).

A more direct means of assessing any radiative impact of increasing aerosol loading is to model the clear-sky SSR given the AERONET observations as the atmospheric state descriptors. We have done this for the Pune AERONET record using the clear-sky Simplified Solis radiative transfer model [35]. AOD at 675 nm is used as proxy for AOD at 700 nm, which is the Solis input variable. We further average the AERONET-Solis record to one-hour temporal resolution and use the timestamps to select corresponding clear-sky measured SSR one-hour averages at Pune between 2004 and 2014. A differential comparison between these data is shown in Figure 8.

The data are variably sparse and therefore we must refrain from any strong conclusions based on them; however, it is apparent that post-2009, Solis-modeled SSR predominantly overestimates the in situ measurement whereas over- and underestimations were much more equally distributed prior to that. In fact, the annual mean biases (relative to in situ SSR measurement) of SARAH-E and Solis-AERONET over Pune are linearly correlated with r² of 0.65. Despite the sparse data, this suggests that the observed decrease in measured SSR at Pune disagrees with both SARAH-E and the AERONET clear-sky measurements.

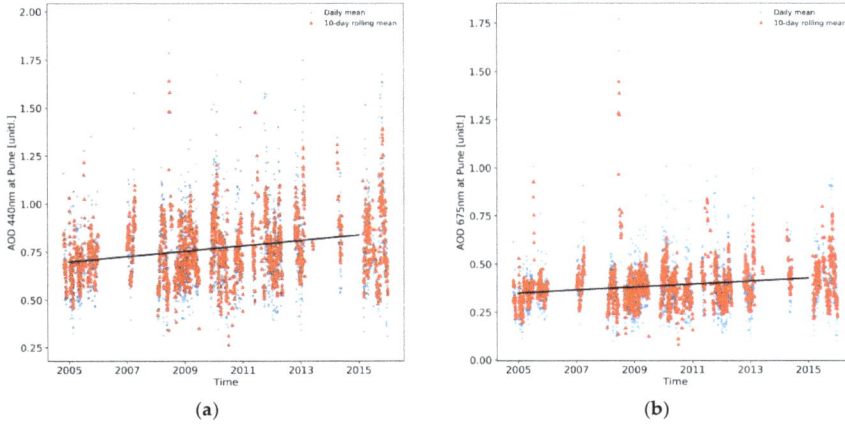

Figure 7. AERONET level 1.5 measurements of Aerosol Optical Depth (AOD) at Pune for: (**a**) 440 nm; and (**b**) 675 nm. Blue circles indicate daily means, red triangles 10-day rolling mean. Black line indicates linear trend for 2005–2015.

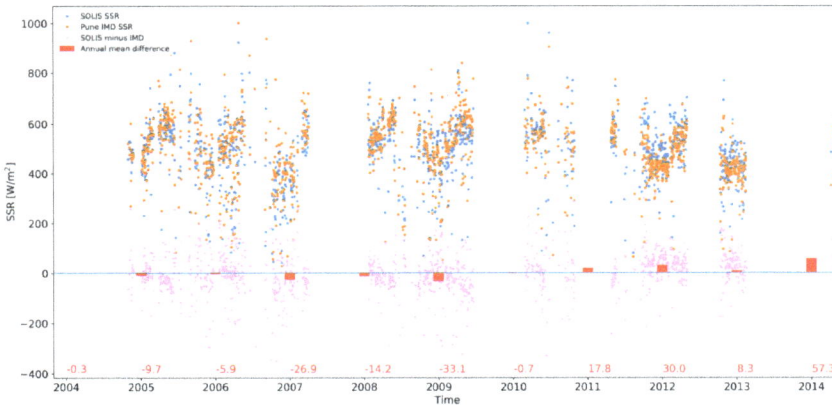

Figure 8. Comparison of daily mean SSR calculated from Simplified Solis with AERONET atmospheric inputs, and in situ measured SSR at Pune. Red bars and numbers at the bottom show their annual mean difference.

Furthermore, a recent validation document with the CM SAF project [38] reports the evaluation of the CM SAF Clouds, Albedo and Radiation dataset—Edition 2 (CLARA-A2) over India. The CLARA-A2 is based on polar-orbiting satellites and its algorithm is independent of SARAH-E; however, the findings on SSR retrieval accuracy over India mainly reflect those reported here: a stable period of retrievals in the 2000s, followed by a sharp increase in retrieval bias associated with an across-the-board drop in in situ measured SSR. The SSR from CLARA-A2 has been validated in the literature against a very large number of in situ measurements from different station networks over

Europe [13] with little evidence of a bias increase post-2009. However, the validity of the aerosol dataset used in CLARA-A2 generation has not been specifically validated over India in the recent years. In addition, it should be kept in mind that both SARAH-E and CLARA-A2 use climatological aerosol backgrounds for their retrievals.

Considering the divergent measurements and retrievals, the question of SSR trend estimation should be approached with utmost care and caution. Because of the varying temporal coverage of the IMD sites, the first precaution is to only include sites which offer the full 1999–2014 coverage, and calculate both in situ and SARAH-E SSR trends based on that subset. The resulting data and its trends are shown in Figure 9. While the overall trend in measured in situ SSR is stronger than in SARAH-E (-1.93 W/m^2/year vs. -0.24 W/m^2/year), the trends diverge strongly only after 2009. Prior to that, both datasets describe a decreasing trend with a comparable magnitude of ~ -0.6 W/m^2/year. All trends described here are statistically significant (i.e., non-zero trend) at the 95% confidence interval when evaluated with either the two-tailed t test or the nonparametric Mann–Kendall test. The diverging trends also remain when the datasets are evaluated through seasonal decomposition (not shown). Again, these findings should be considered preliminary prior to conclusively identifying the cause of the increasing discrepancy between satellite estimates and in situ measurements.

While the period 2010–2014 is too short for any reliable trend estimation, it remains notable that, should present circumstances persist, the IMD measurements would describe a decreasing trend of ~ -7 W/m^2/year, an exceedingly high rate of dimming.

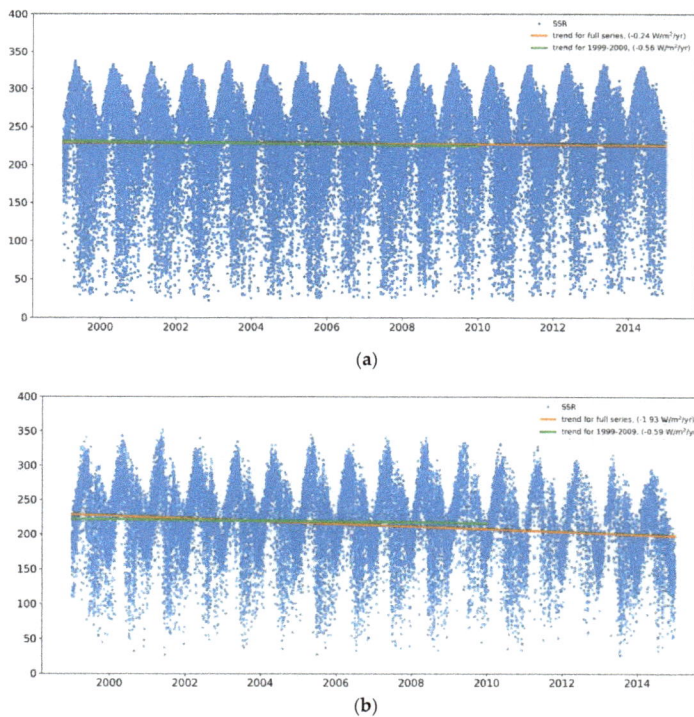

Figure 9. Daily mean SSR trends for 1999–2009 (green) and 1999–2014 (orange): (**a**) from the SARAH-E time series; and (**b**) from IMD in situ measurements. Blue markers indicate available SSR data. SARAH-E and in situ data included only from sites with coverage from 1999 to 2014 (see Table 1).

Overall, we must say that the available data do not allow conclusive findings about the cause of the increasing SARAH-E (or CLARA-A2) bias vs. Indian in situ measurements post-2009. If the sharp drop in measured SSR reflects intensifying aerosol loading, suggested to be caused by anthropogenic sources in the literature [21], this loading is not seen in the aerosol climatologies behind SARAH-E, or CLARA-A2. This is certainly within limits of possibility, especially considering the recent increases in air pollution over India [39]. However, the currently available AERONET measurements at Pune do not describe clear-sky aerosol or water vapor conditions which would agree with a drastically dimming clear-sky global radiation after 2009. This would imply that the cause of the large divergence of this period lies in SSR retrievals under cloudy skies, but there are no Supplementary Materials to test this hypothesis at present.

Contrary to expectations, SARAH-E estimation accuracy appears highest against in situ observations during the monsoon season. This is the period when rainfall scavenging of aerosols is most efficient, although the aerosol loading is reported to be on average higher than the winter period over most of India [40]. In addition, during the October–March (winter/pre-monsoon) season, anthropogenic particles contribute relatively more to the aerosol properties than during the monsoon season when natural particles dominate [40]. In principle, the larger SARAH-E bias during the winter period could therefore be consistent with increasing anthropogenic aerosol loading during that period post-2010 that is not present in the aerosol climatology; however, the sheer magnitude of the negative SSR trend in station observations, present for any month of the year, and our inability to reproduce the decrease with AERONET-measured clear-sky aerosol properties at Pune casts some doubt on that theory.

To summarize, it remains possible that the increasing bias post-2009 results from an undetected issue in the Indian in situ measurement network. On the other hand, deficiencies related to recent changes in aerosol loading or properties unseen by the SARAH-E aerosol climatology over India are also a possible cause. There is also a distinct possibility of a multi-factor explanation, the divergence resulting partly from an instrumentation bias on the ground, and partly from a deficiency in the aerosol background in the last years of this analysis.

4. Conclusions

We have evaluated the surface solar radiation (SSR) retrieval accuracy of the satellite-based SARAH-E dataset against in situ measured SSR over India. The primary findings of this study are:

1. Between 1999 and 2009, SARAH-E consistently overestimates in situ measured SSR over India by 10–30 W/m^2 (~10–20% in relative terms).
2. Between 1999 and 2009, both SARAH-E and the in situ measurements indicate an overall decreasing trend in SSR with a magnitude of ~−0.6 W/m^2/year. The trends are statistically significant at the 95% confidence interval.
3. Post-2009, in situ measured SSR begins to decline sharply whereas SARAH-E retrievals remain stable, leading to a sharply increasing estimation bias.
4. A modeling test of the clear-sky SSR over Pune with AERONET-measured aerosol and water vapor data as input indicates that the relationship between AERONET-based SSR estimate and measured SSR (in clear-sky conditions) is different pre- and post-2009. However, it should also be noted that the SARAH-E aerosol background is based on a 2003–2010 climatology, which may not fully capture recent changes in ambient atmospheric conditions over India.

While the diverging SSR retrievals and measurements are clearly seen post-2009, the cause remains unconfirmed at this stage. Identifying the cause of the rapidly increasing bias would require both radiometric calibration studies on the in situ data as well as decomposition of the satellite retrieval algorithm, both of which are beyond the scope of the present study. Here, we outline potential explanations (in Section 3) and recommend further studies and assessments for both in situ SSR measurements, and the validity of the atmospheric constituents used in the satellite retrievals.

Overall, while the SARAH-E retrieval bias observed in this study during 1999–2009 is higher than reported in general for the dataset [4], the bias for this period is largely stable enough to encourage usage, with due attention to seasonal and geographical limitations. Post 2009, we advise users to caution until the cause of the increasing estimation bias has been ascertained. As stated, we also encourage and highly recommend further studies on the topic, examining the full measurement chain from raw data to temporally averaged SSR means for both satellite and in situ sources.

Supplementary Materials: The following are available online at www.mdpi.com/2072-4292/10/3/392/s1, Figure S1: An illustration of the QA measures taken with the in situ data for one example site (Hyderabad) for one month of data. Circles indicate hourly mean SSR at the site versus measurement time (Indian time, night set to zero SSR). The graphs indicate the maximum allowable SSR by the various QA phases.

Acknowledgments: The research has been funded by the Academy of Finland, decision 284536 and Department of Science & Technology, India, contract number: INT/Fin/P-14, jointly. The authors would like to thank Thomas Huld and Jörg Trentmann for enlightening discussions. The SARAH-E dataset is publicly available from CM SAF: wui.cmsaf.eu. The AERONET data are available through aeronet.gsfc.nasa.gov. We thank the PIs for their efforts in establishing and maintaining the Pune site. IMD can be contacted directly for more information on in situ data availability, calibration, and procurement.

Author Contributions: A.R. co-designed the study, performed the analysis, and wrote the majority of the manuscript. V.K. carried out the in situ data QA testing and analysis, and contributed to the manuscript. S.D. and A.S. procured the SSR and AERONET data, performed an initial QA screening, and contributed to the manuscript. A.V.L. co-designed the study and contributed to the manuscript.

Conflicts of Interest: The authors declare no conflict of interest.

References

1. Bhattacharya, M.; Paramati, S.R.; Ozturk, I.; Bhattacharya, S. The effect of renewable energy consumption on economic growth: Evidence from top 38 countries. *Appl. Energy* **2016**, *162*, 733–741. [CrossRef]
2. Sharma, N.K.; Tiwari, P.K.; Sood, Y.R. Solar energy in India: Strategies, policies, perspectives and future potential. *Renew. Sustain. Energy Rev.* **2012**, *16*, 933–941. [CrossRef]
3. Kar, S.K.; Sharma, A.; Roy, B. Solar energy market developments in India. *Renew. Sustain. Energy Rev.* **2016**, *62*, 121–133. [CrossRef]
4. Amillo, A.G.; Huld, T.; Müller, R. A new database of global and direct solar radiation using the eastern meteosat satellite, models and validation. *Remote Sens.* **2014**, *6*, 8165–8189. [CrossRef]
5. Huld, T.; Müller, R.; Gracia-Amillo, A.; Pfeifroth, U.; Trentmann, J. *Surface Solar Radiation Data Set—Heliosat, Meteosat-East (SARAH-E)*, 1st ed.; Satellite Application Facility on Climate Monitoring: Offenbach, Germany, 2016; Available online: https://doi.org/10.5676/DWD/JECD/SARAH_E/V001 (accessed on 15 December 2017).
6. Rigollier, C.; Lefèvre, M.; Wald, L. The method Heliosat-2 for deriving shortwave solar radiation from satellite images. *Sol. Energy* **2004**, *77*, 159–169. [CrossRef]
7. Huld, T.; Müller, R.; Gambardella, A. A new solar radiation database for estimating PV performance in Europe and Africa. *Sol. Energy* **2012**, *86*, 1803–1815. [CrossRef]
8. Pinker, R.T.; Laszlo, I. Modeling surface solar irradiance for satellite applications on a global scale. *J. Appl. Meteorol.* **1992**, *31*, 194–211. [CrossRef]
9. Lohmann, S.; Schillings, C.; Mayer, B.; Meyer, R. Long-term variability of solar direct and global radiation derived from ISCCP data and comparison with reanalysis data. *Sol. Energy* **2006**, *80*, 1390–1401. [CrossRef]
10. Karlsson, K.G.; Anttila, K.; Trentmann, J.; Stengel, M.; Meirink, J.F.; Devasthale, A.; Benas, N. CLARA-A2: The second edition of the CM SAF cloud and radiation data record from 34 years of global AVHRR data. *Atmos. Chem. Phys.* **2017**, *17*, 5809. [CrossRef]
11. Riihelä, A.; Carlund, T.; Trentmann, J.; Müller, R.; Lindfors, A.V. Validation of CM SAF surface solar radiation datasets over Finland and Sweden. *Remote Sens.* **2015**, *7*, 6663–6682. [CrossRef]
12. Pfenninger, S.; Staffell, I. Long-term patterns of European PV output using 30 years of validated hourly reanalysis and satellite data. *Energy* **2016**, *114*, 1251–1265. [CrossRef]
13. Urraca, R.; Gracia-Amillo, A.M.; Koubli, E.; Huld, T.; Trentmann, J.; Riihelä, A.; Antonanzas-Torres, F. Extensive validation of CM SAF surface radiation products over Europe. *Remote Sens. Environ.* **2017**, *199*, 171–186. [CrossRef] [PubMed]

14. Cano, D.; Monget, J.M.; Albuisson, M.; Guillard, H.; Regas, N.; Wald, L. A method for the determination of the global solar radiation from meteorological satellite data. *Sol. Energy* **1986**, *37*, 31–39. [CrossRef]

15. Beyer, H.G.; Costanzo, C.; Heinemann, D. Modifications of the Heliosat procedure for irradiance estimates from satellite images. *Sol. Energy* **1996**, *56*, 207–212. [CrossRef]

16. Hammer, A.; Heinemann, D.; Hoyer, C.; Kuhlemann, R.; Lorenz, E.; Müller, R.; Beyer, H.G. Solar energy assessment using remote sensing technologies. *Remote Sens. Environ.* **2003**, *86*, 423–432. [CrossRef]

17. Dee, D.P.; Uppala, S.M.; Simmons, A.J.; Berrisford, P.; Poli, P.; Kobayashi, S.; Bechtold, P. The ERA-Interim reanalysis: Configuration and performance of the data assimilation system. *Q. J. R. Meteorol. Soc.* **2011**, *137*, 553–597. [CrossRef]

18. Morcrette, J.J.; Boucher, O.; Jones, L.; Salmond, D.; Bechtold, P.; Beljaars, A.; Schulz, M. Aerosol analysis and forecast in the European Centre for medium-range weather forecasts integrated forecast system: Forward modeling. *J. Geophys. Res. Atmos.* **2009**, *114*, 605–617. [CrossRef]

19. Müller, R.; Pfeifroth, U.; Träger-Chatterjee, C.; Trentmann, J.; Cremer, R. Digging the METEOSAT treasure—3 decades of solar surface radiation. *Remote Sens.* **2015**, *7*, 8067–8101. [CrossRef]

20. Krämer, M.; Müller, R.; Bovensmann, H.; Burrows, J.; Brinkmann, J.; Röth, E.P.; Günther, G. Intercomparison of stratospheric chemistry models under polar vortex conditions. *J. Atmos. Chem.* **2003**, *45*, 51–77. [CrossRef]

21. Soni, V.K.; Pandithurai, G.; Pai, D.S. Evaluation of long-term changes of solar radiation in India. *Int. J. Climatol.* **2012**, *32*, 540–551. [CrossRef]

22. Kumar, V.; Jain, S.K.; Singh, Y. Analysis of long-term rainfall trends in India. *Hydrol. Sci. J.* **2010**, *55*, 484–496. [CrossRef]

23. Tiwari, S.; Dumka, U.C.; Gautam, A.S.; Kaskaoutis, D.G.; Srivastava, A.K.; Bisht, D.S.; Chakrabarty, R.K.; Sumlin, B.J.; Solmon, F. Assessment of PM 2.5 and PM 10 over Guwahati in Brahmaputra River Valley: Temporal evolution, source apportionment and meteorological dependence. *Atmos. Pollut. Res.* **2016**, *8*, 13–28. [CrossRef]

24. Yerramsetti, V.S.; Gauravarapu Navlur, N.; Rapolu, V.; Dhulipala, N.S.K.; Sinha, P.R.; Srinavasan, S.; Anupoju, G.R. Role of nitrogen oxides, black carbon, and meteorological parameters on the variation of surface ozone levels at a tropical urban site–Hyderabad, India. *Clean Soil Air Water* **2013**, *41*, 215–225. [CrossRef]

25. Sreekanth, V. Satellite derived aerosol optical depth climatology over Bangalore, India. *Adv. Space Res.* **2013**, *51*, 2297–2308. [CrossRef]

26. Pani, S.K.; Verma, S. Variability of winter and summertime aerosols over eastern India urban environment. *Atmos. Res.* **2014**, *137*, 112–124. [CrossRef]

27. Bhaskar, V.V.; Safai, P.D.; Raju, M.P. Long term characterization of aerosol optical properties: Implications for radiative forcing over the desert region of Jodhpur, India. *Atmos. Environ.* **2015**, *114*, 66–74. [CrossRef]

28. Biswas, J.; Pathak, B.; Patadia, F.; Bhuyan, P.K.; Gogoi, M.M.; Babu, S.S. Satellite-retrieved direct radiative forcing of aerosols over North-East India and adjoining areas: Climatology and impact assessment. *Int. J. Climatol.* **2017**, *37*, 4756. [CrossRef]

29. Moradi, I. Quality control of global solar radiation using sunshine duration hours. *Energy* **2009**, *34*, 1–6. [CrossRef]

30. Tang, W.; Yang, K.; He, J.; Qin, J. Quality control and estimation of global solar radiation in China. *Sol. Energy* **2010**, *84*, 466–475. [CrossRef]

31. Shi, G.Y.; Hayasaka, T.; Ohmura, A.; Chen, Z.H.; Wang, B.; Zhao, J.Q.; Xu, L. Data quality assessment and the long-term trend of ground solar radiation in China. *J. Appl. Meteorol. Climatol.* **2008**, *47*, 1006–1016. [CrossRef]

32. Long, C.N.; Dutton, E.G. BSRN Global Network Recommended QC Tests, V2. 2010. Available online: https://epic.awi.de/30083/1/BSRN_recommended_QC_tests_V2.pdf (accessed on 15 December 2017).

33. Holmgren, W.F.; Andrews, R.W.; Lorenzo, A.T.; Stein, J.S. Pvlib python 2015. In Proceedings of the IEEE 42nd Photovoltaic Specialist Conference (PVSC), New Orleans, LA, USA, 14–19 June 2015; pp. 1–5.

34. Holben, B.N.; Eck, T.F.; Slutsker, I.; Tanre, D.; Buis, J.P.; Setzer, A.; Lavenu, F. AERONET—A federated instrument network and data archive for aerosol characterization. *Remote Sens. Environ.* **1998**, *66*, 1–16. [CrossRef]

35. Ineichen, P. A broadband simplified version of the Solis clear sky model. *Sol. Energy* **2008**, *82*, 758–762. [CrossRef]

36. Sanchez-Lorenzo, A.; Wild, M.; Trentmann, J. Validation and stability assessment of the monthly mean CM SAF surface solar radiation dataset over Europe against a homogenized surface dataset (1983–2005). *Remote Sens. Environ.* **2013**, *134*, 355–366. [CrossRef]

37. Hakuba, M.Z.; Folini, D.; Sanchez-Lorenzo, A.; Wild, M. Spatial representativeness of ground-based solar radiation measurements. *J. Geophys. Res. Atmos.* **2013**, *118*, 8585–8597. [CrossRef]

38. Sanchez Lorenzo, A. *Stability Assessment and Trends of CM SAF Surface Solar Radiation Data Sets (RadTrend)*; CM SAF Visiting Scientist Report; Satellite Application Facility on Climate Monitoring: Offenbach, Germany, 2017.

39. Gurjar, B.R.; Khaiwal, R.; Nagpure, A.S. Air pollution trends over Indian megacities and their local-to-global implications. *Atmos. Environ.* **2016**, *142*, 475–495. [CrossRef]

40. Dey, S.; Di Girolamo, L. A climatology of aerosol optical and microphysical properties over the Indian subcontinent from 9 years (2000–2008) of Multiangle Imaging Spectroradiometer (MISR) data. *J. Geophys. Res. Atmos.* **2010**, *115*. [CrossRef]

remote sensing

MDPI

Article

Retrieval of Reflected Shortwave Radiation at the Top of the Atmosphere Using Himawari-8/AHI Data

Sang-Ho Lee [1,2], Bu-Yo Kim [1,2], Kyu-Tae Lee [1,2,*], Il-Sung Zo [2], Hyun-Seok Jung [1,2] and Se-Hun Rim [1,2]

[1] Department of Atmospheric and Environmental Sciences, Gangneung-Wonju National University, 7, Jukheon-gil, Gangneung-si, Gangwon-do 25457, Korea; sangho.lee.1990@gmail.com (S.-H.L.); kimbuyo@gwnu.ac.kr (B.-Y.K.); kn_horizon@naver.com (H.-S.J.); shrim789@gmail.com (S.-H.R.)
[2] Research Institute for Radiation-Satellite, Gangneung-Wonju National University, 7, Jukheon-gil, Gangneung-si, Gangwon-do 25457, Korea; zoilsung@gwnu.ac.kr
* Correspondence: ktlee@gwnu.ac.kr; Tel.: +82-33-640-2324

Received: 30 November 2017; Accepted: 25 January 2018; Published: 1 February 2018

Abstract: This study developed a retrieval algorithm for reflected shortwave radiation at the top of the atmosphere (RSR). This algorithm is based on Himawari-8/AHI (Advanced Himawari Imager) whose sensor characteristics and observation area are similar to the next-generation Geostationary Korea Multi-Purpose Satellite/Advanced Meteorological Imager (GK-2A/AMI). This algorithm converts the radiance into reflectance for six shortwave channels and retrieves the RSR with a regression coefficient look-up-table according to geometry of the solar-viewing (solar zenith angle, viewing zenith angle, and relative azimuth angle) and atmospheric conditions (surface type and absence/presence of clouds), and removed sun glint with high uncertainty. The regression coefficients were calculated using numerical experiments from the radiative transfer model (SBDART), and ridge regression for broadband albedo at the top of the atmosphere (TOA albedo) and narrowband reflectance considering anisotropy. The retrieved RSR were validated using Terra, Aqua, and S-NPP/CERES data on the 15th day of every month from July 2015 to February 2017. The coefficient of determination (R^2) between AHI and CERES for scene analysis was higher than 0.867 and the Bias and root mean square error (RMSE) were -21.34–5.52 and 51.74–59.28 Wm^{-2}. The R^2, Bias, and RMSE for the all cases were 0.903, -2.34, and 52.12 Wm^{-2}, respectively.

Keywords: Himawari-8/Advanced Meteorological Imager (Himawari-8/AHI); Geostationary Korea Multi-Purse Satellite/Advanced Meteorological Imager (GK-2A/AMI); broadband albedo at the top of the atmosphere (TOA albedo); reflected shortwave radiation at the top of the atmosphere (RSR); Clouds and the Earth Radiant Energy System (CERES)

1. Introduction

Reflected shortwave radiation at the top of the atmosphere (RSR) is affected by the surface characteristics (15%); atmosphere gases such as aerosols, vapors, etc. (20%); and clouds (65%) [1]. In particular, a clear-sky area is greatly influenced by short-wavelength ultraviolet and near-infrared rays depending on surface characteristics, whereas a cloudy-sky area is affected more by the cloud properties [2]. In addition, aerosols such as particulate matter not only affect cloud distribution and characterization [3], but also increase the planetary albedo in relation to absorption and reflection of RSR [4], causing energy imbalance and global cooling [5,6]. RSR and broadband albedo at the top of the atmosphere (TOA albedo) retrieved from high-resolution satellite data can be used to analyze temporal and spatial changes in the atmosphere due to climate change and aerosols.

Since the 1970s, many studies have been measuring and analyzing RSR using radiative transfer models and satellite-based broadband or narrowband sensor data. However, radiative transfer models require prioritization of numerical experiments based on input data, and they are time consuming because calculation for each lattice is inevitable [7,8]. In addition, studies using broadband sensor data are limited to those on local radiation budget studies because broadband sensors such as National Aeronautics and Space Administration (NASA)'s the Earth Radiation Budget Experiment (ERBE) [9] and Clouds and the Earth Radiant Energy System (CERES) [10], which are mounted on polar orbiting satellites, provide data at resolutions over 20 km. These sensors are not suitable for analyzing the spatial and temporal distribution of RSR [11] because air pollution (natural or anthropogenic), urbanization, and forest fires in small areas cannot be easily detected using their data. In general, the RSR method using a radiation transfer model and broadband sensor data from polar orbit satellites provide higher accuracy than narrowband sensors, but the analysis of spatial and temporal changes is limited. In contrast, narrowband sensor data from geostationary satellites provide superior temporal and spatial continuity. For narrowband sensor data, the reflectance for each channel (CH) is assumed to be isotropic, which means the ratio of shortwave radiation is incident to the atmosphere and irradiance observed from the satellite. However, because isotropy is not satisfied in actual atmospheric conditions (surface type, water vapor, etc.), it is corrected using a radiative transfer model and a broadband sensor [12]. The relationship between the reflectance of a narrowband sensor and TOA albedo [11] and the narrowband reflectance and the RSR are affected [13] by solar zenith angle (SZA), viewing zenith angle (VZA) and relative azimuth angle (RAA) [14], surface type [15], and anisotropy [16,17].

Investigating RSR retrieval using the reflectance of narrowband sensors, Viollier [17] retrieved RSR using the narrowband reflectance of Meteosat data from 20 January to 31 March 1999, assuming that the sun glint (SG) removal method, isotropic only method, and ERBE anisotropy corrected method were comparable. As a result, when the assumptions of isotropy of Meteosat data and SG removal were considered, the coefficient of determination (R^2) with ScaRaB data was 0.880, which was larger than R^2 values of 0.845 and 0.863 in the case of only isotropy and ERBE anisotropy correction. In March, April, June, and September 2014, Vazquez-Navarro et al. [18] retrieved RSR through artificial neural network based on the narrowband reflectance (0.6, 0.8, 1.6 μm) of Meteosat-8/Spinning Enhanced Visible and Infrared Imager (SEVIRI) and input data (SZA, VZA, RAA, land/sea mask). In their study, the Bias compared with CERES Single Scanner Footprint (SSF) data were −15.99, −16.81, −30.88, and −7.48 Wm^{-2}, respectively, and R^2 was over 0.921. To evaluate the accuracy of long-term RSR, Niu and Pinker [12] retrieved RSR using the narrowband reflectance (0.6–0.8 μm) of Meteosat-8/SEVIRI and anisotropy data from April to July 2004. They calculated anisotropy using the radiative transfer model (MODTRAN 3.7) [19] and CERES broadband sensor data. The results of RSR calculation using the reflectance of Meteosat-8/SEVIRI and anisotropy were compared with CERES SRBAVG data in time-space agreement (4 month average, 100 km resolution). Statistical analysis showed that R^2 was 0.960 and the Bias and root mean square error (RMSE) were 2.5 and 5.9 Wm^{-2}, respectively.

In previous studies on retrieving RSR from the radiance of satellites, anisotropy is calculated using a radiative transfer model or broadband sensor data, but input variables required for retrieving the anisotropy of these narrowband reflectances are either missing or inaccurate [2]. Comparing the results of the RSR retrieval with the CERES data, Niu and Pinker [12] used long-term averaged data at 100 km resolution. Their validation results were good, but in evaluating the accuracy of the algorithm, they did not consider problems that may occur in other situations (such as seasons) and specific phenomena (such as precipitation and typhoon) [20]. Vazquez-Navarro et al. [18] produced high spatial-temporal resolution data for a single case over a specific time period and compared it with CERES data, but it is difficult to judge the accuracy of the long-term algorithm.

Considering the above-mentioned issues, this study was conducted to develop an RSR algorithm from the outputs of the next generation Geostationary Korea Multi-Purpose Satellite/Advanced Meteorological Imager (GK-2A/AMI) [21], and is a preliminary study for the RSR retrieval algorithm using Himawari-8/AHI (Advanced Himawari Imager) [22] data, which is similar to GK-2A/AMI data. That is, unlike previous studies using input data of atmospheric elements (clouds, aerosols, water vapor, etc.), Lee et al. [23] retrieved TOA albedos using narrowband reflectance of each CH and the regression coefficient look-up-table (LUT) according to geometry of the solar-viewing (SZA, VZA, and RAA) and atmospheric conditions (surface type and absence/presence of clouds). In this process, the narrowband reflectances was assumed isotropic using radiative transfer model (Santa Barbara DISORT Atmospheric Radiative Transfer, SBDART) [24]. However, in this study, we considered the anisotropy of the geometry of the solar-viewing and atmospheric conditions when retrieving RSR, thereby removing solar reflection regions with high uncertainties, i.e., SG. The retrieved results were compared with those of Terra, Aqua, and Suomi National Polar-orbiting Partnership (S-NPP)/CERES data. Section 2 describes input data and validation data. In Section 3, the theoretical background and retrieval algorithm of RSR are presented. Sections 4 and 5 discusses the output and validation results. Section 6 summarizes the conclusions.

2. Materials

2.1. Input Data

This study is a preliminary investigation of the RSR retrieval algorithm using GK-2A/AMI (128.2°E, 0.0°N), which will be launched in 2018. The algorithm is based on Himawari-8/AHI (140.7°E, 0.0°N) whose sensor characteristics are similar to that of GK-2A/AMI. The Himawari-8 satellite was launched on 7 October 2014 and has 16 CHs [25]. It can retrieve more meteorological factors than the existing MTSAT satellites with five sensors [26]. It has temporal and spatial resolutions of 10 min and 0.5–2.0 km, respectively, as shown in Table 1. In this study, six shortwave CHs of Himawari-8/AHI were used for high-resolution RSR retrieval. Because of the varying spatial resolutions, they were averaged to a spatial distance of 2.0 km. Cao et al. [27] reported that the absolute radiometric calibration accuracy of S-NPP/VIRS is less than 2% for CH1-CH6 and the calibration accuracy of this data and Himawari-8/AHI was reported to be within 6–8% of CH1-CH4 and CH6 except for CH5 (5%) [28].

Table 1. Shortwave channel (CH) data from Himawari-8/AHI for retrieving RSR.

| Channel | Wavelength (μm) | Resolution | | | Main Purpose of Use |
		Spatial	Numbers of Pixels	Temporal	
CH1 (Blue)	0.47 (0.43–0.48)	1.0 km	11,000		
CH2 (Green)	0.51 (0.50–0.52)	1.0 km	11,000		
CH3 (Red)	0.64 (0.63–0.66)	0.5 km	22,000	10-min	Weather forecasting
CH4 (NIR)	0.86 (0.85–0.87)	1.0 km	11,000	Full Disk	Climate modeling
CH5 (NIR)	1.61 (1.60–1.62)	2.0 km	5500		
CH6 (NIR)	2.26 (2.25–2.27)	2.0 km	5500		

2.2. Validation Data

In this study, data from CERES SSF level 2 edition (Terra, Aqua: 4A; S-NPP: 1A) with a resolution of 20 km within the field of view were used for validation. CERES sensors are mounted on low orbiting satellites (TRMM: 1997/12–2015/4, Terra: 1999/12–Present, Aqua: 2002/5–Present, S-NPP: 2011/10–Present) [5,10] and provide long-term radiometric data for shortwave (0.3 to 5.0 μm), atmospheric window (8 to 12 μm), and total-wave (0.3 to 200 μm) region as part of the NASA Earth Observing System (EOS) satellite project [27,29]. Su et al. [30] compared the all-sky flux from the CERES retrieval (along-track observation) with that from the MODIS retrieval and reported uncertainties in the CERES data of 3.3% (9.0 Wm^{-2}) and 2.7 % (8.4 Wm^{-2}) for ocean and land areas, respectively. They further reported that the direct integration test revealed the monthly Bias and RMSE of CERES to be less than 0.5 and 0.8 Wm^{-2}, respectively.

The results of this study, retrieved using AHI, were averaged at intervals of 10 km and coincided with the CERES data because Himawari-8/AHI and CERES have different temporal and spatial resolutions. Based on the results of the study, CERES data within ±5 min were used [20,23]. Then, since the 15th day of each month is the point in time that can represent the whole month in terms of atmospheric conditions (clear fraction, atmospheric transmissivity, etc.) and surface characteristics (albedo, vegetation index, etc.) [31,32], a comparison of Himawari-8/AHI and with Terra, Aqua, and S-NPP/CERES data was conducted on the 15th day of each month from July 2015 to February 2017.

3. Methods

3.1. Theoretical Background

The theoretical background of RSR is shown in Figure 1. In this figure, the solid line is the primary absorption and reflection path of the shortwave radiation (extraterrestrial solar irradiation) incident on the atmosphere, and the dotted line shows the multiple scattering process by the surface and atmosphere [33]. RSR is green-shadowed in this schematic. It was approximated by using an infinite geometric series, as shown in Equation (1), because atmospheric reflection (R) times surface albedo (α) is less than 1 ($\alpha R < 1$) [34]. In Equation (1), the solar constant (=1361 Wm^{-2}) [35] and the SZA [36] are calculated by the theoretical equation in the previous study, however, TOA albedo can be retrieved using the narrowband reflectance of each CH and regression coefficients, as expressed by Equation (2) [11,13–15].

$$
\begin{aligned}
\text{RSR} &= \text{RS} + \alpha S(1-R-A)^2 + \alpha^2 SR(1-R-A)^2 + \cdots \\
&= \text{RS} + S(1-R-A)^2[\alpha + \alpha^2 R + \cdots] \\
&\approx S \times \left(R + \frac{(1-R-A)^2}{1-\alpha R} \right) \\
&\approx S_0 \cos(\text{SZA})\, d_0^2/d^2 \times (\text{TOA albedo})
\end{aligned}
\tag{1}
$$

$$
\begin{aligned}
\text{TOA albedo} &= \sum_{i=1}^{6} c_i(\text{SZA, VZA, RAA, Surface type, Absence/Presence of clouds})\, \rho_i \\
&= c_1\rho_{0.47\,\mu m} + c_2\rho_{0.51\,\mu m} + c_3\rho_{0.64\,\mu m} + c_4\rho_{0.86\,\mu m} + c_5\rho_{1.61\,\mu m} + c_6\rho_{2.26\,\mu m}
\end{aligned}
\tag{2}
$$

In Equations (1) and (2), A, R, and α are atmospheric absorption, scattering, and surface albedo, respectively; S_0, and d_0^2/d^2 are solar constant, and earth-to-sun distance in astronomical units; c_{1-6} is the regression coefficient according to geometry of the solar-viewing and atmospheric conditions; and $\rho_{0.47\,\mu m}$, $\rho_{0.51\,\mu m}$, $\rho_{0.64\,\mu m}$, $\rho_{0.86\,\mu m}$, $\rho_{1.61\,\mu m}$, and $\rho_{2.26\,\mu m}$ are the narrowband reflectance of each CH.

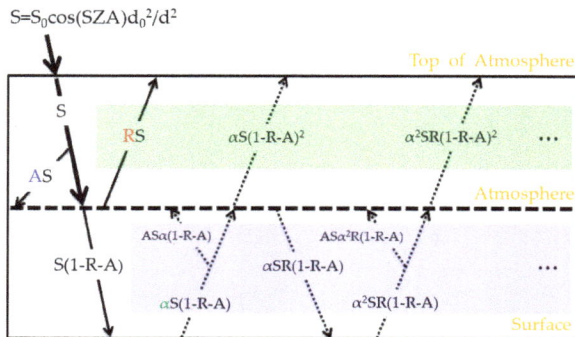

Figure 1. Schematic representing the RSR (green area) in the one-layer solar radiation model. The line at the top of the atmosphere represents the atmospheric contribution (red) associated with cloud reflection, and the dotted line indicates the surface contribution (blue area) associated with surface reflection. The variables A, R, and α are atmospheric absorption and scattering of extraterrestrial solar irradiation, reflection by clouds, and surface albedo, respectively.

3.2. Reflected Shortwave Radiation Retrieval Algorithm

The step-by-step algorithm is shown in Figure 2. As shown in this figure, Himawari-8/AHI was converted into reflectance for each CH using radiance (Process 1). As shown in Equation (2), TOA albedo was retrieved by the narrowband reflectance and the regression coefficient LUT according to geometry of the solar-viewing and atmospheric conditions. Finally, SG was removed and RSR was retrieved using Equation (1). The regression coefficient used in this study was simulated, as shown in Table 2, using the SBDART according to the geometry of the solar-viewing and atmospheric conditions. More specifically, the SBDART simulation was performed under the following conditions: 12 SZAs, 18 VZAs, 19 RAAs, six atmospheric profiles, and five surface types in the 3.3 μm range at 0.2. Additionally, four aerosol types and four aerosol visibilities in the clear-sky area, and eight cloud heights and five cloud optical thicknesses in the cloud-area were included when configuring the simulation conditions. The regression coefficients can be obtained with a multiple linear regression model or ridge regression model, based on the simulation results (independent variables: reflectance of each CH taking into account anisotropy; dependent variable: TOA albedo). When a multiple linear regression model is used to determine the regression coefficient with the least square method [11,13], multicollinearity arises because of the high correlations between the shortwave CHs [37,38], thus lowering the accuracy of the regression coefficient; therefore, we used a ridge regression model [39,40]. Lee et al. [23] compared the accuracy of TOA albedos estimated by a multiple linear regression model and a ridge regression model with the Himawari-8/AHI data of 20 August 2015, by checking their respective results against the corresponding CERES data. The comparison revealed that the ridge regression model outperformed the multiple linear regression model in terms of R^2 and RMSE (0.914 and 0.055 vs. 0.856 and 0.191).

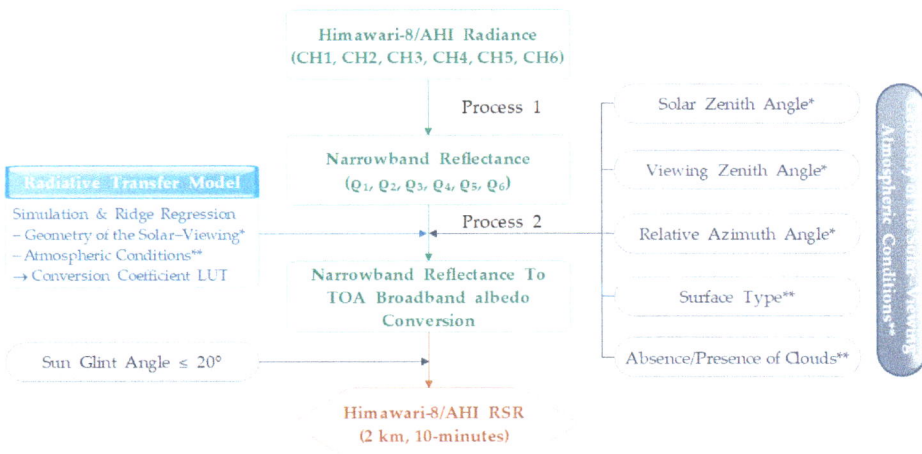

Figure 2. Flow chart of the retrieval algorithm for RSR. Reflectance converted from radiance from each shortwave channel (CH) (Process 1) is used to retrieve RSR using regression coefficients look-up-table (LUT) according to geometry of the solar-viewing (SZA, VZA, and RAA) and atmospheric conditions (surface type and absence/presence of clouds) (Process 2). These regression coefficients were calculated using results from the radiative transfer model (SBDART), which considered geometry of the solar-viewing and atmospheric conditions, and a ridge regression model.

Table 2. Numerical experiments of SBDART for creating regression coefficient look-up-table (LUT).

Parameter	Values Used for Look-Up-Table	Number
Spectral range	0.2 to 3.3 at 0.005 μm	620
Solar zenith angle	0°, 10°, 20°, 30°, 40°, 50°, 60°, 70°, 75°, 80°, and 85°	12
Viewing zenith angle	0° to 85° at 5° increments	18
Relative azimuth angle	0° to 180° at 10° increments	19
Atmospheric profiles	Tropical, Mid-latitude summer, Mid-latitude winter Subarctic summer, Subarctic winter, and US62 standard	6
Surface types	Ocean, Lake, Vegetation, Snow, and Sand	5
Aerosol types	Rural, Urban, Marine, and Tropospheric	4
Aerosol visibilities	5, 10, 15, and 20 km	4
Cloud height	2, 4, 6, 8, 10, 12, 14, and 16 km	8
Cloud optical thickness	8, 16, 32, 64, and 128	5

3.2.1. Anisotropy Consideration

The TOA albedo retrieved by previous studies [23] was calculated assuming the narrowband reflectance to be isotropic. However, in this study, RSR was retrieved considering anisotropy. In other words, in Lee et al.'s [23] algorithm, the radiance of each CH was converted into the reflectance of each CH. As in Equation (2), the regression coefficient LUT were produced according to geometry of the solar-viewing and atmospheric conditions to retrieve the TOA albedo. The regression coefficients used in their study were based on the results of numerical experiments using the SBDART, in which simulations were performed according to atmospheric conditions, and a ridge regression (independent variables: narrowband reflectance of each CH assuming isotropy as in Equation (3); dependent variables: TOA albedo integrated from 0.2 to 3.3 μm using SBDART). In this process, the narrowband reflectance of each CH is assumed to be isotropic as shown in Equation (3). However, to retrieve TOA albedo accurately, the narrowband reflectance should be expressed as anisotropy, as shown in Equations (4) and (5). The anisotropy of the reflectance for each CH varies depending on the atmospheric conditions (surface type, absence/presence of clouds, etc.) [17,41–44]. Moreover, in the process of converting radiance into irradiance using Equation (3), the error in the mean dispersion was found to be higher when the Lambertian assumption (i.e., no anisotropic correction) was adopted than when anisotropy was considered (16.9% vs. 2.2%) [45].

$$\rho_i = \frac{\pi L_i(\text{SZA}, \text{VZA}, \text{SAA}, \text{VAA})}{S_{0,i} \cos(\text{SZA}) \, d_0^2 / d^2}, \, i = \text{CH } 1, 2, 3, 4, 5, 6 \tag{3}$$

$$\rho_i = \frac{F_i}{S_{0,i} \cos(\text{SZA}) \, d_0^2 / d^2}, \, i = \text{CH } 1, 2, 3, 4, 5, 6 \tag{4}$$

$$F_i = \pi L_i(\text{SZA}, \text{VZA}, \text{SAA}, \text{VAA}) / \text{ADM} \tag{5}$$

In Equations (3)–(5), L_i and $S_{0,i}$ are the radiance at the top of the atmosphere and the solar constant of each CH, respectively; SAA, and VAA are solar azimuth angle, and viewing azimuth angle, respectively; and F_i and ADM are irradiance at the top of the atmosphere of each CH [46] and anisotropy [16], respectively.

Considering the abovementioned factors, in this study, the regression coefficient LUT and the TOA albedo were calculated by considering the anisotropy of reflectance for each CH, as shown in Equations (4) and (5). The results were compared with those of Lee et al. [23]. Himawari-8/AHI, Terra, Aqua, and S-NPP/CERES data on the 15th day of each month from July 2015 to February 2017 (all 60 cases) were used. TOA albedo considering isotropy and anisotropy using narrowband reflectance is shown in Table 2.

3.2.2. Sun Glint Removal

SG is a phenomenon in which sunlight reflected from the Earth's surface is incident on the field of view of the satellite sensor depending on the geometry of the solar–viewing. The measured value becomes greater than the originally measured value, and therefore the reflection appears brighter [47]. Previous studies have reported that SG removal or correction is necessary because it can cause serious errors in ocean-related satellite output [48]. In this study, SG was calculated using information of the SZA, VZA, and RAA (Equation (6)), and statistical analysis was performed using Himawari-8/AHI and Terra/CERES data for proper SG removal (Figures 3 and 4).

$$SG = \cos^{-1}[\cos(SZA)\cos(VZA) + \sin(SZA)\sin(VZA)\cos(RAA)] \tag{6}$$

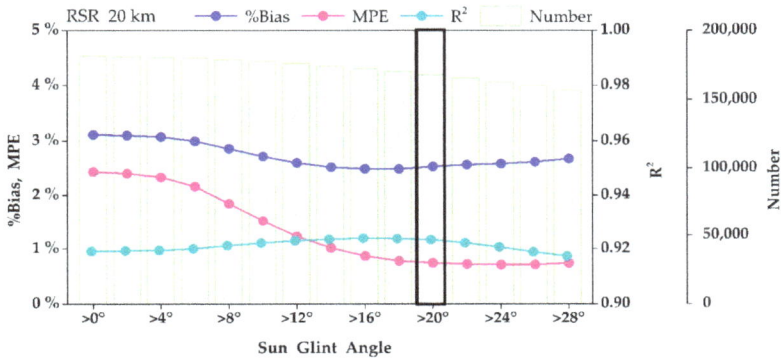

Figure 3. Relative bias (%Bias), mean percentage error (MPE in percent), coefficient of determination (R^2), and number in Himawari-8/AHI and Terra/CERES data sets as a function of sun glint (SG) angle (15 October 2015).

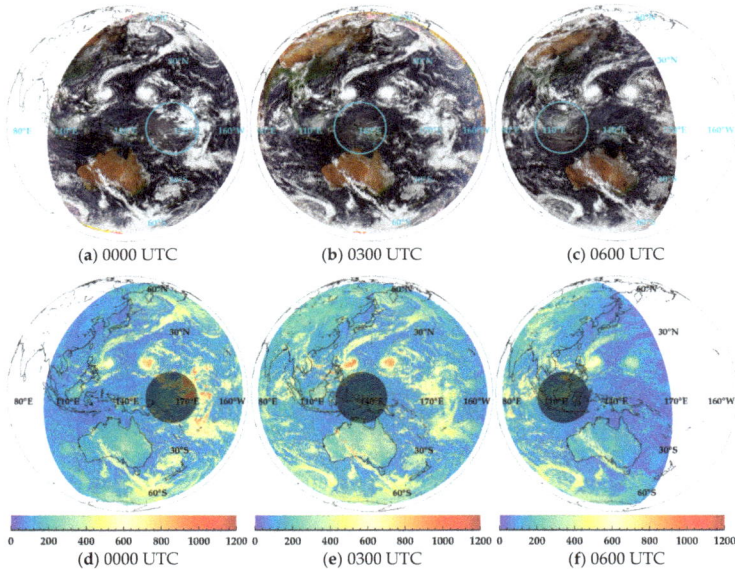

(**a**) 0000 UTC   (**b**) 0300 UTC   (**c**) 0600 UTC

(**d**) 0000 UTC   (**e**) 0300 UTC   (**f**) 0600 UTC

Figure 4. *Cont.*

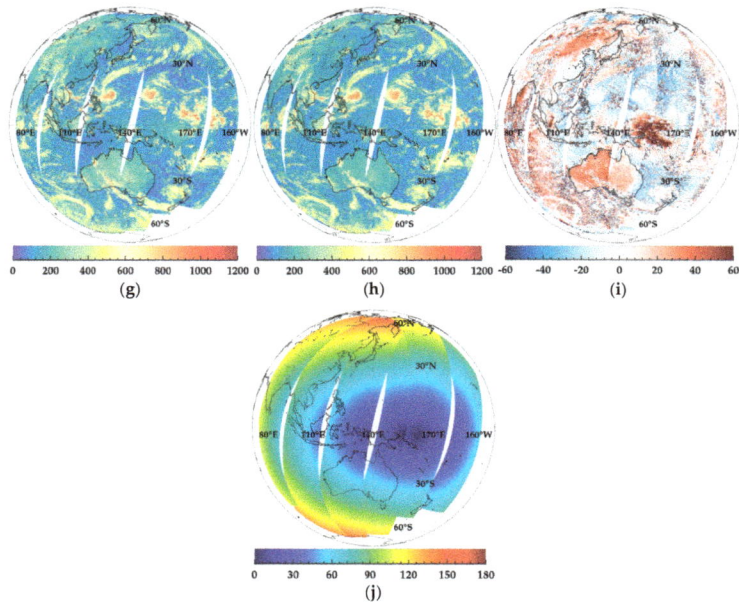

Figure 4. Red–Green–Blue (RGB) composite imagery (**a–c**); and RSR (**d–f**) of Himawari-8/AHI at 0000, 0300, 0600 UTC on 15 October 2015. The cyan circle and black circle (transparency = 50%) represents areas where sun glint (SG) ≤ 20°. Spatio-temporal matched RSR (Wm^{-2}) of: AHI (**g**); and CERES (**h**) on this date; and percentage error (%) of: two dataset sets (**i**); and sun glint (SG) angle (**j**).

4. Results

4.1. Evaluation of the Reflected Shortwave Radiation Algorithm

4.1.1. Anisotropy Consideration

Table 3 compares the results of isotropic and anisotropic assumptions using Himawari-8/AHI and the TOA albedo of Terra, Aqua, and S-NPP/CERES for this study. R^2 values between isotropic and anisotropic Terra, Aqua, and S-NPP/CERES were similar. However, relative Bias (%Bias) and relative RMSE (%RMSE) were −1.14% and 21.04%, respectively, for anisotropic cases, which were improved from the isotropic case (%Bias = 5.16%; %RMSE = 22.67%), and the mean percentage error (MPE) was also improved. The cases of 15 July 2015, 15 May 2016, 15 June 2016, and 15 July 2016 showed better isotropy than the anisotropy, and the cause will be further discussed in Figure 6.

Table 3. Statistical results of TOA albedo using Himawari-8/AHI and Terra, Aqua, and S-NPP/CERES data.

Date	Isotropy				Anisotropy				N
	R^2	%Bias	%RMSE	MPE	R^2	%Bias	%RMSE	MPE	
15 July 2015	0.892	1.07	21.61	2.34	0.892	−4.77	21.78	−3.56	382,868
15 August 2015	0.881	3.99	22.99	4.98	0.882	−2.80	21.98	−1.72	765,902
15 September 2015	0.896	7.54	23.44	7.96	0.897	0.66	20.86	1.25	748,431
15 October 2015	0.894	8.17	24.55	8.65	0.889	1.27	21.94	2.60	769,128
15 November 2015	0.907	7.63	23.23	8.33	0.897	−0.43	21.80	0.95	346,584
15 December 2015	0.894	4.08	21.94	5.37	0.885	−1.74	21.56	0.00	318,051
15 January 2016	0.908	1.76	20.52	3.35	0.905	−0.79	20.47	1.03	209,119
15 February 2016	0.904	8.13	22.09	9.47	0.899	1.39	19.73	3.56	700,569
15 March 2016	0.906	6.93	21.49	7.53	0.902	0.23	19.56	1.53	608,431
15 April 2016	0.900	4.43	22.57	4.63	0.900	−1.18	21.15	−0.54	622,010
15 May 2016	0.910	−0.06	20.18	−0.98	0.910	−4.58	20.24	−5.43	457,664
15 June 2016	0.895	−0.91	20.07	−0.98	0.897	−6.11	20.33	−6.32	414,194
15 July 2016	0.889	0.45	21.73	1.14	0.887	−4.56	21.94	−4.18	350,007
15 August 2016	0.880	4.25	23.00	4.54	0.891	−2.65	20.73	−2.47	673,663
15 September 2016	0.874	5.69	24.20	5.88	0.892	−0.69	20.63	−0.57	729,884
15 October 2016	0.900	5.59	22.04	5.78	0.900	−0.81	20.05	−0.12	769,781
15 November 2016	0.892	4.28	22.79	4.89	0.885	−1.29	22.19	−0.28	585,544
15 December 2016	0.900	3.49	21.33	4.86	0.892	−0.59	21.17	0.92	451,665
15 January 2017	0.900	5.85	22.22	5.99	0.887	−0.77	21.40	0.29	626,594
15 February 2017	0.887	9.59	25.63	8.82	0.878	0.94	22.14	1.36	578,772
All	0.893	5.16	22.67	5.58	0.893	−1.14	21.04	−0.33	11,108,861

Note: R^2 is the coefficient of determination; %Bias is the relative Bias of (Bias/CERES$_{Mean}$) × 100 in percent; %RMSE is the relative RMSE of (RMSE/CERES$_{Mean}$) × 100 in percent; MPE is the mean percentage error of ((AHI-CERES)/CERES) × 100 in percent; and N is the number of pairs.

4.1.2. Sun Glint Removal

Because RSR significantly varies from clear skies to deep convective clouds (0 to 80%) [18], the case of 15 October 2015 was selected, with the accompanying precipitation and typhoons (KOPPU: 16.0°N, 138.8°E, and CHAMPI: 13.3°N, 158.2°E). Figure 3 shows the result of statistical analysis of R^2, %Bias and MPE according to SG as an example of 15 October 2015 for the calculation of a proper angle for SG removal. As shown in this figure, R^2 between Himawari-8/AHI and Terra/CERES was changed from 0.919 to 0.923 and %Bias from 3.11% to 2.53%. The MPE of Himawari-8/AHI and Terra/CERES decreased with increasing sun reflection angle, which was less than 0.74% when the SG was 20°. In this study, sun reflection angles less than 20° were considered as SG.

Figure 4 shows Red–Green–Blue (Blue: CH1, Green: CH2, Red: CH3, RGB) composite images (Figure 4a–c) and RSR (Figure 4d–f) of Himawari-8/AHI analyzing scene error according to SG. In this figure, the black circles (transparency = 50%) represent SG areas (≤20°). Figure 4g–j shows the RSR, percentage error, and SG of Himawari-8/AHI and Terra/CERES, respectively, with space-time agreement. In the RGB composite images in Figure 4a–c, SG appears brighter than the surrounding pixels of the mid-latitude ocean. This corresponds to the RSR pictures in Figure 4d–f. The MPE of Himawari-8/AHI and Terra/CERES was more than 45% (Figure 4i) and the SG angle was less than 20° (Figure 4j). In other words, when the SG area is removed, R^2 between AHI and CERES is 0.919, which is larger than 0.923, and this result is consistent with the result of [17].

4.1.3. Reflected Shortwave Radiation

Lee et al. [23] retrieved TOA albedo assuming narrowband reflectance to be isotropic. In this study, RSR was retrieved by considering anisotropy of the narrowband reflectance by the surface and the atmosphere, and by removing SG. The results are shown in Figure 5. This figure shows the results of RSR of this study. Isotropic, anisotropic, and SG effects of narrowband reflectance were analyzed using the Himawari-8/AHI data (all data: 3,411,624 pixels). Figure 5a is a two-dimensional histogram (2D histogram) of the Terra/CERES data, with RSR retrieved assuming the narrowband reflectance to be isotropic Lee et al. [23] (OLD). Figure 5b,c shows the RSR considering anisotropy of the narrowband reflectance (OLD2)and the case of removing SG (OLD3), respectively. Figure 5d considers Figure 5b,c together; similar to Figure 5a, Figure 5b–d shows scatter plots along with Terra/CERES data. R^2 between RSR retrieved using the Himawari-8/AHI and Terra/CERES data was 0.910 when the narrowband reflectance was assumed to be isotropic as shown in Figure 5a. The Bias and RMSE were 15.13 and 53.94 Wm^{-2}, respectively. However, in this study, R^2 between Himawari-8/AHI and Terra/CERES considering narrowband reflectance as anisotropy was 0.912, which is higher than that in Figure 5a. Furthermore, the Bias and RMSE were 2.80 and 50.14 Wm^{-2}, respectively, showing improvements compared to Figure 5a. Comparing Figure 5a,c, R^2 values are similar, but Bias and RMSE improved slightly. The narrowband reflectance was considered to be anisotropic and SG was removed simultaneously (NEW). Comparing Figure 5a,d, R^2 (0.914), Bias (2.07 Wm^{-2}) and RMSE (49.22 Wm^{-2}) were significantly improved. These results are in good agreement with the results of [17], where R^2 values between Meteosat RSR and ScaRaB RSR were found to be 0.845 and 0.863, respectively, for isotropic and anisotropic calculations.

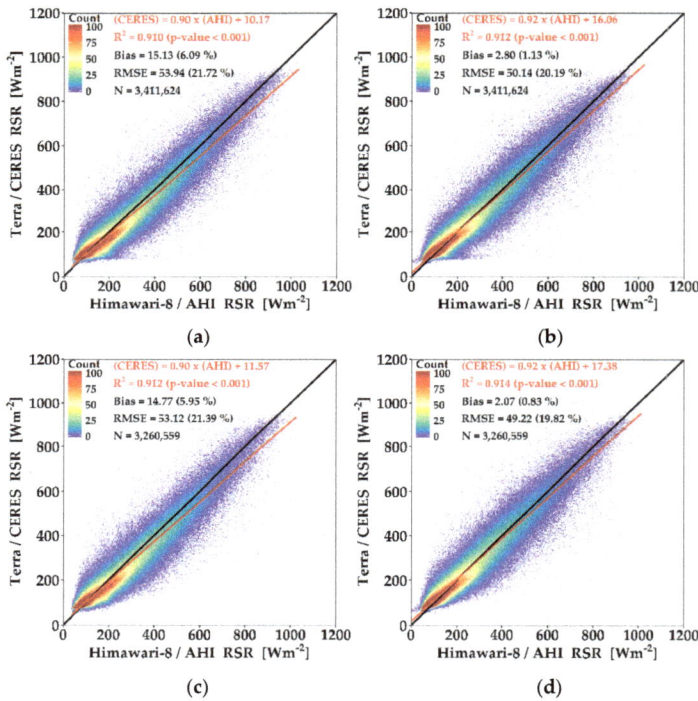

Figure 5. Two-dimensional histograms of RSR from Himawari-8/AHI and Terra/CERES for: OLD (a) OLD2 (+ anisotropy) (b); OLD3 (+ sun glint (SG) $\geq 20°$) (c); and NEW (+ anisotropy, sun glint (SG) $\geq 20°$) (d) on the 15th day of every month from July 2015 to February 2017. The colors represent the 2D histogram (or density) of coincident pairs using a bin size of 1. The solid red line is a linear fit to the data. The black line corresponds to the 1:1 line.

4.2. Validation of Reflected Shortwave Radiation Algorithm Using CERES Data

The narrowband radiance of Himawari-8/AHI and the regression coefficient LUT were applied to the algorithm in Figure 2, and the RSR was retrieved and compared with Terra, Aqua, and S-NPP/CERES data (Figure 6). For validating the method, scalar accuracy measurement and linear relationship analysis were performed. For the scalar accuracy measurement, Bias, relative Bias (%Bias = Bias/CERES$_{Mea}$ × 100%), root mean square error (RMSE), relative RMSE (%RMSE = RMSE/CERES$_{Mean}$ × 100%) and mean percentage error (MPE = ((AHI-CERES)/CERES) × 100%) of Terra, Aqua, and S-NPP/CERES were calculated [49]. In addition, Pearson correlation coefficients [50] and significance level [51] were calculated through correlation analysis and Monte Carlo simulation to determine the linear relationship between the two data sets. For this purpose, SZA and VZAs less than 80° were used and SG angles less than 20° were excluded.

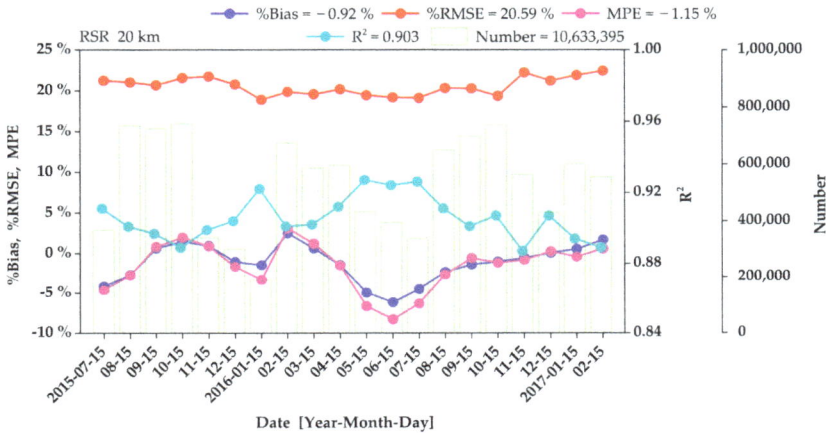

Figure 6. Statistical analysis of RSR using Himawari-8/AHI and Terra, Aqua, and S-NPP/CERES data for the 15th day of each month from July 2015 to February 2017 (%Bias: blue dotted line; %RMSE: red dotted line; MPE: magenta dotted line; R^2: cyan dotted line; Number: green bar chart). The legend shows the statistical results for the all case.

Figure 6 shows a comparison of Terra, Aqua, and S-NPP/CERES with RSR from Himawari-8/AHI. Excluding May–July 2016 cases, %Bias was −4.18–2.49% and MPE was −4.66–3.12%. Between May and July 2016, %Bias and MPE were much lower at −6.20 and −8.31%, respectively, on 15 June 2016, which is consistent with the results of TOA albedo statistical analysis in Table 3 (%Bias, MPE). To clarify the reason for such a result, clouds were subdivided according to the clear fraction (0–100%) used for the RSR retrieval. At this time, overcast was classified as 0–5%, mostly cloudy 5–50%, partly cloudy 50–95%, and clear 95–100% [46].

Figure 7 shows Bias, standard deviation (Stdev), RMSE, and R^2 according to the clear fraction, and the bar chart means the number of data. Except for clear (95–100%) and cloudy fractions (0–5%), R^2 (0.571–0.815) and RMSE (25.09–64.61 Wm^{-2}) of Terra, Aqua, and S-NPP/CERES were not suitable for partly cloudy and mostly cloudy fractions (5–95%). The reason is that averaging the RSR of AHI in the process of spatial resolution matching can generate large errors when the averaged area characteristics (surface or cloud) are different from the CERES observations [20]. In particular, these errors were large in the cloudy area. The number of data in clear area (95–100%)/cloudy area (0–95%) was compared with the all data in each case. On 15 June 2016, the value was 5.94%/94.06% and, compared to other cases (8.71–14.15%/85.85–91.29%), there were less clear areas and too much cloud. Therefore, the results of this study and CERES statistical analysis were unsatisfactory. In the case of RSR retrieval, for error analysis in cloudy areas (<95%), Terra, Aqua, and S-NPP/CERES data were analyzed according to land, ocean, and clear fractions (partly cloudy, mostly cloudy, overcast, and all). Table 4 shows the results. For the ocean area, results of the statistical analysis (R^2, Bias, MPE) were relatively better than that for the land, but both land and ocean area were partly or mostly cloudy, and this tendency was also consistent with MPE.

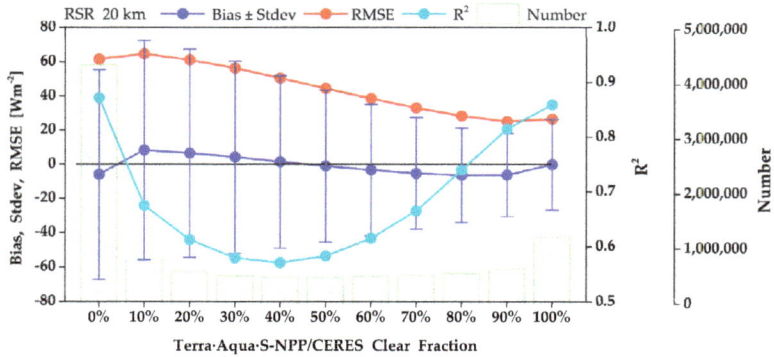

Figure 7. Coefficient of determination (R^2), standard deviation (Stdev), Bias, and RMSE of RSR using Himawari-8/AHI and Terra, Aqua, and S-NPP/CERES for the clear fraction for all cases. The bar graph shows the number of data.

Table 4. Statistical analysis of RSR retrieval results according to land, ocean, and clear fractions (all: 0–100%; partly cloudy: 50–95%; mostly cloudy: 5–50%; overcast: < 5%).

Clear Fraction Land & Ocean	Statistics	R^2	Mean		RMSE (%RMSE)	MPE	N
			AHI	CERES			
Land	Cloudy	0.869	302.70	313.46	56.02 (17.87)	−2.34	2,006,927
	–Partly	0.639	195.27	198.39	38.17 (19.24)	−0.25	574,000
	–Mostly	0.657	270.61	277.79	59.52 (21.42)	−1.66	630,676
	–Overcast	0.861	404.78	423.82	64.4. (14.97)	−4.36	802,251
	All	0.880	274.26	282.03	51.29 (18.29)	−0.51	2,632,865
Ocean	Cloudy	0.902	256.18	256.59	54.14 (21.10)	−0.60	7,417,530
	–Partly	0.429	105.49	111.12	30.62 (27.56)	−5.25	1,864,210
	–Mostly	0.625	192.29	183.33	57.95 (31.61)	3.98	1,999,843
	–Overcast	0.874	371.19	374.14	61.12 (16.34)	−0.74	3,553,477
	All	0.909	243.16	244.24	52.39 (21.45)	−1.37	8,000,530

Note: R^2 is the coefficient of determination; Mean is the average in Wm^{-2}; RMSE is the root mean square error in Wm^{-2}; %RMSE is the relative RMSE of (RMSE/CERES$_{Mean}$) × 100 in percent; MPE is the mean percentage error of ((AHI-CERES)/CERES) × 100 in percent; N is the number of pairs.

To analyze the detailed results according to the validation data of the Figure 6, the statistical analysis results of the three data sets (Terra, Aqua, and S-NPP) are shown in the Figure 8. Bias (%Bias), RMSE (%RMSE), and MPE were improved in the order of Terra, Aqua, and S-NPP. In the case of Terra, %Bias, %RMSE, and MPE were 0.83%, 19.82%, and 0.13%, respectively, while those for Aqua were −1.40%, 20.24%, and −1.59%, and −1.92%, 21.44%, and −1.85% for S-NPP. For further analysis of these results, the data for Terra, Aqua, and S-NPP were compared to those for the overall case according to the clear fraction (Figure 9). As shown in Figure 9, the number of data for clear areas in Terra was approximately 0.24% higher than that of Aqua and S-NPP but about 0.96%, 1.40%, and 1.89% less for cloudy areas (partly cloudy, mostly cloudy, and overcast). In the previous analysis, Terra's error was smaller than that of Aqua and S-NPP because Terra had a slightly larger cloudy area and fewer clear areas than Aqua and S-NPP.

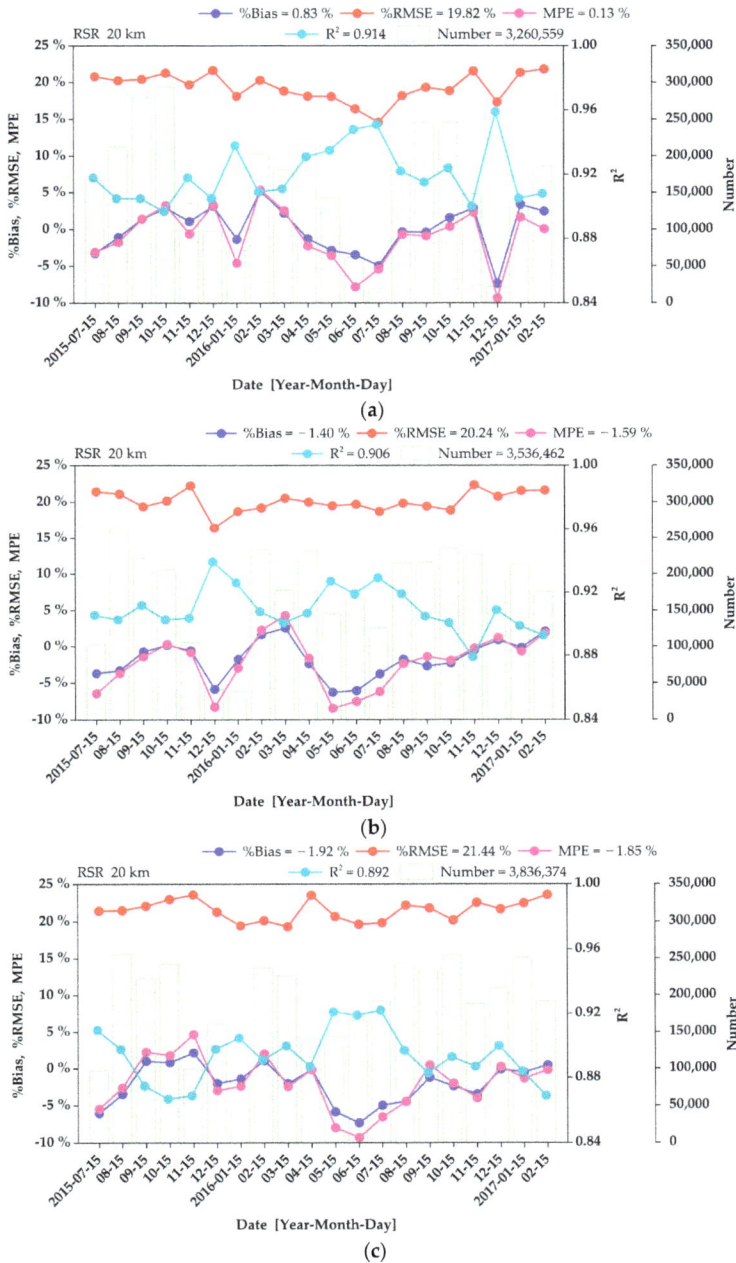

Figure 8. Similar to Figure 6 but for validation data in: Terra (**a**); Aqua (**b**); and S-NPP (**c**).

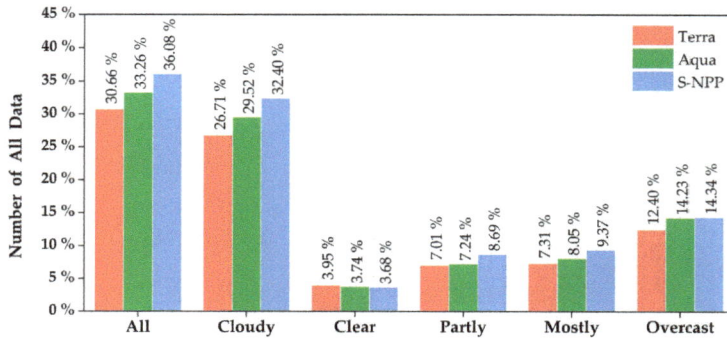

Figure 9. Number of Terra, Aqua and S-NPP data compared to all data according to the clear fraction. Clouds were subdivided according to clear fraction (all: 0–100%; cloudy: 0–95%; clear: ≥ 95%; partly cloudy: 50–95%; mostly cloudy: 5–50%; overcast: < 5%).

Figure 10 shows the results of scene analysis of S-NPP/CERES (36.08%) with the largest number of data among Terra, Aqua, and S-NPP for 20 cases because the S-NPP satellite observes at an altitude of 840 km, which is higher than the Terra and Aqua (705 km) [52]. Figure 10 shows the RSR of Himawari-8/AHI (left) and S-NPP/CERES (middle left), the percentage error of the two data sets (middle right), and the clear fraction of CERES (right). In this figure, RSR differs depending on the absence or presence of clouds and surface types, and the SZA increases as the latitude increases. We analyzed SZA and VZAs of less than 80° with approximately 3–5 scan line for each case. However, some cases had only 1–2 scan line due to the absence of Himawari-8/AHI data. In this figure, R^2 and Bias between Himawari-8/AHI and S-NPP/CERES were 0.867–0.922, and −21.34–5.52 Wm^{-2}, respectively, and the RMSE was 51.74–59.28 Wm^{-2}, similar to the results of [18]. Regarding land surface albedo (0.12–0.36) for the clear area (95–100%), the RMSE of S-NPP/CERES was 14.33–41.65 Wm^{-2}. This value is somewhat less accurate than that for the ocean (0.03–0.06, RMSE = 14.39–24.17 Wm^{-2}), which was relatively small and constant. During the preparation of the regression coefficient LUT, surface was classified into five types (vegetation, desert, snow, ocean, and lake), whereas CERES data were classified into 20 types [46], leading to some inaccuracies. In addition, in the cloud area, the RMSE varied between 54.08 and 61.52 Wm^{-2}, depending on the cloud properties (cloud optical thickness, cloud fraction, cloud type, etc.). Partly and mostly cloudy areas showed larger errors compared with overcast areas, particularly with increasing SZA and VZA. This can be attributed to the generation of the regression coefficient LUT using SBDART [11,53] and the actual atmospheric error assuming plane-parallel atmospheres [18]. In addition, the inability to refine the LUT according to the atmospheric conditions was considered to lead to an error; SZA (12 including 0°, 10°, 20°, 30°, 40°, 50°, 60°, 65°, 70°, 75°, 80°, and 85°), VZA (18 from 0–85° at 5° intervals), and RAA (19 from 0–180° at 10° intervals) [53]. Vermote and kotchenova [54] reported an error of 0.002 for 22 SZA and VZAs and 73 RAAs.

Date (Statistical analysis results)

Himawari-8/AHI RSR   S-NPP/CERES RSR   Percentage Error   Clear Fraction

15 July 2015 (R^2 = 0.910, Bias = −15.87 Wm^{-2}, RMSE = 56.20 Wm^{-2})

15 August 2015 (R^2 = 0.898, Bias = −8.93 Wm^{-2}, RMSE = 55.33 Wm^{-2})

15 September 2015 (R^2 = 0.875, Bias = 2.51 Wm^{-2}, RMSE = 54.49 Wm^{-2})

15 October 2015 (R^2 = 0.867, Bias = 1.97 Wm^{-2}, RMSE = 54.27 Wm^{-2})

15 November 2015 (R^2 = 0.869, Bias = 5.52 Wm^{-2}, RMSE = 59.28 Wm^{-2})

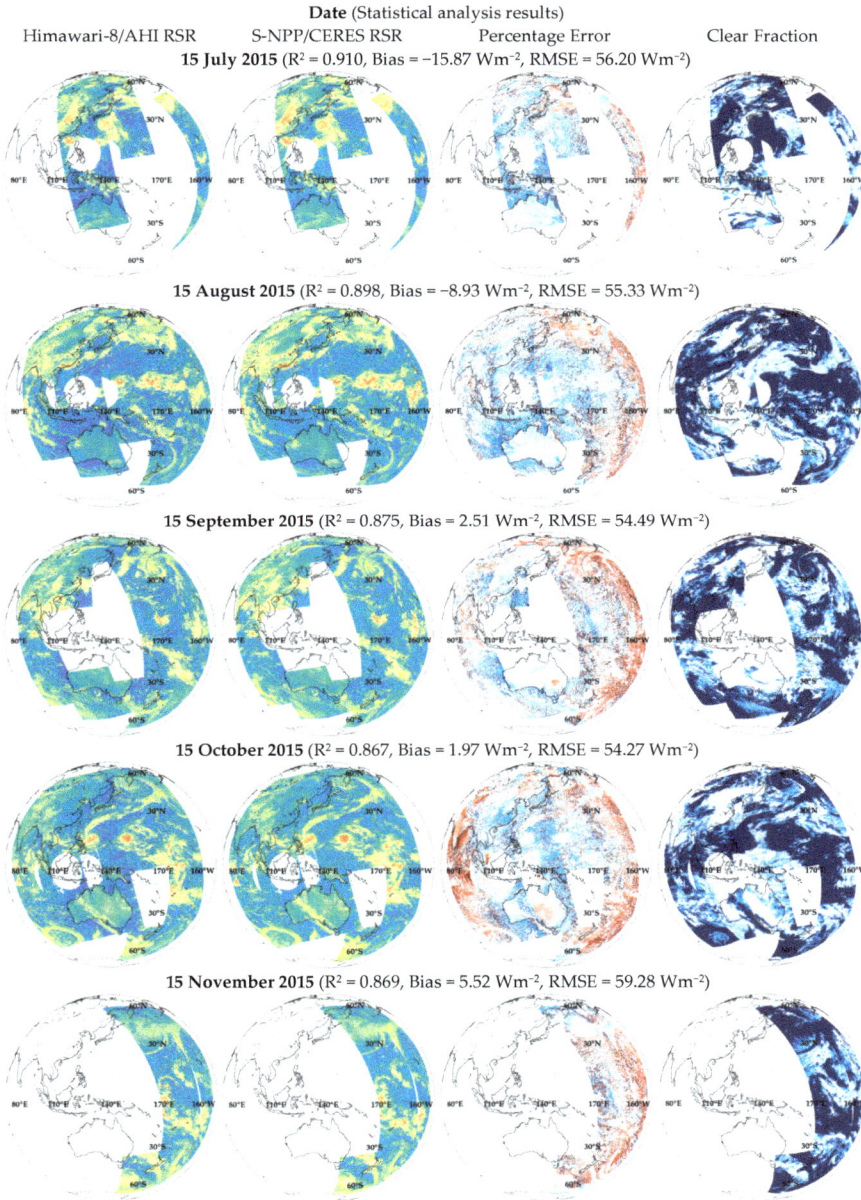

Figure 10. *Cont.*

15 December 2015 (R^2 = 0.898, Bias = −5.50 Wm^{-2}, RMSE = 58.02 Wm^{-2})

15 January 2016 (R^2 = 0.905, Bias = −3.73 Wm^{-2}, RMSE = 51.87 Wm^{-2})

15 February 2016 (R^2 = 0.891, Bias = 2.97 Wm^{-2}, RMSE = 55.56 Wm^{-2})

15 March 2016 (R^2 = 0.900, Bias = −5.75 Wm^{-2}, RMSE = 54.29 Wm^{-2})

15 April 2016 (R^2 = 0.887, Bias = −0.38 Wm^{-2}, RMSE = 56.69 Wm^{-2})

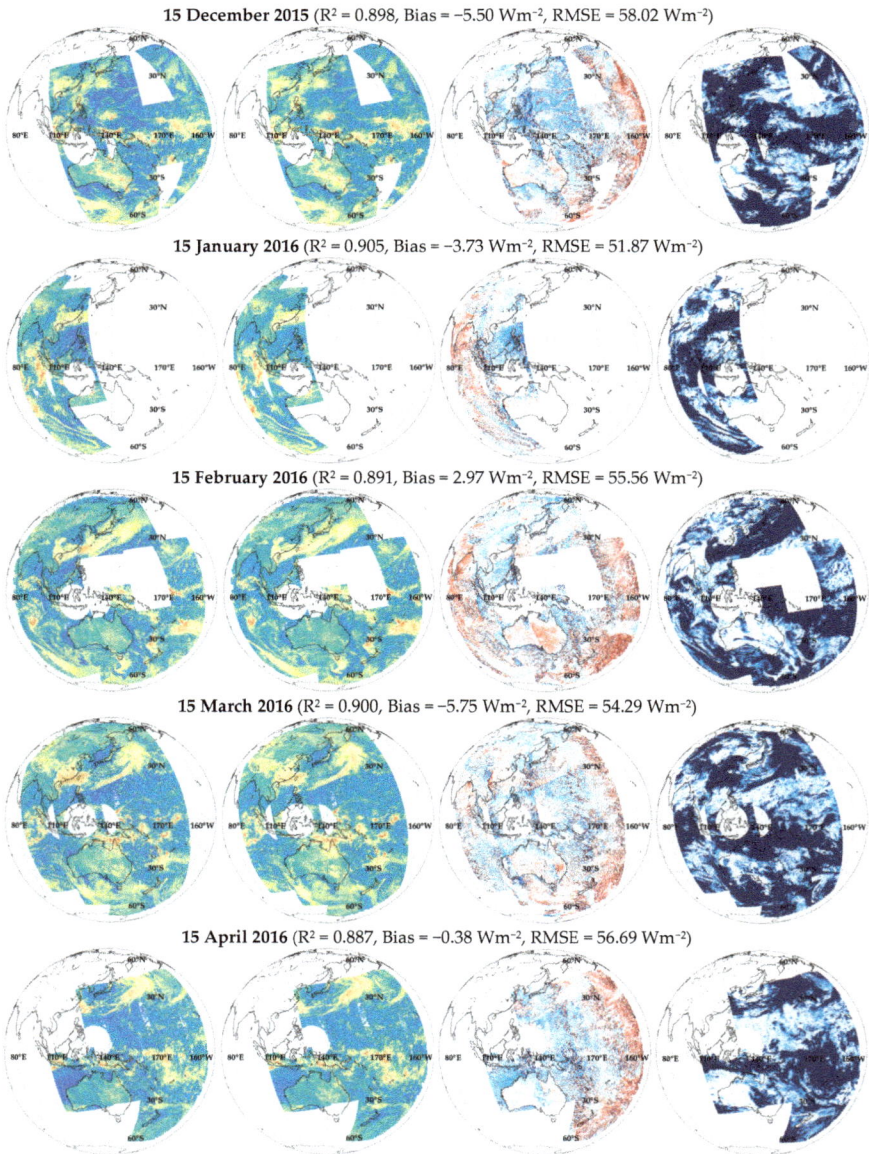

Figure 10. *Cont.*

15 May 2016 (R^2 = 0.921, Bias = −14.96 Wm^{-2}, RMSE = 52.79 Wm^{-2})

15 June 2016 (R^2 = 0.919, Bias = −21.34 Wm^{-2}, RMSE = 57.07 Wm^{-2})

15 July 2016 (R^2 = 0.922, Bias = −12.97 Wm^{-2}, RMSE = 51.74 Wm^{-2})

15 August 2016 (R^2 = 0.897, Bias = −11.30 Wm^{-2}, RMSE = 56.35 Wm^{-2})

15 September 2016 (R^2 = 0.883, Bias = −3.15 Wm^{-2}, RMSE = 55.73 Wm^{-2})

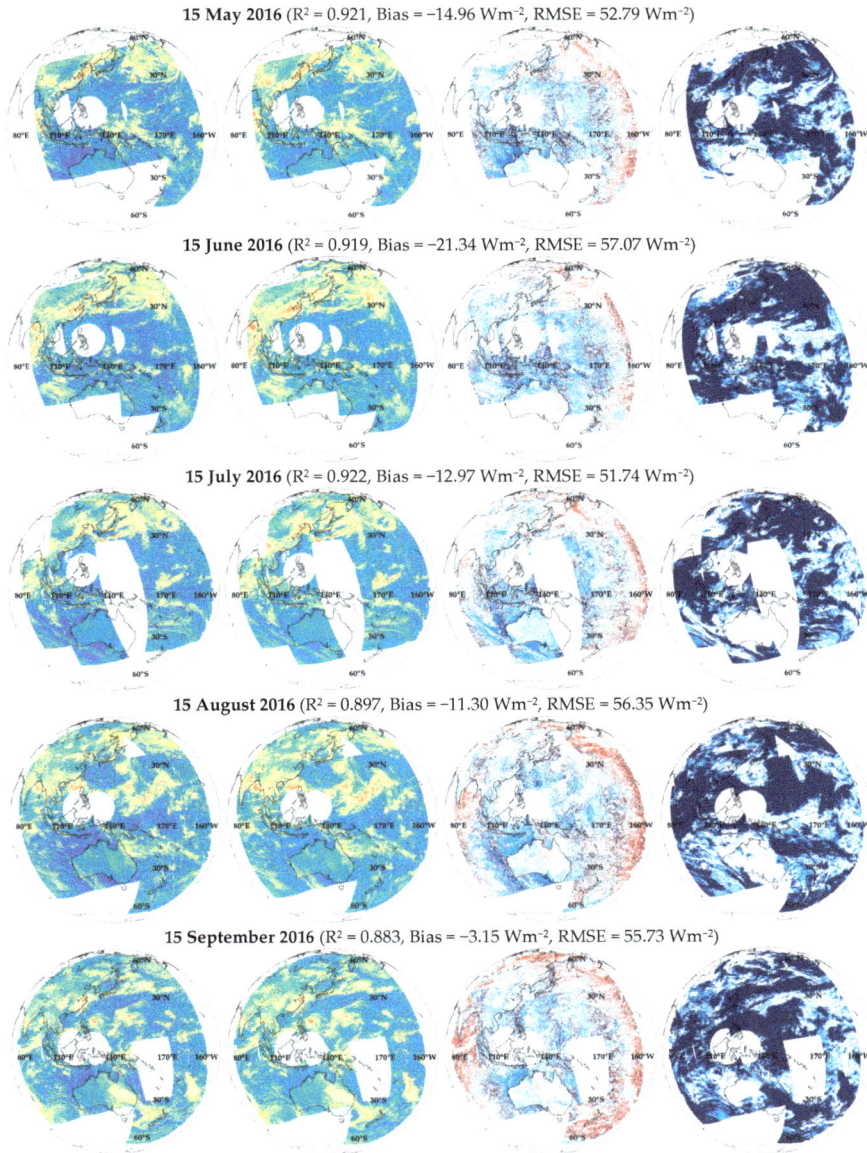

Figure 10. *Cont.*

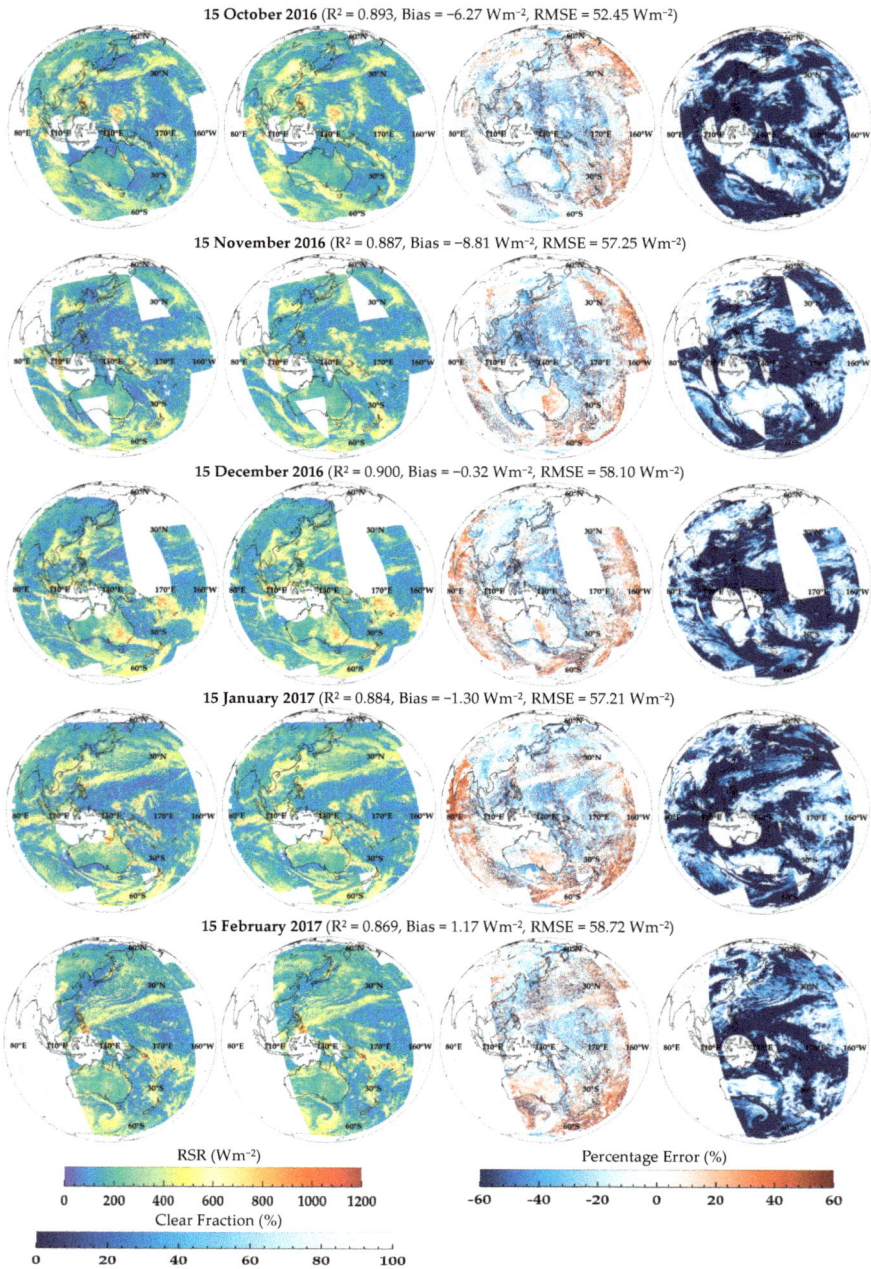

Figure 10. RSR (Wm^{-2}) of Himwari-8/AHI (**left**) and S-NPP/CERES (**middle left**) retrieved using data from the 15th day of every month between July 2015 and February 2017; percentage error (%) (**middle right**) and clear fraction (%) (**right**) between Himawari-8/AHI and S-NPP/CERES data sets; R^2, Bias, and RMSE (top) between the two data sets.

Figure 11 shows the results of averaging the data of Terra, Aqua, and S-NPP in the same manner averaging the RSR retrieved for these 20 cases with a resolution of 20 km for CERES. Figure 11 shows the averaged RSR (20 km) of AHI (Figure 11a) and CERES (Figure 11b) from 15 July 2015 to 15 February 2017, and the percentage error (Figure 11c) and the clear fraction (Figure 11d) of the two data sets. Figure 11e,f shows averaged results with respect to latitude and longitude, and RSR of AHI and CERES and the percentage error of both data sets, respectively. The results of this study, as shown in Figure 11a, are similar to the CERES average, as shown in Figure 11b. However, as shown in Figure 11c, the SZA and VZAs increases as sunrise or sunset approaches. In Figure 11e, the error between AHI and CERES for the latitude averages were within ±4%. However, the average longitude of Figure 11f had larger inaccuracies as bifurcations increased to 95 and 180° based on 140°. Then, in Figure 11d, the RSR of CERES above the Australian desert area shows a lower value than that of the AHI. This may be explained by the fact that RSR is greatly reduced by vapors, although the effects can vary in clear-sky areas depending on the surface type and absorption gases [55]. Since the CERES observes in the broadband shortwave range, RSR is greatly reduced in the range sensitive to vapor absorption (0.89–0.97 μm), whereas less attenuation occurs in the shortwave CHs of the Himawari-8/AHI used in this study because they are relatively transparent, compared to CERES. Similarly, the total precipitable water observed in this area during the study period was approximately 2.00 cm, indicating a MPE of approximately 3.81% in comparison to CERES. Therefore, the attenuation effect of vapors must be considered when retrieving RSR using a narrowband sensor. Figure 12 shows the results of this study (AHI) and Terra, Aqua, and S-NPP/CERES from 60 studies. In this figure, R^2 is 0.903 and Bias and RMSE are −2.34 and 52.12 Wm^{-2}, respectively.

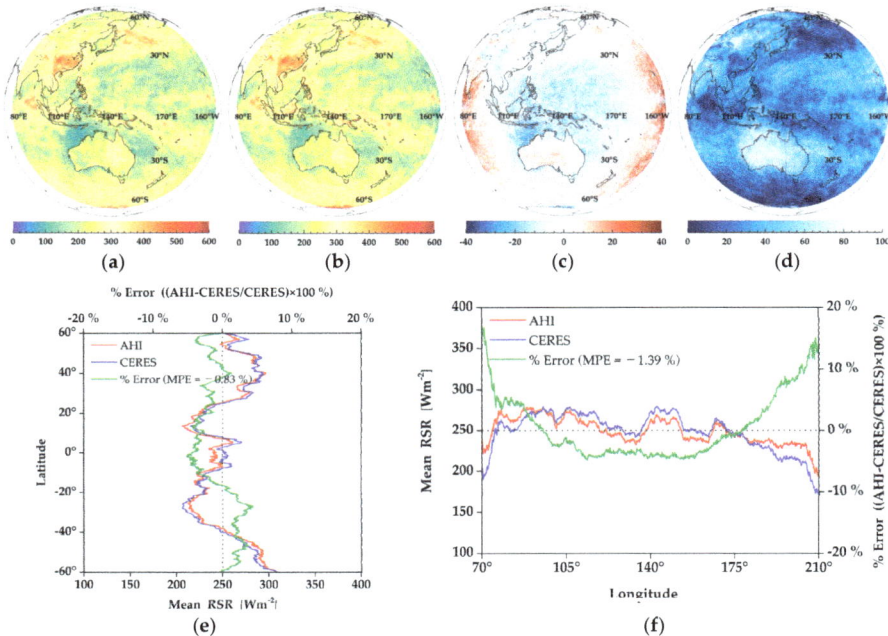

Figure 11. Similar to Figure 10, but for 20 km gridded mean (15 July 2015–15 February 2017) spatial distribution of RSR (Wm^{-2}) from: AHI (**a**); and CERES (**b**); percentage error (**c**); and clear fraction (**d**). Mean RSR from AHI and CERES averaged along each latitude and percentage error on double y axis (**e**); and similar to but along each longitude and percentage error on double x axis (**f**).

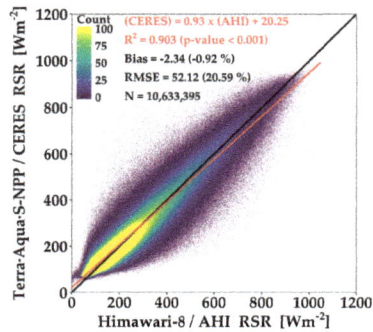

Figure 12. Similar to Figure 5, but for Table 4.

5. Discussion

In this study, the narrowband reflectance that had been assumed to be isotropic in Lee et al. [23] was considered to be anisotropic and the RSR was retrieved by removing the sun reflection region that caused a large error. Because the anisotropy of the atmosphere differs depending on the geometry of the solar-viewing, as well as the characteristics of the Earth's surface, the results improved in comparison with the CERES data when anisotropy was considered; the errors are listed in Table 3. Furthermore, SG below 20° was removed through statistical analysis (Figure 3) because strong sun reflection appears in a specific direction compared to the theoretical value and causes large errors in ocean areas [47,56]. After the SG was removed, the R^2 between the AHI and CERES was 0.923, which is higher than the value when SG was not removed (R^2 = 0.919). Furthermore, the Bias and MPE decreased from 3.11% and 2.43% to 2.53% and 0.74%, respectively. Based on these results, either the narrowband reflectance was assumed to be anisotropic or the SG was removed, as shown in Figure 5. The case where the narrowband reflectance was considered to be anisotropic and the SG was removed showed better results than the other cases, i.e., the case where the narrowband reflectance was assumed to be isotropic or considered anisotropic, or SG removed. In other words, consideration of anisotropy and removal of SG at the time of retrieving the RSR can reduce the error, as suggested by [17].

We compared the RSR retrieved using Himawari-8/AHI with the RSR observed in Terra, Aqua, and S-NPP/CERES. In this process, errors were generated by differences in spatial resolution and discrepancies in observation time (validation of ±5 min data, i.e., 0010 UTC vs. 0005-0015 UTC) between AHI (2 km) and CERES (20 km) (Figure 7). To solve this problem, it is desirable to select and analyze a spatiotemporal resolution that shows good results by performing statistical analysis according to spatial resolution (0.5°, 1°, 2°, 5°, and 10°) for each daily or monthly average, as suggested by [2]. In addition, instead of analyzing data from only one satellite, all the Terra, Aqua and S-NPP data should be used together to carry out relatively continuous validation because polar-orbiting satellites equipped with broadband sensors perform observations only for specific time periods. Furthermore, when all the 60 cases and satellite data used in this study were averaged, relatively large errors were generated at SZA and VZAs above 70°, as shown in Figure 11. This is because several satellite-based data are error-prone due to the lack of accuracy in sunrise and sunset times [57]. Furthermore, the refraction effect at a SZA of 85° or less is negligible between the apparent SZA and the real SZA but produces significant errors at sunrise or sunset [58].

Even though errors were due to the difference between the spatiotemporal resolution and the SZA and VZAs, the validation results in Figure 12 show a similar trend to results presented in [1,18]. However, since the spatial resolution of the averaged satellite data was greater and more long-term analysis was performed in the current study, the R^2 [59], Bias and RMSE [2,60] all improved. Therefore, the results of this study are improved in comparison to those of other studies [2,12,60].

6. Conclusions

This study used Himawari-8/AHI data to develop an algorithm for retrieving RSR. Himawari-8/AHI data have characteristics similar to those of the next-generation geostationary orbit meteorological satellite (GK-2A/AMI), which is to be launched in 2018. This algorithm converts the radiance of narrowband sensor into reflectance and then into TOA albedo through a regression coefficient LUT according to geometry of the solar-viewing (SZA, VZA, and RAA) and atmospheric conditions (surface type and absence/presence of clouds). The regression coefficients used in this study were calculated through numerical experimental using the SBDART and ridge regression for the TOA albedo and narrowband reflectance of sensor channels considering anisotropy. In addition, statistical analysis was performed to remove SG because SG can cause serious errors. For this purpose, reflection angles less than 20° were removed. Terra, Aqua, and S-NPP/CERES data were used to validate this RSR.

The results of this study (Himawari-8/AHI) showed improved performance of the algorithm, and the validation with Terra, Aqua, and S-NPP/CERES data also showed good results, except from 15 May to 15 July 2016. In particular, on 15 June 2016, comparing all data, the number of data in the clear areas and the cloudy area was influenced, and therefore the statistical analysis results were unsatisfactory. R^2 (0.571–0.815) and RMSE (25.09–64.61 Wm^{-2}) of the partly and mostly cloudy areas were large, which could be attributed to a space-time mismatch between AHI and CERES in the spatial resolution matching process. To further analyze these results, we divided the data into three sets (Terra, Aqua, and S-NPP). The results were better in the order of Terra, Aqua, and S-NPP, because Terra accounted for a larger number of data in the clear area and less cloud area than Aqua and S-NPP. This implies that the validation results are affected by the number of data in the clear and cloudy areas.

The analysis of the scan line showed an RMSE of 14.33–41.65 Wm^{-2} on the land for the clear area, which slightly differs from that of the ocean (RMSE = 14.39–24.17 Wm^{-2}) due to the classification difference of the surface type of AHI and CERES. In addition, because SBDART assumes a plane-parallel atmosphere in the cloud region, the error increases as the SZA and VZAs increase in the real atmosphere and the plane-parallel atmosphere. Nevertheless, R^2 in this study was higher than 0.867 and Bias and RMSE were −21.34–5.52 and 51.74–59.28 Wm^{-2} for scan line, respectively. R^2 between AHI and Terra, Aqua, and S-NPP/CERES for all case was 0.903, showing significance of approximately 0.001, and Bias and RMSE were −2.34 and 52.12 Wm^{-2}, respectively.

High temporal and spatial resolution data, such as the next-generation geostationary meteorological satellites (Himawari-8/AHI, GK-2A/AMI, etc.), are used to conduct research on understanding the effects of aerosols [61] and clouds [62], which have large temporal and spatial variabilities, on radiation [18]. It can also be applied as basic input for numerical prediction and climate models, which will contribute to improving the accuracy of the calculation [63] and further monitoring of radiation balance to understand the mechanism of climate change [64]. As a preliminary study for the radiative element output of GK-2A/AMI, the successor satellite of the Republic of Korea, this study will guide future research and development of the sensor. Nevertheless, the algorithm cannot classify surface types [65] in clear areas and the RSR error is generated because cloudy areas do not reflect cloud properties (cloud optical thickness, cloud fraction, cloud type, etc.) in detail. Future studies will require a recalculation LUT that reflects the characteristics of the surface and clouds and should be periodically corrected by long-term observed broadband radiation [5,66].

Acknowledgments: This work was supported by "Development of Radiation/Aerosol Algorithms" project, funded by ETRI, which is a subproject of "Development of Geostationary Meteorological Satellite Ground Segment (NMSC-2016-01)" program funded by NMSC (National Meteorological Satellite Center) of KMA (Korea Meteorological Administration).

Author Contributions: Sang-Ho Lee led manuscript writing and contributed to the data analysis and research design. Kyu-Tae Lee and Bu-Yo Kim supervised this study, contributed to the research design and manuscript writing, and served as the corresponding authors. Il-Sung Zo, Hyun-Seok Jung and Se-Hun Rim contributed to the discussion of the results and manuscript writing.

Conflicts of Interest: The authors declare no conflict of interest.

References

1. AWG Radiation Budget Application Team. GOES-R Advanced Baseline Imager (ABI) Algorithm Theoretical Basis Document for Downward Shortwave Radiation (Surface), and Reflected Shortwave Radiation (TOA), NOAA NESDIS Center for Satellite Applications and Research, 27 September 2010. Available online: https://www.goes-r.gov/products/ATBDs/baseline/baseline-DSR-v2.0.pdf (accessed on 28 November 2017).
2. Bhartia, P.K. Top-of-the-atmosphere shortwave flux estimation from satellite observations: An empirical neural network approach applied with data from the a-train constellation. *Atmos. Meas. Tech.* **2016**, *9*, 2813–2826.
3. Kim, Y.-J.; Kim, B.-G.; Miller, M.; Min, Q.; Song, C.-K. Enhanced aerosol-cloud relationships in more stable and adiabatic clouds. *Asia-Pac. J. Atmos. Sci.* **2012**, *48*, 283–293. [CrossRef]
4. Trenberth, K.E.; Dai, A.; Van Der Schrier, G.; Jones, P.D.; Barichivich, J.; Briffa, K.R.; Sheffield, J. Global warming and changes in drought. *Nat. Clim. Chang.* **2014**, *4*, 17–22. [CrossRef]
5. Doelling, D.R.; Loeb, N.G.; Keyes, D.F.; Nordeen, M.L.; Morstad, D.; Nguyen, C.; Wielicki, B.A.; Young, D.F.; Sun, M. Geostationary enhanced temporal interpolation for CERES flux products. *J. Atmos. Ocean. Technol.* **2013**, *30*, 1072–1090. [CrossRef]
6. Stephens, G.L.; O'Brien, D.; Webster, P.J.; Pilewski, P.; Kato, S.; Li, J.L. The albedo of earth. *Rev. Geophys.* **2015**, *53*, 141–163. [CrossRef]
7. Hatzianastassiou, N.; Matsoukas, C.; Fotiadi, A.; Pavlakis, K.; Drakakis, E.; Hatzidimitriou, D.; Vardavas, I. Global distribution of earth's surface shortwave radiation budget. *Atmos. Chem. Phys.* **2005**, *5*, 2847–2867. [CrossRef]
8. Stubenrauch, C.; Rossow, W.; Kinne, S. Assessment of global cloud data sets from satellites a project of the world climate research programme global energy and water cycle experiment (GEWEX) radiation panel lead authors. *Am. Meteorol. Soc.* **2012**. [CrossRef]
9. Luther, M.; Cooper, J.; Taylor, G. The earth radiation budget experiment nonscanner instrument. *Rev. Geophys.* **1986**, *24*, 391–399. [CrossRef]
10. Wielicki, B.A.; Barkstrom, B.R.; Harrison, E.F.; Lee, R.B., III; Louis Smith, G.; Cooper, J.E. Clouds and the earth's radiant energy system (CERES): An earth observing system experiment. *Bull. Am. Meteorol. Soc.* **1996**, *77*, 853–868. [CrossRef]
11. Wang, D.; Liang, S. Estimating high-resolution top of atmosphere albedo from moderate resolution imaging spectroradiometer data. *Remote Sens. Environ.* **2016**, *178*, 93–103. [CrossRef]
12. Niu, X.; Pinker, R.T. Revisiting satellite radiative flux computations at the top of the atmosphere. *Int. J. Remote Sens.* **2012**, *33*, 1383–1399. [CrossRef]
13. Buriez, J.C.; Parol, F.; Poussi, Z.; Viollier, M. An improved derivation of the top-of-atmosphere albedo from POLDER/ADEOS-2: 2. Broadband albedo. *J. Geophys. Res. Atmos.* **2007**, *112*. [CrossRef]
14. Laszlo, I.; Gruber, A.; Jacobowitz, H. The relative merits of narrowband channels for estimating broadband albedos. *J. Atmos. Ocean. Technol.* **1988**, *5*, 757–773. [CrossRef]
15. Wydick, J.E.; Davis, P.A.; Gruber, A. *Estimation of Broadband Planetary Albedo from Operational Narrowband Satellite Measurements*; National Oceanic and Atmospheric Administration: Washington, DC, USA, 1987.
16. Loeb, N.G.; Manalo-Smith, N.; Kato, S.; Miller, W.F.; Gupta, S.K.; Minnis, P.; Wielicki, B.A. Angular distribution models for top-of-atmosphere radiative flux estimation from the clouds and the earth's radiant energy system instrument on the tropical rainfall measuring mission satellite. Part I: Methodology. *J. Appl. Meteorol.* **2003**, *42*, 240–265. [CrossRef]
17. Viollier, M. Restitution of longwave and shortwave radiative fluxes at the top of the atmosphere from combination of scarab and meteosat data. In Proceedings of the Megha-Tropiques 2nd Scientific Workshop, Paris, France, 2–6 July 2001.
18. Vazquez-Navarro, M.; Mayer, B.; Mannstein, H. A fast method for the retrieval of integrated longwave and shortwave top-of-atmosphere upwelling irradiances from MSG/SEVIRI (RRUMS). *Atmos. Meas. Tech.* **2013**, *6*, 2627–2640. [CrossRef]
19. Berk, A.; Bernstein, L.; Robertson, D. *Modtran: A Moderate Resolution Model for LOWTRAN 7*; Rep. AFGL-TR-83-0187; Air Force Geophysical Laboratory, Hanscom Air Force Base: Hanscom, MA, USA, 1983.

20. Kim, B.-Y.; Lee, K.-T.; Jee, J.-B.; Zo, I.-S. Retrieval of outgoing longwave radiation at top-of-atmosphere using himawari-8 AHI data. *Remote Sens. Environ.* **2018**, *204*, 498–508. [CrossRef]
21. Choi, Y.-S.; Ho, C.-H. Earth and environmental remote sensing community in South Korea: A review. *Remote Sens. Appl. Soc. Environ.* **2015**, *2*, 66–76. [CrossRef]
22. Murata, H.; Takahashi, M.; Kosaka, Y. Vis and IR bands of Himawari-8/AHI compatible with those of MTSAT-2/Imager. 2015. Available online: www.data.jma.go.jp/mscweb/technotes/msctechrep60.pdf (accessed on 30 January 2018).
23. Lee, S.-H.; Lee, K.-T.; Kim, B.-Y.; Zo, I.-S.; Jung, H.-S.; Rim, S.-H. Retrieval Algorithm for Broadband Albedo at the Top of the Atmosphere. *Asia Pac. J. Atmos. Sci.* **2017**, accepted.
24. Ricchiazzi, P.; Yang, S.; Gautier, C.; Sowle, D. Sbdart: A research and teaching software tool for plane-parallel radiative transfer in the earth's atmosphere. *Bull. Am. Meteorol. Soc.* **1998**, *79*, 2101–2114. [CrossRef]
25. Bessho, K.; Date, K.; Hayashi, M.; Ikeda, A.; Imai, T.; Inoue, H.; Kumagai, Y.; Miyakawa, T.; Murata, H.; Ohno, T. An introduction to himawari-8/9—Japan's new-generation geostationary meteorological satellites. *J. Meteorol. Soc. Jpn. Ser. II* **2016**, *94*, 151–183. [CrossRef]
26. Puschell, J.J.; Lowe, H.A.; Jeter, J.W.; Kus, S.M.; Hurt, W.T.; Gilman, D.; Rogers, D.L.; Hoelter, R.L.; Ravella, R. Japanese Advanced Meteorological Imager: A next-generation geo imager for MTSAT-1R. In Proceedings of the Earth Observing Systems VII, Seattle, WA, USA, 7–11 July 2002; International Society for Optics and Photonics: Bellingham, WA, USA, 2002; pp. 152–162.
27. Cao, C.; De Luccia, F.J.; Xiong, X.; Wolfe, R.; Weng, F. Early on-orbit performance of the visible infrared imaging radiometer suite onboard the Suomi National Polar-Orbiting Partnership (S-NPP) satellite. *IEEE Trans. Geosci. Remote Sens.* **2014**, *52*, 1142–1156. [CrossRef]
28. Yu, F.; Wu, X. Radiometric inter-calibration between Himawari-8 AHI and S-NPP VIIRS for the solar reflective bands. *Remote Sens.* **2016**, *8*, 165. [CrossRef]
29. Paden, J.; Smith, G.L.; Lee, R.B.; Pandey, D.K.; Thomas, S. Reality check: A point response function (PRF) comparison of theory to measurements for the clouds and the earth's radiant energy system (CERES) tropical rainfall measuring mission (TRMM) instrument. In Proceedings of the Visual Information Processing VI, Orlando, FL, USA, 21–25 April 1997; International Society for Optics and Photonics: Bellingham, WA, USA, 1997; pp. 109–118.
30. Su, W.; Corbett, J.; Eitzen, Z.; Liang, L. Next-generation angular distribution models for top-of-atmosphere radiative flux calculation from CERES instruments: Validation. *Atmos. Meas. Tech.* **2015**, *8*, 3297–3313. [CrossRef]
31. Zerefos, C.S.; Isaksen, I.S.; Ziomas, I. *Chemistry and Radiation Changes in the Ozone Layer*; Springer Science & Business Media: Berlin/Heidelberg, Germany, 2012; Volume 557.
32. Blanc, P.; Gschwind, B.; Lefevre, M.; Wald, L. Twelve monthly maps of ground albedo parameters derived from MODIS data sets. In Proceedings of the IEEE International Geoscience and Remote Sensing Symposium (IGARSS), Quebec City, QC, Canada, 13–18 July 2014; pp. 3270–3272.
33. Liang, S. *Quantitative Remote Sensing of Land Surfaces*; John Wiley & Sons: Hoboken, NJ, USA, 2005; Volume 30.
34. Qu, X.; Hall, A. Surface contribution to planetary albedo variability in cryosphere regions. *J. Clim.* **2005**, *18*, 5239–5252. [CrossRef]
35. Kopp, G.; Lean, J.L. A new, lower value of total solar irradiance: Evidence and climate significance. *Geophys. Res. Lett.* **2011**, *38*. [CrossRef]
36. Michalsky, J.J. The astronomical almanac's algorithm for approximate solar position (1950–2050). *Sol. Energy* **1988**, *40*, 227–235. [CrossRef]
37. Nanni, M.R.; Demattê, J.A.M. Spectral reflectance methodology in comparison to traditional soil analysis. *Soil Sci. Soc. Am. J.* **2006**, *70*, 393–407. [CrossRef]
38. Mokhtari, M.H.; Busu, I. Downscaling albedo from moderate-resolution imaging spectroradiometer (MODIS) to advanced space-borne thermal emission and reflection radiometer (ASTER) over an agricultural area utilizing aster visible-near infrared spectral bands. *Int. J. Phys. Sci.* **2011**, *6*, 5804–5821.
39. Draper, N.R.; Smith, H.; Pownell, E. *Applied Regression Analysis*; Wiley: New York, NY, USA, 1966; Volume 3.
40. Kleinbaum, D.; Kupper, L.; Nizam, A.; Rosenberg, E. *Applied Regression Analysis and Other Multivariable Methods*; Nelson Education: Scarborough, ON, Canada, 2013.
41. Loeb, N.G.; Hinton, P.O.R.; Green, R.N. Top-of-atmosphere albedo estimation from angular distribution models: A comparison between two approaches. *J. Geophys. Res. Atmos.* **1999**, *104*, 31255–31260. [CrossRef]

42. Loeb, N.G.; Parol, F.; Buriez, J.-C.; Vanbauce, C. Top-of-atmosphere albedo estimation from angular distribution models using scene identification from satellite cloud property retrievals. *J. Clim.* **2000**, *13*, 1269–1285. [CrossRef]

43. Kato, S.; Marshak, A. Solar zenith and viewing geometry-dependent errors in satellite retrieved cloud optical thickness: Marine stratocumulus case. *J. Geophys. Res. Atmos.* **2009**, *114*. [CrossRef]

44. Gardner, A.S.; Sharp, M.J. A review of snow and ice albedo and the development of a new physically based broadband albedo parameterization. *J. Geophys. Res. Earth Surf.* **2010**, *115*. [CrossRef]

45. Loeb, N.G.; Kato, S. Top-of-atmosphere direct radiative effect of aerosols over the tropical oceans from the clouds and the earth's radiant energy system (CERES) satellite instrument. *J. Clim.* **2002**, *15*, 1474–1484. [CrossRef]

46. Geier, E.; Green, R.; Kratz, D.; Minnis, P.; Miller, W.; Nolan, S.; Franklin, C. Clouds and the Earth's Radiant Energy System (CERES). Data Management System: Single Satellite Footprint TOA/Surface Fluxes and Clouds (SSF) Collection Document. Release 2 Version 1; 2003. Available online: https://ceres.larc.nasa.gov/documents/collect_guide/pdf/SSF_CG.pdf (accessed on 28 November 2017).

47. Zhang, H.; Wang, M. Evaluation of sun glint models using MODIS measurements. *J. Quant. Spectrosc. Radiat. Transf.* **2010**, *111*, 492–506. [CrossRef]

48. Kay, S.; Hedley, J.D.; Lavender, S. Sun glint correction of high and low spatial resolution images of aquatic scenes: A review of methods for visible and near-infrared wavelengths. *Remote Sens.* **2009**, *1*, 697–730. [CrossRef]

49. Wilks, D.S. *Statistical Methods in the Atmospheric Sciences*; Academic Press: Cambridge, MA, USA, 2011; Volume 100.

50. Benesty, J.; Chen, J.; Huang, Y.; Cohen, I. Pearson correlation coefficient. In *Noise Reduction in Speech Processing*; Springer: New York, NY, USA, 2009; pp. 1–4.

51. Livezey, R.E.; Chen, W. Statistical field significance and its determination by Monte Carlo techniques. *Mon. Weather Rev.* **1983**, *111*, 46–59. [CrossRef]

52. Thomas, S.; Priestley, K.; Shankar, M.; Smith, N.; Timcoe, M. Pre-launch sensor characterization of the CERES flight model 5 (FM5) instrument on NPP mission. *Proc. SPIE* **2011**. [CrossRef]

53. Lu, N.; Liu, R.; Liu, J.; Liang, S. An algorithm for estimating downward shortwave radiation from GMS 5 visible imagery and its evaluation over china. *J. Geophys. Res. Atmos.* **2010**, *115*. [CrossRef]

54. Vermote, E.F.; Kotchenova, S. Atmospheric correction for the monitoring of land surfaces. *J. Geophys. Res. Atmos.* **2008**, *113*. [CrossRef]

55. Sena, E.; Artaxo, P.; Correia, A. The effects of smoke aerosols, land-use change and water vapor reduction on the shortwave radiative budget over the Amazônia. In Proceedings of the EGU General Assembly Conference Abstracts, Vienna, Austria, 27 April–2 May 2014; Volume 16.

56. Bertrand, C.; Clerbaux, N.; Ipe, A.; Dewitte, S.; Gonzalez, L. Angular distribution models anisotropic correction factors and sun glint: A sensitivity study. *Int. J. Remote Sens.* **2006**, *27*, 1741–1757. [CrossRef]

57. Urraca, R.; Gracia-Amillo, A.M.; Koubli, E.; Huld, T.; Trentmann, J.; Riihelä, A.; Lindfors, A.V.; Palmer, D.; Gottschalg, R.; Antonanzas-Torres, F. Extensive validation of CM SAF surface radiation products over europe. *Remote Sens. Environ.* **2017**, *199*, 171–186. [CrossRef] [PubMed]

58. Madhavan, B. Interactive comment on "shortwave surface radiation budget network for observing small-scale cloud inhomogeneity fields" by B.L. Madhavan et al. *Atmos. Meas. Tech. Discuss.* **2015**, *8*, C2233–C2250. [CrossRef]

59. Li, Z.; Cribb, M.; Chang, F.L.; Trishchenko, A.; Luo, Y. Natural variability and sampling errors in solar radiation measurements for model validation over the atmospheric radiation measurement southern great plains region. *J. Geophys. Res. Atmos.* **2005**, *110*. [CrossRef]

60. Wang, D.; Liang, S. Estimating top-of-atmosphere daily reflected shortwave radiation flux over land from modis data. *IEEE Trans. Geosci. Remote Sens.* **2017**, *55*, 4022–4031. [CrossRef]

61. He, L.; Wang, L.; Lin, A.; Zhang, M.; Bilal, M.; Tao, M. Aerosol optical properties and associated direct radiative forcing over the Yangtze River basin during 2001–2015. *Remote Sens.* **2017**, *9*, 746. [CrossRef]

62. Katagiri, S.; Kikuchi, N.; Nakajima, T.Y.; Higurashi, A.; Shimizu, A.; Matsui, I.; Hayasaka, T.; Sugimoto, N.; Takamura, T.; Nakajima, T. Cirrus cloud radiative forcing derived from synergetic use of MODIS analyses and ground-based observations. *Sola* **2010**, *6*, 25–28. [CrossRef]

63. Allan, R.P.; Slingo, A.; Milton, S.F.; Culverwell, I. Exploitation of geostationary earth radiation budget data using simulations from a numerical weather prediction model: Methodology and data validation. *J. Geophys. Res. Atmos.* **2005**, *110*. [CrossRef]

64. Urbain, M.; Clerbaux, N.; Ipe, A.; Tornow, F.; Hollmann, R.; Baudrez, E.; Velazquez Blazquez, A.; Moreels, J. The CM SAF TOA radiation data record using MVIRI and SEVIRI. *Remote Sens.* **2017**, *9*, 466. [CrossRef]

65. Congalton, R.G.; Gu, J.; Yadav, K.; Thenkabail, P.; Ozdogan, M. Global land cover mapping: A review and uncertainty analysis. *Remote Sens.* **2014**, *6*, 12070–12093. [CrossRef]

66. Doelling, D.R.; Sun, M.; Nguyen, L.T.; Nordeen, M.L.; Haney, C.O.; Keyes, D.F.; Mlynczak, P.E. Advances in geostationary-derived longwave fluxes for the CERES synoptic (SYN1deg) product. *J. Atmos. Ocean. Technol.* **2016**, *33*, 503–521. [CrossRef]

remote sensing

MDPI

Review

Modeling Solar Radiation in the Forest Using Remote Sensing Data: A Review of Approaches and Opportunities

Alex S. Olpenda [1,*], Krzysztof Stereńczak [2] and Krzysztof Będkowski [3]

[1] Department of Geomatics and Spatial Planning, Faculty of Forestry, Warsaw University of Life Sciences, 02-787 Warsaw, Poland

[2] Laboratory of Geomatics, Forest Research Institute, Sękocin Stary, 05-090 Raszyn, Poland; K.Sterenczak@ibles.waw.pl

[3] Department of Geoinformation, Institute of Urban Geography and Tourism, Faculty of Geographical Sciences, University of Lodz, 90-137 Lodz, Poland; krzysztof.bedkowski@geo.uni.lodz.pl

* Correspondence: alecsolpenda@mail.sggw.pl; Tel.: +48-579-113-977

Received: 22 March 2018; Accepted: 26 April 2018; Published: 1 May 2018

Abstract: Solar radiation, the radiant energy from the sun, is a driving variable for numerous ecological, physiological, and other life-sustaining processes in the environment. Traditional methods to quantify solar radiation are done either directly (e.g., quantum sensors), or indirectly (e.g., hemispherical photography). This study, however, evaluates literature which utilized remote sensing (RS) technologies to estimate various forms of solar radiation or components, thereof under or within forest canopies. Based on the review, light detection and ranging (LiDAR) has, so far, been preferably used for modeling light under tree canopies. Laser system's capability of generating 3D canopy structure at high spatial resolution makes it a reasonable choice as a source of spatial information about light condition in various parts of forest ecosystem. The majority of those using airborne laser system (ALS) commonly adopted the volumetric-pixel (voxel) method or the laser penetration index (LPI) for modeling the radiation, while terrestrial laser system (TLS) is preferred for canopy reconstruction and simulation. Furthermore, most of the studies focused only on global radiation, and very few on the diffuse fraction. It was also found out that most of these analyses were performed in the temperate zone, with a smaller number of studies made in tropical areas. Nonetheless, with the continuous advancement of technology and the RS datasets becoming more accessible and less expensive, these shortcomings and other difficulties of estimating the spatial variation of light in the forest are expected to diminish.

Keywords: solar radiation; understory light condition; forest canopy; subcanopy light regime; PAR; shortwave radiation; light attenuation; remote sensing

1. Introduction

The solar radiation in the form of light is a driving variable for many biological, ecological, physiological, and hydrological processes in the environment [1–5]. Aside from photosynthesis and transpiration, radiation in the forest also affects vegetation patterns [5,6], stand development [7], forest growth [5,8], and production efficiency [9,10]. By penetrating through the canopy, the influence of solar radiation can also be seen down to the forest floor, as it is also closely related to germination and understory growth [8,11], regeneration and succession [12–14], soil conditions [13,15,16], and even biodiversity [17,18]. It is, therefore, essential to have an understanding of the light condition if one wishes to study forest ecology [19].

Light intensity in the forest is mainly affected by canopy structure, site characteristics, atmospheric conditions and solar elevation [13,19–23]. These factors produce complex understory light patterns that express not only horizontal heterogeneity, but also the vertical variation at any given time [6].

Both solar elevation and atmospheric conditions are primarily dependent on the geographical location of the specified site. Thus, the discussions of these factors were deliberately excluded. The remaining factors: canopy structure and site characteristics are often represented and described with the use of remote sensing (RS) data [21,22,24,25]. Among others, canopy closure, canopy cover or canopy height are proxies of structural and geometrical information of leaves and branches, which are important variables characterizing light conditions inside the stand [6,9,19,26–29]. Such information carries critical parameters that can only be described effectively in a 3D space. In addition, site characteristics, such as micro relief, aspect, slope, and height above sea level, can be derived from an accurate high-resolution digital elevation model (DEM) acquired with the use of an active system, such as the light detection and ranging (LiDAR) [30,31]. Clearly, the remote sensing technology can be a very useful and important source of information for light condition inside the forest [32–34].

Existing comprehensive reviews on quantifying forest light environment have been published so far by Lieffers et al. [28], Comeau [35], and Promis [36]. The authors showed the nature and properties of the different instruments used, methods applied, the accuracies, as well as the associated costs. The use of handheld instruments, called ceptometers, and other quantum-equipped sensors as part of direct measurements, was widely mentioned, while hemispherical photography is considered the most common indirect way. Nonetheless, what was not considered in the scope of the three reviews, was the utilization of RS technology and other allied sciences.

Combined with field data, RS provides a handful of benefits, such as the generation of continuous spatial information with an effective cost on a wider scale. Modeling with the use of geographic information system (GIS), for example, can be used in lieu of field-recorded data for areas where field data is not available, or would be too expensive to record [37]. With this strategy, a continuous map of light conditions can be delivered as a final result.

While a passive optical sensor in two-dimensional format proved to be useful in various forestry applications, three-dimensional data may offer even more canopy details and a better understanding of the forest's structure [38–40]. Aside from that, it overcomes some of the disadvantages of passive remote sensing, such as cloud-cover issues and vegetation index saturation problems [41]. A wide-ranging description of available airborne and satellite sensors and their capabilities, in general, could be referred to in the publications by Wang et al. [42], Brewer et al. [43] and, more recently, by Toth and Jozkow [44], with the latest discussions not only on sensors, but on platforms as well.

This paper, however, evaluates published articles that used or have at least a component of remote sensing in the context of solar radiation at forest stands in various scales. The aim of this study is to (i) synthesize the studies to determine how and what has been done with this state-of-the-art technology, and (ii) detect gaps in methodology or specific technology used to address improvements in the future. The first section of the review presents the physical concepts of solar radiation, and how it transmits through the canopy. The second section focuses on the different techniques and modeling approaches done by researchers on analyzing subcanopy solar radiation with the use of RS data. The next section discusses what we think are the critical issues that came out after synthesizing the articles, and selects papers that stand out and demonstrate novel approaches that only RS technology could offer. Lastly, conclusions and recommendations are laid down on how to improve future directions of similar studies.

2. Overview of Concepts

Scientists estimate that roughly 1368 watts m^{-2}, averaged over the globe and over several years, illuminates the outermost atmosphere of the Earth [45]. This value is known as solar constant or the total solar irradiance (TSI), which is the maximum possible power that the sun can deliver to the Earth at the mean distance between them [23,46]. Only about $\frac{1}{4}$ of the TSI that is considered the incoming solar radiation, collectively called shortwave radiation, enters the top of Earth's atmosphere [46]. Out of this proportion, approximately 20% and 30% is absorbed by the atmosphere and reflected back to space, respectively [1]. The remaining 50% penetrates the atmosphere, and is taken in by the land and the oceans [45]. Gibson [47] said that about half of the shortwave radiation is in the visible region (0.4–0.7 µm) of the electromagnetic spectrum, and the other half is mostly in the near-infrared (0.7 µm–100 µm). Ultraviolet radiation (0.01–0.4 µm) makes up only a little over 8% of the total [47]. The entire spectrum of the absorbed radiation drives photosynthesis, fuels evaporation, melts snow and ice, and warms the Earth [46].

Once the shortwave radiation touches the Earth's surface, there are three forms of interactions that take place—absorption, transmission, and reflection [48]. Their proportions depend on the wavelength of the energy, and the material and condition of the feature [48]. According to Brown and Gillespie [49], a single layer of leaf will generally absorb 80% of incoming visible radiation, whilst reflecting 10% and transmitting the remaining 10%. Approximately 20% of infrared is absorbed, with 50% reflected and 30% transmitted. These interactions may be modified considering the heterogeneous spatiotemporal characteristics of canopy based on the type of leaf, arrangements, density, and the angle of incidence which determines the projected ("shadowed") leaf area in the direction of the radiation [8,50,51].

Furthermore, in a subcanopy environment, the direct component of radiation is more heterogeneous than the diffused light. A predominantly direct-beam radiation that passes through openings in the forest canopy is called sunfleck [52]. The amount of sunfleck in the understory depends on different, often interacting factors: the coincidence of solar path with a canopy opening, the movement of clouds that obscure or reveal the sun, and the wind-induced movement of foliage and branches [53]. On the other hand, the penetration of the diffuse light is less variable, as it depends on the level of sky brightness, and the number, size, and spatial distribution of canopy openings, the canopy geometry, and the spatial distribution and optical characteristics of the forest biomass [54].

Generally, solar radiation below the canopy and on the forest floor has been expressed as transmittance, which depends on (1) the density and thickness of the vegetation layer, (2) the terrain, and (3) the position of the sun [36,55]. Comeau [35] defined transmittance as the ratio of solar radiation that reaches a sampling point within a forest to the incident radiation measured in the open or over the canopy at the same time. A widely accepted theory on light transmission in the forest is used by treating the canopy as a turbid medium [34,35]. This equation is called the Beer–Lambert–Bouguer law, or simply, Beer's law [6,56]:

$$I = I_0 exp^{-kL} \tag{1}$$

where: I = below-canopy light intensity

I_0 = incident radiation at the canopy top

k = extinction coefficient

L = leaf area index (LAI)

The extinction coefficient which corresponds to an optical depth per unit leaf area [51] is determined by a number of factors, such as leaf angle distribution, canopy structure, and clumping level [57]. Depending on the type of vegetation, it usually varies between 0.3 and 0.6 [57–59]. For an assumed spherical leaf angle distribution, k is approximated as a function of the solar zenith angle θ [23,58]:

$$k = \frac{0.5}{\cos \theta} \tag{2}$$

The LAI which represents a structure parameter in the vertical direction is defined as the total one-sided area of leaf tissue per unit ground surface area (m^2 m^{-2}) [51,57]. Nearly all vegetation and land-surface models include parameterizations of LAI, as it characterizes the canopy–atmosphere interface, where most of the energy fluxes exchange [57,60]. Estimation of the LAI can be derived from light transmission applying Beer's law, where the fraction of light that is intercepted (*T*) changes exponentially as the layer of leaves increases [61]:

$$T = 1 - exp^{-kL} \tag{3}$$

So, with a theoretical value of 0.5 for k, an LAI of 1, 3, 6, and 9 intercepts, respectively, 39, 78, 95, and 99% of the visible light [61].

In addition, Nyman et al. [55] describes another way to measure transmittance as a function of the path length (P), based on the earlier work of Seyednasrollah and Kumar [62]. In this approach, the density of the vegetation and the extinction coefficients are combined into a single parameter which describes the rate of absorption per unit path length. Path length of the direct beam is computed as the ratio of canopy height and cosine of solar zenith angle (Figure 1). The probability of attenuation of an incident radiation is directly proportional to the path length itself, leaf density (ratio of LAI and canopy height) and the leaf area facing the beam of light, or the leaf normal oriented in the light's direction [63].

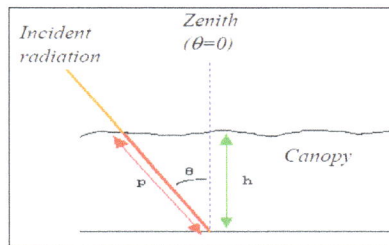

Figure 1. A diagram showing the path length of ray (P) travelling through the canopy at an angle between the zenith and that of the incident radiation (θ), h—canopy height. Image source [63].

Solar radiation energy is often measured as an energy flux density (watts per square meter), and is appropriate for energy balance studies because watt is a unit of power [23,64]. But for studies of light interception in relation to plant's health and growth, the photon density is better suited because the rate of photosynthesis depends on the number of photons received, rather than photon energy [23]. The specific range of 400 to 700 nm wavelength is where photosynthesis is active, and so it is called the photosynthetically active radiation or PAR region [64,65]. Owing to its similarity to spectral range, PAR is often used synonymously with visible light [28,66]. Energy in the PAR region is referred to as photosynthetic photon flux density (PPFD), with units in $\mu mol\ m^{-2}\ s^{-1}$ [65].

3. Modelling Approaches Using RS Technology

Based on the selected scientific literatures, a schematic diagram (Figure 2) was created to illustrate how the authors made use of RS technology in building their models. Airborne laser scanning (ALS) and terrestrial laser scanning (TLS) are the two commonly used systems, while additional information was taken from optical images either from satellites or UAV-based digital cameras. Point clouds from either ALS or TLS were processed and transformed into voxels, where transmission model was applied. Most of the voxel modeling relied on Beer–Lambert–Bouguer law, where attenuation of sunlight is calculated as a function of vegetation structure. Generation of LiDAR metrics, such as canopy height or canopy density, was also observed, then used as inputs for further analysis. Others used it in ray-tracing model or as a substitute to LAI, particularly in Equation (1). For simulation of light regime, reconstruction of the canopy was performed especially from the TLS dataset. Letters (a) and (b) of Figure 2 illustrate how laser pulses best represent the direct beam of sunlight hitting various parts of the canopy, as well as the ground. Detailed discussions on how these techniques were executed are given in the subsequent section.

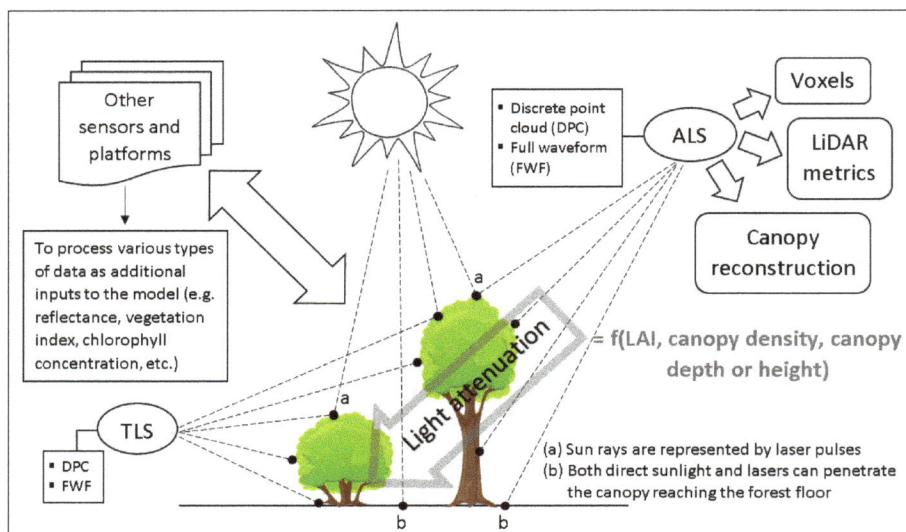

Figure 2. Schematic diagram of how RS technology is used to construct different models and approaches in estimating solar radiation below forest canopy. ALS and TLS are the two commonly used systems. Light attenuation is a function of various vegetation structures.

3.1. Physical Models and Techniques

Most of the authors depended on the capability of laser technology where 3D generation of the data is possible. It is probably one of the major advantages of LiDAR over other sensors. The usage of voxel or volumetric pixel modeling is on top of the list where the structure of the canopy is represented by cubical volume elements that are either filled with elements of the vegetation, or not [12,67,68]. The biophysical and structural property of a forest canopy can be derived from LiDAR technique with high detail [16,67].

Another technique of 3D model generation is the explicit geometric reconstruction of the canopy itself as "seen" by the sensor, and then simulation of the radiation regime within it. TLS has been tested successfully as a tool to reconstruct canopy structures, tree trunk, and diameter at breast height (DBH) to estimate LAI, stem volume, and aboveground biomass [38,39,69–71]. Van Leeuwen et al. [34] used this kind of technique with good results, utilizing the Echidna Validation Instrument (EVI). EVI is a ground-based laser waveform-recording, multiview angle-scanning LiDAR system designed to capture the complete upper hemisphere, and down to a zenith angle of just under 130° [72]. EVI was demonstrated to retrieve forest stand structural parameters, such as DBH, stand height, stem density, LAI, and stand foliage profile, with impressive accuracy [73]. The data used by Van Leeuwen et al. [34] for stem and canopy reconstruction pipeline was discrete, but the point clouds were derived from full waveform data. Single and last returns were used for creating virtual geometric models of the forest plots. These returns were obtained from the full waveform information using methods described by Yang et al. [71]. Jupp et al. [72] were able to process EVI data to generate canopy gap probabilities at different zenith ring ranges to infer radiation transmission. The canopy gap probability is equivalent to the probability that the ground surface is directly visible from airborne and spaceborne remote sensing platforms [74]. Both Van Leeuwen et al. [34] and Yang et al. [75] found strong relationships between the transmission properties of their respective radiation models and EVI-derived gap probabilities.

Other light transmission models, that are not in the reviewed articles but worth mentioning, are LITE (Light Interception and Transmittance Estimator) [76], SLIM (Spot Light Interception Model) [77], and DART (Discrete Anisotropic Radiative Transfer) [78]. The first two models are separate executable programs that provide a flexible association of utilities for entering light data from several sources and for outputting 3D estimates of leaf area and both instantaneous and seasonal PPFD beneath tree canopies [77]. SLIM prepares data for entry into LITE, but SLIM is also a "stand-alone" utility designed to estimate leaf area index (LAI), gap fraction, and transmittance from hemispherical photographs or ceptometer data. LITE provides maps of LAI and PPFD across the study site, and provides information on the light regime at selected locations [79]. However, we believe that there has not been any attempt yet to integrate RS data to either of the models.

DART [78] is probably one of the most popular tools for simulating remote sensing images for conducting physical models [80,81]. Developed by the Center for the Study of the Biosphere from Space (CESBIO) laboratory in France, DART is a comprehensive physically-based 3D model for Earth–atmosphere radiation interaction, from visible to thermal infrared. It simulates measurements of passive and active satellite/plane sensors for urban and natural landscapes. These simulations are computed in any experimental (atmosphere, terrain, forest, date) and instrumental (spatial and spectral resolutions, viewing direction) configuration [78]. DART successfully simulated reflectance of forests at different spectral domains [82]. The model just recently introduced a method to simulate LiDAR signals, and showed potential for forestry applications [83], but we have not found any actual research applying the model to a specific site.

Whether it is a simple equation-type or computer-based light model, various inputs are needed, which LiDAR metrics can provide. At least four of the articles [16,84–86] generated LiDAR-derived canopy or tree metrics before executing their models. Some of these metrics are canopy height and canopy density, essential for voxel modeling. As emphasized by Mucke and Hollaus [13], knowledge about the canopy architecture, meaning the spatial composition of trees or bushes, and the arrangement of their branches and leaves or needles, is of critical importance for the modeling of light transmission, and no other technique, except LiDAR, is well-suited for the derivation of canopy's geometric information. Moreover, there is a different kind of LiDAR format, called full waveform, which is designed to digitize and record the entire backscattered signal of each emitted laser pulse [87]. Full-waveform data is rich in information, and is particularly useful for mapping within a canopy structure [88]. Unlike the basic metrics from the more popular discrete point clouds, however, this type of system is rarely used, due to acquisition costs, data storage limitations, and the lack of physically-based processing methods for their interpretation [88,89].

Laser beams could act as sunrays (a and b of Figure 2), which are either intercepted by leaves and branches, or penetrate through the holes. The latter is analogous to what a laser penetration index (LPI) does, which, according to Barilotti et al. [90], explains the ratio of the points that reach the ground over the total points in an area. In contrast, Kwak and Lee [91] exploited those point clouds intercepted by the crown, and called it the laser interception index (LII). LPI and LII were used separately to predict LAI, and both produced high accuracy results [90,91]. For radiation studies, LPI was used by Yamamoto et al. [84], and successfully estimated the transmittance, even with low pulse density (e.g., 2.5 pulses m^{-2}). They utilized a modified LPI to automatically and easily separate the laser pulses into those of canopy and below-canopy returns. The system was called Automated LiDAR Data Processing Procedure (ALPP) [92] and worked by integrating LiDAR Data Analysis System [93] and GRASS software. In ALPP, a top surface model was first created from the locations of treetops extracted from the LiDAR data within a plot. Then, the return pulses within the plot were converted to vertical distance (VD) from which the frequency distribution of 0.5 m class VD was statistically separated into canopy and below-canopy classes. Finally, linear regression between the LPI and diffuse transmittance taken from hemispherical photographs was performed. The highest coefficient of determination (R^2 = 0.95) of the regression line was seen at 12.5 m radius LPI cylindrical plots. The accuracy of the methodology may be put to test upon applying to a different forest or a species-type other than cypress in Japan. Moreover, from a logical point of view, the analogy of LPI seemed to be more appropriate for direct component rather than the diffuse one. Musselman et al. [16] utilized LPI to infer solely the direct beam transmittance inside a conifer forest for various types of land surface modeling, such as melting snow. Interestingly, the most effective radius they discovered for LPI extraction, to estimate LAI, was 35 m; almost 3× that of the findings by Yamamoto et al. [84]. Although Musselman et al. [16] used a Beer's type model specific to direct beam for LAI estimation, this may still be debatable, because like diffuse transmittance, they are both closely related.

Alexander et al. [21] used ALS as a proxy to a hemispherical camera by transforming point clouds into canopy cover and canopy closure. The former was made possible by generating Thiessen polygons from the ALS points, while the latter was done by plotting the same points in the polar coordinate system, with the azimuth angle as the angular coordinate, and the zenith angle as the radial coordinate. Moreover, these synthetic images mimicking hemispherical photos were correlated to the species classified according to Ellenberg values [94]. These values, ranging from 1 to 9, represent varying shade tolerance with 1 denoting species preferring deep shade, and 9 for species preferring full sunlight. It revealed that ALS-based canopy closure is a reasonable indicator of understory light availability, and has the advantage over field-based methods that it can be rapidly estimated for extensive areas.

A convenient way to generate spatial information is to have a working environment in a GIS platform. GIS software that comes with a solar radiation analysis package that could estimate irradiance at a watershed scale has been used by a number of researchers [22,95]. GIS solar models operate on digital terrain model (DTM) raster layers, allowing them to accurately estimate insolation reduction due to slope, aspect, and topographic shading across watersheds [96]. Both Solar Analyst from ESRI's ArcGIS, and from that of Helios Environmental Modeling Institute, adapt the sky region technique used in hemispherical photos for use with raster grids [97]. On the other hand, the r.sun module of the GRASS software (see sample map in Figure 3) is a clear sky solar model designed to take topographic angles and shading into account [98].

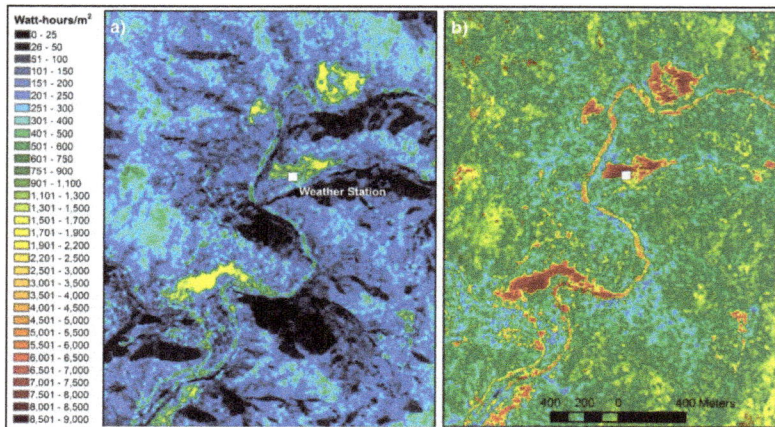

Figure 3. Light condition below canopy at landscape scale made possible by geographic information system (GIS)-based solar radiation tool (left: winter solstice; right: summer solstice) [22].

3.2. Summary of the Studies

Table 1 summarizes the literature subjected for review. Each paper's salient features, some of them unique, are described briefly.

Table 1. Studies on solar radiation below canopy utilizing remote sensing or allied sciences.

References	Salient Methodology Features
Nyman et al., 2017 [55]	• Compared various transmission models including a light penetration index (LPI) with a weighing factor to account for the path length
Tymen et al., 2017 [95]	• Developed light transmission model using voxels generated from point clouds • Used LPI
Cifuentes et al., 2017 [99]	• Voxel-based canopy modeling generated from terrestrial laser system (TLS) • Classified point clouds into leaves and non-leaves, then assigned properties before conducting light simulation
Yamamoto et al., 2015 [84]	• Utilized owned version of LPI then correlated with relative illuminance
Bode et al., 2014 [22]	• Used LPI and solar radiation module of GRASS software (r.sun)
Peng et al., 2014 [6]	• Generated 3D canopy structure from point clouds • Implemented Beer's Law through voxel models with ray trace method
Moeser et al., 2014 [29]	• Improvised synthetic hemispherical photos generated from point clouds
Widlowski et al., 2014 [25]	• Voxel-based canopy reconstruction from TLS • Bidirectonal reflectance factor (BRF) simulation in virtual environment
van Leeuwen et al., 2013 [34]	• Used ground-based laser scanner Echidna Validation Instrument (EVI) to • reconstruct geometric explicit models of canopy • characterize radiation transmission properties from LiDAR full waveform
Musellman et al., 2013 [16]	• Used Beer's-type transmittance model based on LiDAR-derived LAI • Developed solar raytrace model applied to 3D canopy derived from multiple LiDAR flights
Alexander et al., 2013 [21]	• Estimated canopy cover by producing Thiessen polygons from point clouds • Calculated canopy closure by transforming point clouds from Cartesian to spherical coordinates
Bittner et al., 2012 [67]	• 3D voxel representation of the canopy architecture derived from TLS • Different attributes of light assigned to voxels of stem, leaf or air
Guillen-Climent et al., 2012 [100]	• Used a 3D radiative transfer model called forest light interaction model (FLIGHT) • Mapped with high-resolution imagery from unmanned aerial vehicle (UAV) with multispectral camera
Kobayashi et al., 2011 [101]	• Generated canopy height model, tree and crown segmentation from point clouds as inputs to CANOAK-FLiES (forest light environmental simulator) • Derived canopy reflectance from airborne airborne visible/infrared imaging spectroradiometer (AVIRIS)
Van der Zande et al., 2011 [3]	• Voxel-based representation of trees derived from TLS • Light simulation using voxel-based light interception model (VLIM)
Van der Zande et al., 2010 [12]	• Generate 3D representations of the forest stands, enabling structure feature extraction and light interception modeling, using the voxel-based light interception model (VLIM)
Yang et al., 2010 [75]	• Estimated canopy gap probability from ground-based LiDAR (EVI)
Lee et al., 2009 [102]	• Defined a conical field-of-view (scope) function between observer points just above the forest floor and the sun, which relates PAR to the LiDAR data
Essery et al., 2008 [85]	• Colored orthophotograph and laser scanning were used to map out tree locations, heights, and crown diameter as inputs to mathematical radiation modeling and simulation
Thomas et al., 2006 [86]	• LiDAR metrics generated from airborne laser system (ALS) to determine spatial variability of canopy structure • Canopy chlorophyll concentration was derived from airborne hyperspectral imagery
Todd et al., 2003 [103]	• Analyzed foliage distribution from LiDAR observation
Parker et al., 2001 [32]	• Estimated canopy transmittance using the scanning LiDAR imager of canopies by echo recovery (SLICER), a waveform-sampling laser altimeter
Kucharik et al., 1998 [104]	• Captured visible and NIR images of canopies from 16-bit charge-coupled device (CCD) multiband camera from the ground looking vertically upward to estimate sunlit and shaded foliage

The reviewed studies were generally classified based on the type of sensor the authors preferred to use, either as active or passive. The latter is a type of system that uses the sun as the source of electromagnetic radiation, while the former pertains to sensors that supply their own source of energy to illuminate features of interest [47,105]. Digital cameras on board UAVs and satellite optical imagery sensors are passive types, while radar and LiDAR technologies belong to active sensors. It was found out that the majority of the studies conducted are in the active domain of RS (Figure 4a). Out of the 81% bulk, almost two-thirds are airborne-based (Figure 4b,c).

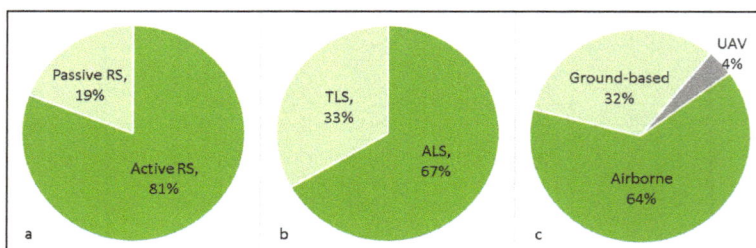

Figure 4. Reviewed articles based on remote sensing (RS) classification (**a**), sensors used (**b**), and platforms (**c**).

Only a few solar radiation studies in forest that utilized RS technology were conducted in the late 1990s up to mid-2000 (Figure 5). Although, Brunner [106] cited numerous research studies concerning forest canopy with RS component between 1983 and 1996. However, those articles were of limited use for modeling of light underneath canopy, because the focus of the study was reflection, which is considered a minor part of total irradiation. When the study of Essery et al. [85] came out, they mentioned that airborne remote sensing has not yet been widely used in radiation modeling. The reason probably is that the available RS tools during those years were just inappropriate to apply to such kinds of research. On the other hand, steady research interest on the subject matter started in the last eight years or so. In fact, an increasing trend can be noted, with a peak in the year 2014. It can be observed that most of these papers relied on laser technology. In 1996, there was only one company selling commercial ALS systems, and the service providers could only be counted on the fingers of one hand [107]. Three years later, manufacturers of major ALS components increased, while the number of companies providing services have jumped to forty, worldwide [107]. Furthermore, early development and usage of the laser systems were mainly focused on topographic and bathymetric surveys.

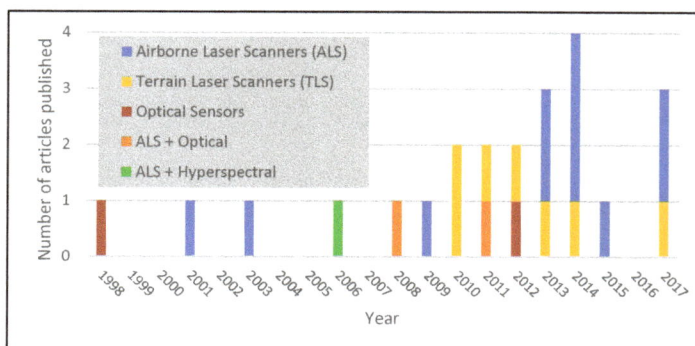

Figure 5. Summary of studies per year on forest canopy in relation to solar radiation utilizing remote sensing or allied sciences.

4. Integration of RS Data and Models

Integrating RS datasets of different types or from various sensors has the advantage of increasing the accuracy of the model. Based on literature, multiple RS data, sometimes with field-measured variables, are combined as inputs to a radiative transfer model, or used as predictors in a regression model. Kobayashi et al. [101] used the optical sensor, AVIRIS, to capture canopy reflectance as an additional parameter to perform sunlight simulation within a forest.

On the other hand, the 3D forest light interaction model (FLIGHT) [100] requires tree characteristics, like the crown's height, radius, and shape as vital inputs for their model. A spectrometer was also used to acquire leaf reflectance and transmittance, which were additional requirements to run the model. Other specific physiological characteristics, such as soil reflectance and vegetation indices, were obtained from a 6-band multispectral camera on board a UAV.

Thomas et al. [86] made ALS-derived canopy variables and physiological tree property from an airborne hyperspectral sensor as independent variables to spatially estimate the fraction of PAR (fPAR). First, mean LiDAR heights were derived from a theoretical cone that originated from below canopy PAR sensor and extended skywards. The angle of the cone was tested with multiple values to determine an optimal one. Accordingly, it is possible to define whether there exists a volume of point clouds located above the PAR sensor that is most closely related to fPAR. Meanwhile, chlorophyll map was derived from the compact airborne spectrographic imager (CASI) sensor having 72 channels with an approximate bandwidth of 7.5 nm over the wavelength range of 400–940 nm. The average values of the maximum first derivative over the red edge were then extracted from the sampling plots, and then compared to the field measurements of average chlorophyll concentration. The linear regression model generated from these two variables was applied to the entire image to produce a map of total chlorophyll for the study site. One of their findings revealed that for a theoretical LiDAR cone angle of 55°, linear regression models were developed, with 90% and 82% of the variances explained for diffuse and direct sunlight conditions, respectively. In addition, fPAR x chlorophyll has stronger correlations with LiDAR than fPAR alone. Probably the only major downsides of this method are the additional cost and other resources necessary to acquire multiple datasets.

Bode et al. [22] have shown the integration of LPI, a LiDAR metric approach, and a GIS-based solar radiation module to produce a solar radiation map (Figure 3). Meanwhile, Nyman et al. [55] modified the LPI introduced by the former, by integrating a weighting factor to account for path length. The weighing factor is 1 when the incident angle is 0 (sun is directly overhead). According to their results, the models that include path length in the transmission term are more flexible in terms of reproducing subdaily and seasonal variations. Although the variant LPI model performed well at their three study sites, it displayed some systematic bias in dense forests, resulting in lower performance. A higher resolution LiDAR data would result in more representative metrics of vegetation structure, and therefore, may improve performance of the model [55].

5. Critical Issues for Future Research Perspective

5.1. The Diffuse Component of Radiation

Ecologists, botanists, and the like, have proven the claim that diffuse component of sunlight is preferred by plants for higher productivity and efficiency [108–111]. In spite of its role and significance to forest ecosystem and being directly related to PAR [28], there seems to be little interest amongst the RS specialist to focus on the diffuse component exclusively. The majority of the articles opted to incorporate diffuse component into the global radiation, which means it would be difficult to determine exactly the percentage of diffuse, as well as direct, that is involved. Figure 6 illustrates the discrepancy between global radiation and diffuse component in terms of analysis. All those studies with diffuse components were done using ALS. Instruments, like quantum sensors and hemispherical photography, are capable of measuring, separately, the diffuse and direct radiation effectively, but it still seems to be a demanding task when using RS technology.

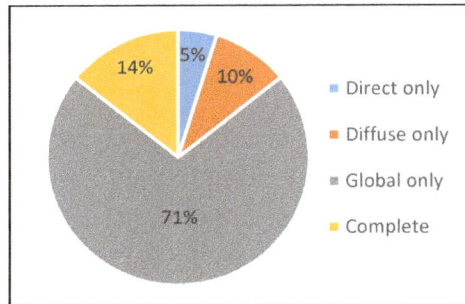

Figure 6. Components of solar radiation modeled below canopy using RS technology.

Bode et al. [22] were able to compute, separately, the direct and diffuse components, then later on, summed up to produce an insolation map beneath the canopy. The authors assessed canopy openness by using laser beams from ALS as a proxy to direct sunlight hitting the forest floor. Using this concept, they calculated LPI by simply dividing the ground hits over the total hits which can be translated as light's probability of reaching the ground. Also, ALS-derived DEM (or bare earth) was applied to a GIS-based solar radiation tool to generate insolation underneath forest vegetation. After which, the generated insolation was multiplied by LPI to produce the direct component. For the diffuse radiation, however, a simple linear regression was used based on the light from above canopy. Upon validation of selected points with a pyranometer, total and direct solar radiation had a relatively precise estimation ($R^2 = 0.92$ and $R^2 = 0.90$, respectively), but the diffuse component was quite weak ($R^2 = 0.30$). While the overall performance of the model is promising, translating the result to a continuous spatial information (i.e., map) proved to be difficult. As observed from previous studies, an underestimation occurs between map values of diffuse radiation when compared to point measurements, and thus, upscaling it to watershed scale is a significant source of uncertainty [22,86].

5.2. Dataset Fusion

While most of the integration happened by treating multiple data as separate inputs, it might as well be worthy to fuse ALS and TLS datasets to complement each other. Peng et al. [6], as mentioned above, needed to acquire the underbranch height of the forest using a handheld laser rangefinder, not only to describe the canopy structure, but as an input to their ray trace model. While such a method worked well, it could be improved for finer resolution with the use of ground-based LiDAR. Furthermore, to illustrate the fusion advantage, Chasmer et al. [112] collected LiDAR data both airborne and ground-based for a single forest site, then graphed the point cloud frequency distribution (Figure 7). It is apparent that most of the laser pulses from ALS are concentrated on top of the canopy, while the TLS collected a much higher percentage in the understory. Hopkinson et al. [113] integrated terrestrial and airborne LiDAR to calibrate a canopy model for mapping effective LAI. However, co-registering the point clouds from the two data sources presented a challenge. Firstly, the point clouds from the two sources needed to be horizontally and vertically co-registered to ensure that 3D attributes are directly comparable. This was done by surveying the TLS instrument's location using single frequency rapid static differential GPS to within ~1 m absolute accuracy. After knowing the coordinates of the scanner, a manual interpretative approach was used to translate and rotate the TLS datasets, until they visually matched the ALS data. Such an approach must be performed with extra attention if applied to a more complex and heterogeneous forest. Furthermore, they found out that a radius of ~11.3 m was appropriate for the integration of ALS data captured from overhead, with hemispherical TLS data captured below the canopy. Accordingly, using a radius that corresponds to a plot of 400 m^2 is small enough to contain unique tree crown attributes from a single or small number of trees (the study site was a Eucalyptus forest) while being large enough to ensure mitigations of misalignments and spatial uncertainties. It should be noted, however, that their results might not hold to another site with different data acquisition, canopy height, or foliage density conditions.

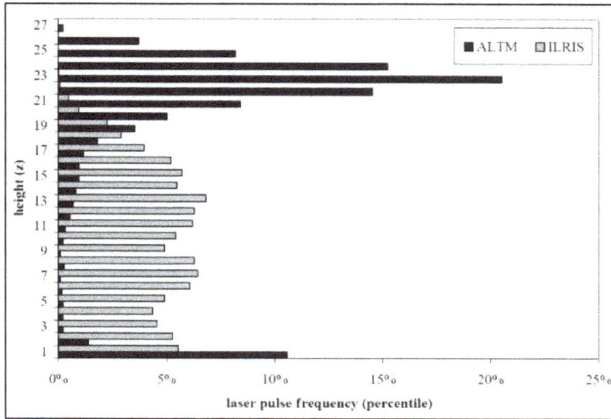

Figure 7. Average laser pulse frequency (percentile) distributions for airborne laser terrain mapper (ALTM) and intelligent laser ranging and imaging system (ILRIS), a ground-based LiDAR developed by Optech® obtained for the mixed deciduous forest plot [113].

5.3. Light along Vertical Gradients

There are about three papers in this review [6,32,103] that were able to characterize light intensities at different heights along the vertical gradient of the canopy. Of course, this is something a conventional measurement method could possibly do, but may demand additional resources. The camera assembly for taking hemispherical photographs could be placed on a monopod, a folding step, or climbing ladders that can reach from 3 m up to 12 m in height [65]. Aside from being time consuming, the method is also prone to errors.

Peng et al. [6] were able to describe the spatiotemporal patterns of understory light in the forest floor and along its vertical gradient. The authors not only illustrated the spatial patterns of the light intensity at various heights (0.5, 2, 5, 10, 15, and 20 m, based on the forest floor), but also showed the strong variations at different times of the day. The model was built based on voxels derived from ALS and field-measured data. Working in a GIS environment, a canopy height model (CHM) was generated, first from point clouds, then coupled with another raster of the underbranch height. Both of these were superimposed, and then divided into grids, thus producing the voxels. The authors then modeled the distance travelled by the solar ray in the crowns (the voxels), with consideration of the solar position, which means the ray could travel in many directions. This calculated distance was used to replace a particular parameter, the LAI, in Beer's Law. In the original equation, LAI can only estimate the light intensity of the forest floor in the vertical direction. The computed distance and transmittance showed exponential relationship, with $R^2 = 0.94$ and $p < 0.05$. Estimated and observed understory light intensities were obviously positively correlated, with $R^2 = 0.92$ and $p < 0.01$. As the study introduced the elevation as a parameter to determine if the solar rays reached the ground, mapping the light intensity along the vertical gradient by altering the DEM values was convenient.

5.4. Limited Representation in Terms of Biomes

Figure 8 displays the location of the study sites for each publication gathered. There are areas in the list that are not purely forest, or not a physical forest at all. Van der Zande et al. [3] created a virtual stand of trees to simulate light conditions in a forest environment while the paper by Guillen-Climent et al. [100] was done on a mixed matured orchard. Many authors relied directly on permanent plots inside national parks of experimental forested areas as large as 2 km², likely because the stand profiles were already available for use. Authors of two papers [6,22] were able to produce a continuous map of solar radiation reaching up to 3.8 km². The number of sample plots ranges from 1 to 81 for rectangular plots and 29 to 96 for circular plots, with 15 m being the most preferable radius. The average area of each plot is 1642 m², with a median value of 707 m². Aside from species name, among the consistently-measured biophysical tree properties were diameter at breast height (DBH), total height, canopy cover, canopy closure, and crown radius. Furthermore, coniferous and deciduous trees comprise 31% and 42%, respectively, as the existing primary species in their corresponding study areas. The remaining 27% were a mixture of both. Species composition is a significant factor, as shown by Cifuentes et al. [99], in that the mean differences between observed and simulated light values in a heterogeneous forest are larger than in a pure forest. While the articles reviewed captured a diverse type of important species and major groups of trees (e.g., deciduous and coniferous), there is a limited number of forest-type biomes covered by existing works. All but one were conducted in temperate regions, mostly in the northern hemisphere, and concentrated in Europe, North America, and Canada. The study site of the research by Tymen et al. [95] was the only one done in a tropical mixed species forest in French Guiana, South America. The same author suggested that the distinction of forest types is meaningful, because differences in microenvironmental conditions, including light, will potentially impact forest dynamics, ecosystem processes, and composition in habitat.

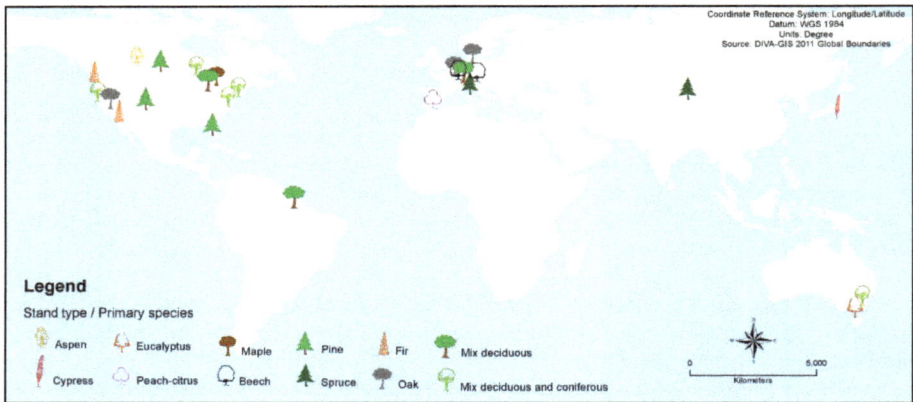

Figure 8. Global locations of the study areas of the reviewed articles.

5.5. Cost and Time Consideration

The cost requirements in considering which approach to apply, lies largely on the dataset or the type of sensors to use. Almost all the articles in Table 1 utilized active remote sensors, particularly laser scanners. After a thorough online search, we have found several countries that are making their LiDAR archives accessible to the public. A compilation of these countries with their respective links for accessing LiDAR data is given in Table 2. It must be noted though that most of these are available only in selected states or regions per country. Not all of the surveys made for ALS and the distributions of the same were done by the state or any government organizations, but rather, of academic and research institutes. The readers are encouraged to check the information on the open data policy of each website as they may have different levels of requirements before anyone can download. The OpenTopography facility (http://www.opentopography.org) which is based in

University of California, USA, also houses selected datasets from certain areas of Antarctica, Brazil, Canada, China, Mexico, New Zealand, and Puerto Rico.

Table 2. Selected countries with airborne LiDAR dataset available for public use.

Country	Uniform Resource Locator (URL)/Helpful Links
Australia	http://www.ga.gov.au/elvis/ http://www.opentopography.org/index.php
Denmark	https://download.kortforsyningen.dk/
Finland	https://tiedostopalvelu.maanmittauslaitos.fi/tp/kartta?lang=en
Germany	https://open.nrw/
Luxembourg	https://data.public.lu/en/datasets/LiDAR-projet-pilote-dun-releve-3d-du-territoire-luxembourgeois/
Netherlands	https://www.arcgis.com/home/webmap/viewer.html?useExisting=1&layers=9039d4ec38ed444587c46f8689f0435e
Norway	https://hoydedata.no/LaserInnsyn/
Italy	http://www.pcn.minambiente.it/mattm/en/online-the-new-procedure-for-the-request-of-LiDAR-data-and_or-interferometric-ps/
Philippines	https://lipad.dream.upd.edu.ph/
Scotland	https://remotesensingdata.gov.scot/
Slovenia	http://evode.arso.gov.si/indexd022.html?q=node/12
Spain	https://b5m.gipuzkoa.eus/url5000/es/G_22485/PUBLI&consulta=HAZLIDAR http://www.murcianatural.carm.es/natmur08/descarga.html http://centrodedescargas.cnig.es/CentroDescargas/buscadorCatalogo.do?codFamilia=LIDAR http://www.icgc.cat/en/
Switzerland	https://geoweb.so.ch/map/LiDAR
United Kingdom	http://environment.data.gov.uk/ds/survey/index.jsp#/survey http://www.ceda.ac.uk/
USA	http://www.opentopography.org/index.php https://coast.noaa.gov/inventory/ https://en.wikipedia.org/wiki/National_LiDAR_Dataset_(United_States)

An initiative called GlobALS (Global ALS Data Providers Database) is being conceptualized, aiming to establish a network of possible providers of ALS data. A worldwide geodatabase could be found in this online map [114]. On the other hand, there are also countries, such as Poland, that are still generous enough to make their remote sensing data generally available for free, as long as it is for research purposes. Apart from acquisition date and other technical issues that might affect a study, the fact that a certain amount of ALS data does exist is not bad at all, to start conducting research-related or forest management activities.

Notwithstanding these initiatives from government agencies and academic communities, many institutions still rely on private companies for data collection. There are "off-the-shelf" datasets available for selected areas, but the price could go even higher for sites not previously flown for acquisition. Although, according to a recent study, LiDAR became affordable in the past 17 years [44]. McGaughey et al. [115] said that while LiDAR system's capabilities have dramatically increased over the last decade, its data acquisition costs have correspondingly decreased. LiDAR acquisition in the years 2007 and 2015 were USD 3.34 and 2.00 per hectare, respectively [88,116]. Furthermore, a twin otter plane and the associated crew will cost somewhere in the region of USD 4000/h before any data processing. Thus, prices are low per unit area if you have a lot of area to cover, but if sites are small and isolated, costs could be a lot higher [88].

Many TLS systems are commercially available for purchase or rental, and can be easily operated. Static terrestrial LiDAR instruments cost between USD 40,000 and 200,000, including a range of optional features [88]. Due to its limited spatial coverage, this tool requires extensive fieldwork and demands a lot of time and manpower to cover a wide area. Forest type, sample design, scanner specifications, instrument settings, and weather conditions can all significantly extend the time required to complete a campaign [117]. Data handling is also technically demanding, and entails additional processing time. As a guide, scanning a 1 ha plot typically takes a team of three people between 3 and 8 days, dependent upon topography and understory conditions [117].

6. Summary

Treating a forest canopy in voxel elements, which is in 3D form, can ensure spatial distribution of the understory light in forest floor or along the vertical gradient of the forest stand [6]. Although this approach requires more computation time and canopy structural variables than one dimensional models, it is expected to give more reliable energy and carbon fluxes when the canopy structural variables are available [101]. The 3D model can fill the theoretical gap between 1D models and actual ecosystems, and can be used to investigate where and when the simplified models give a large error to simulate the radiation energy and carbon fluxes [101,118]. Nevertheless, most voxel-based calculations focus on the canopy alone, and do not account for the radiation that reaches the forest floor [22]. It was demonstrated by Bode et al. [22] that by using ALS, there are ground hits that should be considered, because it verifies that light penetrates the vegetation, reaching the understory floor. The same author also added that for dense vegetation, voxel shading will overestimate light penetration at low angles, due to an absence of hits near the ground.

Applying ray trace model with Beer's Law is a widely accepted theory on radiation transmission, as it assumes that the forest canopy is a turbid medium [34,101]. This means that leaves are randomly distributed, and a homogeneous layering of foliage, such as in the case of even-aged mono-tree plantations. However, it is impractical in natural, heterogeneous systems, especially those with extensive riparian areas where species diversity is high [22]. In fact, the radiation transfer model intercomparison [119] found that such models compare well for homogeneous canopies, but still have large discrepancies for complex heterogeneous canopies [85].

Because of improvements in the LiDAR technique in recent years, TLS makes it possible to retrieve the structural data of forests in high detail [67]. The high level of structural detail of these data provides an important opportunity to parameterize geometrically explicit radiative transfer models [34]. TLS has potential use within canopy light environment studies, as well as linking structure with function [88]. What is more, is the advantage of using either ALS or TLS is that they are not affected by light conditions, compared to hemispherical photos [16,21]. Furthermore, integration of ALS and TLS would be something new in this line of research, and may provide a higher degree of accuracy or may fill some gaps. There are differences between the parts of the canopy captured by the two laser scanners, but combining point clouds from these two platforms would result in a higher point density, and thus, we expect it to bear more robust information. However, the algorithm for fusing the two datasets are not yet fully standardized, and therefore, needs further study. In addition, laser sensors on board UAVs is a possible solution to have a uniform distribution of the point cloud coverage, from the top to the bottom of the canopy.

On the other hand, solar radiation software packages can provide rapid, cost-efficient estimates of solar radiation at a subcompartment scale (e.g., stands <10 ha) [120]. Kumar [37] conducted a study to compare the accuracy of such a tool to ground-recorded meteorological data for several locations, and highlight the strong correlation between the two sets. A low error rate (2%) suggests that modelled data can be utilized in micrometeorology and evapotranspiration-related studies with a high degree of confidence. The choice of software is crucial though, especially when working with large scales. Bode et al. [22] tested ESRI's ArcGIS and its Solar Analyst on raster layers of up to 4 million cells, before failing. He then chose the open-source GRASS for GIS processing, because it is designed to handle large datasets. It could be manifested in the works of Szymanowski et al. [121] and Alvarez et al. [24], who applied it in Poland and Chile, respectively, at a regional scale encompassing various landscape surfaces. Moreover, a number of studies have been relying also on the radiation module of the open-source software System for Automated Geoscientific Analyses (SAGA) [122]. However, to the best of our knowledge, the module has not been implemented in any research relating to canopy light transmission.

7. Conclusions

Understory sunlight condition is absolutely essential in understanding forest dynamics, and is a critical parameter to any modeling in a forested environment condition. For decades, researchers and scientists have developed various approaches to measure it as accurately as possible. At this age, where RS technology offers new sensors and platforms, modeling solar radiation has taken to a new level. Based on this review, the following main conclusions are stated:

- As far the type of sensors is concerned, the active domain, particularly laser technology, rules the choice in analyzing light conditions below or within the forest canopy.
- Not a single set of data derived from a passive sensor inferring spatial solar radiation was used in the reviewed studies.
- Aside from high 3D spatial resolution, airborne laser scanner's ability to penetrate the canopy through the gap openings is also an advantage, as it takes account of the forest floor. Those studies that utilized laser scanning mostly applied voxel models or the laser penetration index (LPI).
- The latter may exhibit varying performance and accuracy, depending on the forest type and consequently, the canopy structure.
- The use of UAVs for future research is also an interesting prospect, as it gives flexibility in terms of the coverage, and can address the gaps by both ALS and TLS.
- Lastly, as evident in the market where LiDAR technology is getting less expensive and more countries are opening their databases for public access, various entities are encouraged to take advantage and the initiative to expand their research efforts for a more science-based monitoring and management of our resources.

Author Contributions: A.O. and K.S. conceived the ideas and designed the methodology; A.O. collected and analysed the data and led the writing of the manuscript. K.S. was responsible for gaining financial support for the project leading to this publication. Both K.S. and K.B. contributed critically to the drafts and gave final approval for publication.

Acknowledgments: This work was supported financially by the Project LIFE+ ForBioSensing PL "Comprehensive monitoring of stand dynamics in the Białowieża Forest", as supported by remote sensing techniques. The Project has been co-funded by Life Plus (contract number LIFE13 ENV/PL/000048) and Poland's National Fund for Environmental Protection and Water Management (contract number 485/2014/WN10/OP-NM-LF/D). The authors would like to express their gratitude to the two anonymous reviewers for their insightful comments which improved the quality of this paper.

Conflicts of Interest: The authors declare no conflict of interest.

References

1. Graham, C.P. The Water Cycle: Feature Articles. Available online: https://earthobservatory.nasa.gov/Features/Water/page2.php (accessed on 2 October 2017).
2. Martens, S.N.; Breshears, D.D.; Meyer, C.W. Spatial distributions of understory light along the grassland/forest continuum: Effects of cover, height, and spatial pattern of tree canopies. *Ecol. Model.* **2000**, *126*, 79–93. [CrossRef]
3. Van der Zande, D.; Stuckens, J.; Verstraeten, W.W.; Mereu, S.; Muys, B.; Coppin, P. 3D modeling of light interception in heterogeneous forest canopies using ground-based LiDAR data. *Int. J. Appl. Earth Obs. Geoinf.* **2011**, *13*, 792–800. [CrossRef]

4. Riebeek, H. Water Watchers: Feature Articles. Available online: https://earthobservatory.nasa.gov/Features/WaterWatchers/printall.php (accessed on 22 November 2017).

5. Leuchner, M.; Hertel, C.; Rötzer, T.; Seifert, T.; Weigt, R.; Werner, H.; Menzel, A. *Solar Radiation as a Driver for Growth and Competition in Forest Stands*; Springer: Berlin/Heidelberg, Germany, 2012; pp. 175–191.

6. Peng, S.; Zhao, C.; Xu, Z. Modeling spatiotemporal patterns of understory light intensity using airborne laser scanner (LiDAR). *ISPRS J. Photogramm. Remote Sens.* **2014**, *97*, 195–203. [CrossRef]

7. Oliver, C.; Larson, B. *Forest Stand Dynamics*, update ed; John Wiley and Sons Inc.: New York, NY, USA; ISBN 0-471-13833-9.

8. Grant, R. Partitioning of biologically active radiation in plant canopies. *Int. J. Biometeorol.* **1997**, *40*, 26–40. [CrossRef]

9. Englund, S.R.; O'brien, J.J.; Clark, D.B. Evaluation of digital and film hemispherical photography and spherical densiometry for measuring forest light environments. *Can. J. Forest Res.* **2000**, *30*, 1999–2005. [CrossRef]

10. Zavitkovski, J. Ground vegetation biomass, production, and efficiency of energy utilization in some northern Wisconsin forest ecosystems. *Ecology* **1976**, *57*, 694–706. [CrossRef]

11. Anderson, M.; Denhead, O. Shortwave radiation on inclined surfaces in model plant communities. *Agron. J.* **1969**, *61*, 867–872. [CrossRef]

12. Van der Zande, D.; Stuckens, J.; Verstraeten, W.W.; Muys, B.; Coppin, P. Assessment of Light Environment Variability in Broadleaved Forest Canopies Using Terrestrial Laser Scanning. *Remote Sens.* **2010**, *2*, 1564–1574. [CrossRef]

13. Mücke, W.; Hollaus, M. Modelling light conditions in forests using airborne laser scanning data. In Proceedings of the SilviLaser 2011, 11th International Conference on LiDAR Applications for Assessing Forest Ecosystems, University of Tasmania, Hobart, TAS, Australia, 16–20 October 2011; Volume 2011.

14. Sakai, T.; Akiyama, T. Quantifying the spatio-temporal variability of net primary production of the understory species, Sasa senanensis, using multipoint measuring techniques. *Agric. Forest Meteorol.* **2005**, *134*, 60–69. [CrossRef]

15. von Arx, G.; Dobbertin, M.; Rebetez, M. Spatio-temporal effects of forest canopy on understory microclimate in a long-term experiment in Switzerland. *Agric. Forest Meteorol.* **2012**, *166–167*, 144–155. [CrossRef]

16. Musselman, K.N.; Margulis, S.A.; Molotch, N.P. Estimation of solar direct beam transmittance of conifer canopies from airborne LiDAR. *Remote Sens. Environ.* **2013**, *136*, 402–415. [CrossRef]

17. Théry, M. Forest light and its influence on habitat selection. In *Tropical Forest Canopies: Ecology and Management*; Springer: Berlin/Heidelberg, Germany, 2001; pp. 251–261.

18. Battisti, A.; Marini, L.; Pitacco, A.; Larsson, S. Solar radiation directly affects larval performance of a forest insect: Effects of solar radiation on larval performance. *Ecol. Entomol.* **2013**, *38*, 553–559. [CrossRef]

19. Jennings, S.B.; Brown, A.G.; Sheil, D. Assessing forest canopies and understorey illumination: Canopy closure, canopy cover and other measures. *Forestry* **1999**, *72*, 59–74. [CrossRef]

20. Anderson, M. Studies of the woodland light climate I. The photographic computation of light conditions. *J. Ecol.* **1964**, *52*, 27–41. [CrossRef]

21. Alexander, C.; Moeslund, J.E.; Bøcher, P.K.; Arge, L.; Svenning, J.-C. Airborne laser scanner (LiDAR) proxies for understory light conditions. *Remote Sens. Environ.* **2013**, *134*, 152–161. [CrossRef]

22. Bode, C.A.; Limm, M.P.; Power, M.E.; Finlay, J.C. Subcanopy Solar Radiation model: Predicting solar radiation across a heavily vegetated landscape using LiDAR and GIS solar radiation models. *Remote Sens. Environ.* **2014**, *154*, 387–397. [CrossRef]

23. Jones, H.G.; Archer, N.; Rotenberg, E.; Casa, R. Radiation measurement for plant ecophysiology. *J. Exp. Bot.* **2003**, *54*, 879–889. [CrossRef] [PubMed]

24. Álvarez, J.; Mitasova, H.; Allen, H.L. Estimating monthly solar radiation in South-Central Chile. *Chil. J. Agric. Res.* **2011**, *71*, 601–609. [CrossRef]

25. Widlowski, J.-L.; Côté, J.-F.; Béland, M. Abstract tree crowns in 3D radiative transfer models: Impact on simulated open-canopy reflectances. *Remote Sens. Environ.* **2014**, *142*, 155–175. [CrossRef]

26. Welles, J.M.; Cohen, S. Canopy structure measurement by gap fraction analysis using commercial instrumentation. *J. Exp. Bot.* **1996**, *47*, 1335–1342. [CrossRef]

27. Angelini, A.; Corona, P.; Chianucci, F.; Portoghesi, L. Structural attributes of stand overstory and light under the canopy. *CRA J.* **2015**, *39*, 23–31.

28. Lieffers, V.J.; Messier, C.; Stadt, K.J.; Gendron, F.; Comeau, P.G. Predicting and managing light in the understory of boreal forests. *Can. J. Forest Res.* **1999**, *29*, 796–811. [CrossRef]

29. Moeser, D.; Roubinek, J.; Schleppi, P.; Morsdorf, F.; Jonas, T. Canopy closure, LAI and radiation transfer from airborne LiDAR synthetic images. *Agric. Forest Meteorol.* **2014**, *197*, 158–168. [CrossRef]

30. Stereńczak, K.; Ciesielski, M.; Balazy, R.; Zawiła-Niedźwiecki, T. Comparison of various algorithms for DTM interpolation from LIDAR data in dense mountain forests. *Eur. J. Remote Sens.* **2016**, *49*, 599–621. [CrossRef]

31. Sterenczak, K.; Zasada, M.; Brach, M. The accuracy assessment of DTM generated from LIDAR data for forest area—A case study for scots pine stands in Poland. *Balt. For.* **2013**, *19*, 252–262.

32. Parker, G.G.; Lefsky, M.A.; Harding, D.J. *PAR Transmittance in Forest Canopies Determined Using Airborne Laser Altimetry and In-Canopy Quantum Measurements*; SERC: London, UK, 2001.

33. Lefsky, M.A.; Cohen, W.B.; Parker, G.G.; Harding, D.J. Lidar Remote Sensing for Ecosystem Studies. *BioScience* **2002**, *52*, 19–30. [CrossRef]

34. van Leeuwen, M.; Coops, N.C.; Hilker, T.; Wulder, M.A.; Newnham, G.J.; Culvenor, D.S. Automated reconstruction of tree and canopy structure for modeling the internal canopy radiation regime. *Remote Sens. Environ.* **2013**, *136*, 286–300. [CrossRef]

35. Comeau, P. *Measuring Light in the Forest*; Technical Report; Ministry of Forests: Victoria, BC, Canada, 2000. [CrossRef]

36. Promis, Á. Measuring and estimating the below-canopy light environment in a forest: A Review. In *Revista Chapingo Serie Ciencias Forestales y del Ambiente*; Universidad Autónoma Chapingo: Chapingo, Mexico, 2013; Volume XIX, pp. 139–146. [CrossRef]

37. Kumar, L. Reliability of GIS-based solar radiation models and their utilisation in agro-meteorological research. In Proceedings of the 34th International Symposium on Remote Sensing of Environment—The GEOSS Era: Towards Operational Environmental Monitoring, Sydney, Australia, 10–15 April 2011.

38. Moskal, L.M.; Erdody, T.; Kato, A.; Richardson, J.; Zheng, G.; Briggs, D. Lidar applications in precision forestry. In Proceedings of the SilviLaser 2009, College Station, TX, USA, 14–16 October 2009; pp. 154–163.

39. Seidel, D.; Fleck, S.; Leuschner, C. Analyzing forest canopies with ground-based laser scanning: A comparison with hemispherical photography. *Agric. Forest Meteorol.* **2012**, *154–155*, 1–8. [CrossRef]

40. Magney, T.S.; Eitel, J.U.H.; Griffin, K.L.; Boelman, N.T.; Greaves, H.E.; Prager, C.M.; Logan, B.A.; Zheng, G.; Ma, L.; Fortin, E.A.; et al. LiDAR canopy radiation model reveals patterns of photosynthetic partitioning in an Arctic shrub. *Agric. Forest Meteorol.* **2016**, *221*, 78–93. [CrossRef]

41. Zheng, G.; Moskal, L.M. Retrieving Leaf Area Index (LAI) Using Remote Sensing: Theories, Methods and Sensors. *Sensors* **2009**, *9*, 2719–2745. [CrossRef] [PubMed]

42. Wang, J.; Sammis, T.W.; Gutschick, V.P.; Gebremichael, M.; Dennis, S.O.; Harrison, R.E. Review of Satellite Remote Sensing Use in Forest Health Studies. *Open Geogr. J.* **2010**, *3*, 28–42. [CrossRef]

43. Brewer, C.; Monty, J.; Johnson, A.; Evans, D.; Fisk, H. *Forest Carbon Monitoring: A Review of Selected Remote Sensing and Carbon Measurement Tools for REDD+; RSAC-10018-RPT1*; Department of Agriculture, Forest Service, Remote Sensing Applications Center: Salt Lake City, UT, USA, 2011; p. 35.

44. Toth, C.; Jóźków, G. Remote sensing platforms and sensors: A survey. *ISPRS J. Photogramm. Remote Sens.* **2016**, *115*, 22–36. [CrossRef]

45. John Weier, R.C. Solar Radiation and Climate Experiment (SORCE) Fact Sheet: Feature Articles. Available online: https://earthobservatory.nasa.gov/Features/SORCE/sorce.php (accessed on 15 April 2018).

46. Lindsey, R. Climate and Earth's Energy Budget: Feature Articles. Available online: https://earthobservatory.nasa.gov/Features/EnergyBalance/ (accessed on 22 November 2017).

47. Gibson, J. UVB Radiation, Definitions and Characteristics. Available online: http://uvb.nrel.colostate.edu/UVB/publications/uvb_primer.pdf (accessed on 10 September 2017).

48. Canada Center for Remote Sensing Fundamentals of Remote Sensing: A Tutorial. Available online: https://www.nrcan.gc.ca/sites/www.nrcan.gc.ca/files/earthsciences/pdf/resource/tutor/fundam/pdf/fundamentals_e.pdf (accessed on 20 March 2018).

49. Brown, R.D.; Gillespie, T.J. *Microclimatic Landscape Design*; Wiley: New York, NY, USA, 1995.

50. Shahidan, M.F.; Shariff, M.K.M.; Jones, P.; Salleh, E.; Abdullah, A.M. A comparison of Mesua ferrea L. and Hura crepitans L. for shade creation and radiation modification in improving thermal comfort. *Landsc. Urban Plan.* **2010**, *97*, 168–181. [CrossRef]

51. Schleppi, P.; Paquette, A. Solar Radiation in Forests: Theory for Hemispherical Photography. In *Hemispherical Photography in Forest Science: Theory, Methods, Applications*; Fournier, R.A., Hall, R.J., Eds.; Springer: Dordrecht, The Netherlands, 2017; Volume 28, pp. 15–52. ISBN 978-94-024-1096-9.

52. Chazdon, R. Sunflecks and Their Importance to Forest Understorey Plants. *Adv. Ecol. Res.* **1998**, *18*, 1–63.

53. Chazdon, R.L.; Pearcy, R.W. The Importance of Sunflecks for Forest Understory Plants. *BioScience* **1991**, *41*, 760–766. [CrossRef]

54. Promis, A.; Schindler, D.; Reif, A.; Cruz, G. Solar radiation transmission in and around canopy gaps in an uneven-aged Nothofagus betuloides forest. *Int. J. Biometeorol.* **2009**, *53*, 355–367. [CrossRef] [PubMed]

55. Nyman, P.; Metzen, D.; Hawthorne, S.N.D.; Duff, T.J.; Inbar, A.; Lane, P.N.J.; Sheridan, G.J. Evaluating models of shortwave radiation below Eucalyptus canopies in SE Australia. *Agric. Forest Meteorol.* **2017**, *246*, 51–63. [CrossRef]

56. Monsi, M.; Saeki, T. On the Factor Light in Plant Communities and its Importance for Matter Production. *Ann. Bot.* **2005**, *95*, 549–567. [CrossRef] [PubMed]

57. Breda, N.J.J. Ground-based measurements of leaf area index: A review of methods, instruments and current controversies. *J. Exp. Bot.* **2003**, *54*, 2403–2417. [CrossRef] [PubMed]

58. Macfarlane, C.; Hoffman, M.; Eamus, D.; Kerp, N.; Higginson, S.; McMurtrie, R.; Adams, M. Estimation of leaf area index in eucalypt forest using digital photography. *Agric. Forest Meteorol.* **2007**, *143*, 176–188. [CrossRef]

59. Solberg, S.; Brunner, A.; Hanssen, K.H.; Lange, H.; Næsset, E.; Rautiainen, M.; Stenberg, P. Mapping LAI in a Norway spruce forest using airborne laser scanning. *Remote Sens. Environ.* **2009**, *113*, 2317–2327. [CrossRef]

60. Cowling, S.A.; Field, C.B. Environmental control of leaf area production: Implications for vegetation and land-surface modeling: Environmental controls of leaf area production. *Glob. Biogeochem. Cycles* **2003**, *17*. [CrossRef]

61. Waring, R.H.; Running, S.W. CHAPTER 2—Water Cycle. In *Forest Ecosystems*, 3rd ed.; Academic Press: San Diego, CA, USA, 2007; pp. 19–57, ISBN 978-0-12-370605-8.

62. Seyednasrollah, B.; Kumar, M. Effects of tree morphometry on net snow cover radiation on forest floor for varying vegetation densities: Tree morphometry effects on radiation. *J. Geophys. Res. Atmos.* **2013**, *118*, 12508–12521. [CrossRef]

63. Regent Instrument Canada. *WinSCANOPY Technical Manual for Canopy Analysis*; Regent Instrument Canada: Quebec City, QC, Canada, 2014.

64. Skye Instrument Ltd Light Guidance Notes. Available online: http://www.skyeinstruments.com/wp-content/uploads/LightGuidanceNotes.pdf (accessed on 30 January 2018).

65. Rich, P.M. *A Manual for Analysis of Hemispherical Canopy Photography*; Los Alamos National Laboratory, New Mexico: Los Alamos, NM, USA, 1989; Volume 92.

66. Alados, I.; Foyo-Moreno, I.; Alados-Arboledas, L. Photosynthetically active radiation: Measurements and modelling. *Agric. Forest Meteorol.* **1996**, *78*, 121–131. [CrossRef]

67. Bittner, S.; Gayler, S.; Biernath, C.; Winkler, J.B.; Seifert, S.; Pretzsch, H.; Priesack, E. Evaluation of a ray-tracing canopy light model based on terrestrial laser scans. *Can. J. Remote Sens.* **2012**, *38*, 619–628. [CrossRef]

68. Gastellu-Etchegorry, J.P.; Martin, E.; Gascon, F. DART: A 3D model for simulating satellite images and studying surface radiation budget. *Int. J. Remote Sens.* **2004**, *25*, 73–96. [CrossRef]

69. Seidel, D.; Beyer, F.; Hertel, D.; Fleck, S.; Leuschner, C. 3D-laser scanning: A non-destructive method for studying above- ground biomass and growth of juvenile trees. *Agric. Forest Meteorol.* **2011**, *151*, 1305–1311. [CrossRef]

70. Omasa, K.; Hosoi, F.; Uenishi, T.M.; Shimizu, Y.; Akiyama, Y. Three-Dimensional Modeling of an Urban Park and Trees by Combined Airborne and Portable On-Ground Scanning LIDAR Remote Sensing. *Environ. Model. Assess.* **2008**, *13*, 473–481. [CrossRef]

71. Yang, X.; Strahler, A.H.; Schaaf, C.B.; Jupp, D.L.B.; Yao, T.; Zhao, F.; Wang, Z.; Culvenor, D.S.; Newnham, G.J.; Lovell, J.L.; et al. Three-dimensional forest reconstruction and structural parameter retrievals using a terrestrial full-waveform lidar instrument (Echidna®). *Remote Sens. Environ.* **2013**, *135*, 36–51. [CrossRef]

72. Jupp, D.L.B.; Culvenor, D.S.; Lovell, J.L.; Newnham, G.J.; Strahler, A.H.; Woodcock, C.E. Estimating forest LAI profiles and structural parameters using a ground-based laser called 'Echidna(R). *Tree Physiol.* **2008**, *29*, 171–181. [CrossRef] [PubMed]

73. Strahler, A.; Jupp, D.; Woodcock, C.; Schaaf, C.; Yao, T.; Zhao, F.; Yang, X.; Lovell, J.; Culvenor, D.; Newnham, G.; et al. Retrieval of forest structural parameters using a ground-based LiDAR instrument (Echidna †). *Can. J. Remote Sens.* **2014**, *34*. [CrossRef]

74. Armston, J.; Disney, M.; Lewis, P.; Scarth, P.; Phinn, S.; Lucas, R.; Bunting, P.; Goodwin, N. Direct retrieval of canopy gap probability using airborne waveform Lidar. *Remote Sens. Environ.* **2013**, *134*, 24–38. [CrossRef]

75. Yang, W.; Ni-Meister, W.; Kiang, N.Y.; Moorcroft, P.R.; Strahler, A.H.; Oliphant, A. A clumped-foliage canopy radiative transfer model for a Global Dynamic Terrestrial Ecosystem Model II: Comparison to measurements. *Agric. Forest Meteorol.* **2010**, *150*, 895–907. [CrossRef]

76. Comeau, P.; Macdonald, R.; Bryce, R.; Groves, B. *Lite: A Model for Estimating Light Interception and Transmission Through Forest Canopies, User's Manual and Program Documentation*; Working Paper 35/1998; Research Branch, Ministry of Forests: Victoria, BC, Canada, 1998.

77. Comeau, P. Modeling Light Using SLIM & LITE. Available online: https://sites.ualberta.ca/~pcomeau/Light_Modeling/lightusingSLIM_and_LITE.htm (accessed on 20 July 2017).

78. Gastellu-Etchegorry, J.-P.; Grau, E.; Lauret, N. DART: A 3D model for remote sensing images and radiative budget of earth surfaces. In *Modeling and Simulation in Engineering*; John Wiley & Sons: Hoboken, NJ, USA, 2012; ISBN 978-953-307-959-2.

79. Estimating Light Beneath Forest Canopies with LITE and SLIM—Ministry of Forests and Range—Research Branch. Available online: https://www.for.gov.bc.ca/hre/StandDevMod/LiteSlim/ (accessed on 24 April 2018).

80. Malenovskỳ, Z.; Martin, E.; Homolová, L.; Gastellu-Etchegorry, J.-P.; Zurita-Milla, R.; Schaepman, M.E.; Pokornỳ, R.; Clevers, J.G.; Cudlín, P. Influence of woody elements of a Norway spruce canopy on nadir reflectance simulated by the DART model at very high spatial resolution. *Remote Sens. Environ.* **2008**, *112*, 1–18. [CrossRef]

81. Sobrino, J.A.; Mattar, C.; Gastellu-Etchegorry, J.P.; Jimenez-Munoz, J.C.; Grau, E. Evaluation of the DART 3D model in the thermal domain using satellite/airborne imagery and ground-based measurements. *Int. J. Remote Sens.* **2011**, 1–25. [CrossRef]

82. Guillevic, P.; Gastellu-Etchegorry, J.-P. Modeling BRF and Radiation Regime of Boreal and Tropical Forest. *Remote Sens. Environ.* **1999**, *68*, 281–316. [CrossRef]

83. Gastellu-Etchegorry, J.-P.; Yin, T.; Lauret, N.; Cajgfinger, T.; Gregoire, T.; Grau, E.; Feret, J.-B.; Lopes, M.; Guilleux, J.; Dedieu, G. Discrete Anisotropic Radiative Transfer (DART 5) for modeling airborne and satellite spectroradiometer and LIDAR acquisitions of natural and urban landscapes. *Remote Sens.* **2015**, *7*, 1667–1701. [CrossRef]

84. Yamamoto, K.; Murase, Y.; Etou, C.; Shibuya, K. Estimation of relative illuminance within forests using small-footprint airborne LiDAR. *J. Forest Res.* **2015**, *20*, 321–327. [CrossRef]

85. Essery, R.; Bunting, P.; Rowlands, A.; Rutter, N.; Hardy, J.; Melloh, R.; Link, T.; Marks, D.; Pomeroy, J. Radiative Transfer Modeling of a Coniferous Canopy Characterized by Airborne Remote Sensing. *J. Hydrometeorol.* **2008**, *9*, 228–241. [CrossRef]

86. Thomas, V.; Finch, D.A.; McCaughey, J.H.; Noland, T.; Rich, L.; Treitz, P. Spatial modelling of the fraction of photosynthetically active radiation absorbed by a boreal mixed wood forest using a lidar–hyperspectral approach. *Agric. Forest Meteorol.* **2006**, *140*, 287–307. [CrossRef]

87. Mallet, C.; Bretar, F. Full-waveform topographic lidar: State-of-the-art. *ISPRS J. Photogramm. Remote Sens.* **2009**, *64*, 1–16. [CrossRef]

88. Beland, M.; Parker, G.; Harding, D.; Hopkinson, C.; Chasmer, L.; Antonarakis, A. *White Paper–On the Use of LiDAR Data at AmeriFlux Sites*; Ameriflux Network, Berkeley Lab: Berkeley, CA, USA, 2015.

89. Flood, M. Laser altimetry: From science to commercial lidar mapping. *Photogramm. Eng. Remote Sens.* **2001**, *67*, 1209–1217.

90. Barilotti, A.; Turco, S.; Alberti, G. LAI determination in forestry ecosystem by LiDAR data analysis. In Proceedings of the Workshop 3D Remote Sensing in Forestry, Vienna, Austria, 14–15 February 2006; Volume 1415.

91. Kwak, D.-A.; Lee, W.-K.; Cho, H.-K. Estimation of LAI Using Lidar Remote Sensing in Forest. In Proceedings of the ISPRS Workshop on Laser Scanning and SilviLaser, Espoo, Finland, 12–14 September 2007; Volume 6.

92. Yamamoto, K.; Takahashi, T.; Miyachi, Y.; Kondo, N.; Morita, S.; Nakao, M.; Shibayama, T.; Takaichi, Y.; Tsuzuku, M.; Murate, N. Estimation of mean tree height using small-footprint airborne LiDAR without a digital terrain model. *J. Forest Res.* **2011**, *16*, 425–431. [CrossRef]

93. Takahashi, T.; Yamamoto, K.; Senda, Y.; Tsuzuku, M. Predicting individual stem volumes of sugi (Cryptomeria japonica) plantations in mountainous areas using small-footprint airborne LiDAR. *J. Forest Res.* **2005**, *10*, 305–312. [CrossRef]

94. Ellenberg, H. *Vegetation Ecology of Central Europe*, 4th ed.; Cambridge University Press: Cambridge, UK, 1988.

95. Tymen, B.; Vincent, G.; Courtois, E.A.; Heurtebize, J.; Dauzat, J.; Marechaux, I.; Chave, J. Quantifying micro-environmental variation in tropical rainforest understory at landscape scale by combining airborne LiDAR scanning and a sensor network. *Ann. Forest Sci.* **2017**, *74*. [CrossRef]

96. Dozier, J.; Frew, J. Rapid calculation of terrain parameters for radiation modeling from digital elevation data. *IEEE Trans. Geosci. Remote Sens.* **1990**, *28*, 963–969. [CrossRef]

97. Fu, P.; Rich, P.M. Design and implementation of the Solar Analyst: An ArcView extension for modeling solar radiation at landscape scales. In Proceedings of the Nineteenth Annual ESRI User Conference, San Diego, CA, USA, 1999; Volume 1, pp. 1–31.

98. Hofierka, J.; Suri, M.; Šúri, M. The solar radiation model for Open source GIS: Implementation and applications. In Proceedings of the Open source GIS—GRASS Users Conference, Trento, Italy, 11–13 September 2002; pp. 1–19.

99. Cifuentes, R.; Van der Zande, D.; Salas, C.; Tits, L.; Farifteh, J.; Coppin, P. Modeling 3D Canopy Structure and Transmitted PAR Using Terrestrial LiDAR. *Can. J. Remote Sens.* **2017**, *43*, 124–139. [CrossRef]

100. Guillen-Climent, M.L.; Zarco-Tejada, P.J.; Berni, J.A.J.; North, P.R.J.; Villalobos, F.J. Mapping radiation interception in row-structured orchards using 3D simulation and high-resolution airborne imagery acquired from a UAV. *Precis. Agric.* **2012**, *13*, 473–500. [CrossRef]

101. Kobayashi, H.; Baldocchi, D.D.; Ryu, Y.; Chen, Q.; Ma, S.; Osuna, J.L.; Ustin, S.L. Modeling energy and carbon fluxes in a heterogeneous oak woodland: A three-dimensional approach. *Agric. Forest Meteorol.* **2012**, *152*, 83–100. [CrossRef]

102. Lee, H.; Slatton, K.C.; Roth, B.E.; Cropper, W.P. Prediction of forest canopy light interception using three-dimensional airborne LiDAR data. *Int. J. Remote Sens.* **2009**, *30*, 189–207. [CrossRef]

103. Todd, K.W.; Csillag, F.; Atkinson, P.M. Three-dimensional mapping of light transmittance and foliage distribution using lidar. *Can. J. Remote Sens.* **2003**, *29*, 544–555. [CrossRef]

104. Kucharik, C.J.; Norman, J.M.; Gower, S.T. Measurements of leaf orientation, light distribution and sunlit leaf area in a boreal aspen forest. *Agric. Forest Meteorol.* **1998**, *91*, 127–148. [CrossRef]

105. Lillesand, T.M.; Kieffer, R.W. *Remote Sensing and Image Interpretation*, 2nd ed.; John Wiley and Sons, Inc.: Toronto, ON, Canada, 1987.

106. Brunner, A. A light model for spatially explicit forest stand models. *Forest Ecol. Manag.* **1998**, *107*, 19–46. [CrossRef]

107. Baltsavias, E.P. Airborne laser scanning: Existing systems and firms and other resources. *ISPRS J. Photogramm. Remote Sens.* **1999**, *54*, 164–198. [CrossRef]

108. Gu, L.; Baldocchi, D.; Verma, S.B.; Black, T.A.; Vesala, T.; Falge, E.M.; Dowty, P.R. Advantages of diffuse radiation for terrestrial ecosystem productivity. *J. Geophys. Res. Atmos.* **2002**, *107*. [CrossRef]

109. Cavazzoni, J.; Volk, T.; Tubiello, F.; Monje, O. Modelling the effect of diffuse light on canopy photosynthesis in controlled environments. *Acta Hortic.* **2002**, *593*, 39–45. [CrossRef] [PubMed]

110. Li, T.; Heuvelink, E.; Dueck, T.A.; Janse, J.; Gort, G.; Marcelis, L.F.M. Enhancement of crop photosynthesis by diffuse light: Quantifying the contributing factors. *Ann. Bot.* **2014**, *114*, 145–156. [CrossRef] [PubMed]

111. Li, T.; Yang, Q. Advantages of diffuse light for horticultural production and perspectives for further research. *Front. Plant Sci.* **2015**, *6*. [CrossRef] [PubMed]

112. Chasmer, L.; Hopkinson, C.; Treitz, P. Assessing the three-dimensional frequency distribution of airborne and ground-based lidar data for red pine and mixed deciduous forest plots. *Int. Arch. Photogramm. Remote Sens. Spat. Inf. Sci.* **2004**, *36*, W2.

113. Hopkinson, C.; Lovell, J.; Chasmer, L.; Jupp, D.; Kljun, N.; van Gorsel, E. Integrating terrestrial and airborne lidar to calibrate a 3D canopy model of effective leaf area index. *Remote Sens. Environ.* **2013**, *136*, 301–314. [CrossRef]

114. GlobALS: Global ALS Data Providers Database. Interactive Map. Available online: https://www.google.com/maps/d/viewer?mid=1-K-a1MvbjFRE19i8YzOvgkAfEwQ2SGBU&ll=38.48478719903015%2C53.557913999999926&z=3 (accessed on 26 February 2018).

115. McGaughey, R.J.; Andersen, H.-E.; Reutebuch, S.E. Considerations for planning, acquiring, and processing LiDAR data for forestry applications. In Proceedings of the 11th Biennial USDA Forest Service Remote Sensing Applications Conference, Salt Lake City, Utah, USA, 24–28 April 2006.

116. Hummel, S.; Hudak, A.T.; Uebler, E.H.; Falkowski, M.J.; Megown, K.A. A comparison of accuracy and cost of LiDAR versus stand exam data for landscape management on the Malheur National Forest. *J. For.* **2011**, *109*, 267–273.

117. Wilkes, P.; Lau, A.; Disney, M.; Calders, K.; Burt, A.; Gonzalez de Tanago, J.; Bartholomeus, H.; Brede, B.; Herold, M. Data acquisition considerations for Terrestrial Laser Scanning of forest plots. *Remote Sens. Environ.* **2017**, *196*, 140–153. [CrossRef]

118. Widlowski, J.-L.; Pinty, B.; Clerici, M.; Dai, Y.; De Kauwe, M.; de Ridder, K.; Kallel, A.; Kobayashi, H.; Lavergne, T.; Ni-Meister, W.; et al. RAMI4PILPS: An intercomparison of formulations for the partitioning of solar radiation in land surface models. *J. Geophys. Res. Biogeosci.* **2011**, *116*. [CrossRef]

119. Pinty, B.; Gobron, N.; Widlowski, J.-L.; Gerstl, S.A.; Verstraete, M.M.; Antunes, M.; Bacour, C.; Gascon, F.; Gastellu, J.-P.; Goel, N. Radiation transfer model intercomparison (RAMI) exercise. *J. Geophys. Res. Atmos.* **2001**, *106*, 11937–11956. [CrossRef]

120. Saremi, H.; Kumar, L.; Turner, R.; Stone, C.; Melville, G. DBH and height show significant correlation with incoming solar radiation: A case study of a radiata pine (Pinus radiata D. Don) plantation in New South Wales, Australia. *GISci. Remote Sens.* **2014**, *51*, 427–444. [CrossRef]

121. Szymanowski, M.; Kryza, M.; Miga, K.; Sobolewski, P.; Kolondra, L. Modelowanie. Modelling and validation of the potential solar radiation for the hornsund region—Application of the r.sun model. *Rocz. Geomatyki Ann. Geomat.* **2008**, *6*, 107–112.

122. Conrad, O.; Bechtel, B.; Bock, M.; Dietrich, H.; Fischer, E.; Gerlitz, L.; Wehberg, J.; Wichmann, V.; Böhner, J. System for Automated Geoscientific Analyses (SAGA) v. 2.1.4. *Geosci. Model Dev.* **2015**, *8*, 1991–2007. [CrossRef]

remote sensing

MDPI

Article

Modeling Photosynthetically Active Radiation from Satellite-Derived Estimations over Mainland Spain

Jose M. Vindel [1],*, Rita X. Valenzuela [1], Ana A. Navarro [1], Luis F. Zarzalejo [1], Abel Paz-Gallardo [2], José A. Souto [3], Ramón Méndez-Gómez [3], David Cartelle [3] and Juan J. Casares [3]

[1] Renewable Energy Division–CIEMAT, Madrid 28040, Spain; r.valenzuela@ciemat.es (R.X.V.);
    a.navarro@ciemat.es (A.A.N.); lf.zarzalejo@ciemat.es (L.F.Z.)
[2] Extremadura Research Center for Advanced Technologies (CETA–CIEMAT), Trujillo 10200, Spain;
    abelfrancisco.paz@ciemat.es
[3] Chemistry Engineering Department, University of Santiago de Compostela, Santiago de Compostela 15782,
    Spain; ja.souto@usc.es (J.A.S.); ramon.mendez.gomez@gmail.com (R.M.-G.);
    david.cartelle@troposfera.es (D.C.); juanjose.casares@usc.es (J.J.C.)
* Correspondence: josemaria.vindel@ciemat.es; Tel.: +34-91-346-6496

Received: 4 April 2018; Accepted: 24 May 2018; Published: 30 May 2018

Abstract: A model based on the known high correlation between photosynthetically active radiation (PAR) and global horizontal irradiance (GHI) was implemented to estimate PAR from GHI measurements in this present study. The model has been developed using satellite-derived GHI and PAR estimations. Both variables can be estimated using Kato bands, provided by Satellite Application Facility on Climate Monitoring (CM-SAF), and its ratio may be used as the variable of interest in order to obtain the model. The study area, which was located in mainland Spain, has been split by cluster analysis into regions with similar behavior, according to this ratio. In each of these regions, a regression model estimating PAR from GHI has been developed. According to the analysis, two regions are distinguished in the study area. These regions belong to the two climates dominating the territory: an Oceanic climate on the northern edge; and a Mediterranean climate with hot summer in the rest of the study area. The models obtained for each region have been checked against the ground measurements, providing correlograms with determination coefficients higher than 0.99.

Keywords: photosynthetically active radiation; global horizontal irradiance; clustering analysis; Kato bands

1. Introduction

Photosynthetically active radiation (PAR) is radiation with wavelengths of 400–700 nm in the solar spectrum. Biomass and algae production, plant physiology, energy balance in ecosystems, natural illumination of greenhouses, etc., require knowledge of this part of the solar spectrum. Despite its importance, PAR measurement stations are very scarce and, thus, usually it is estimated from empirical expressions relating it to solar global irradiance [1–7], which is measured more frequently. PAR estimations can also be obtained from satellites. Liang et al. [8] developed a method based on the look-up table approach for estimating PAR from Moderate-Resolution Imaging Spectrometer (MODIS) data. Similarly, other authors [9] have derived PAR using Geostationary Operational Environmental Satellite (GOES) data. On the other hand, Rubio et al. [10] estimated global horizontal irradiance (GHI) from a satellite, before PAR was obtained using an empirical model proposed by Alados-Arboledas et al. [11], which was developed from a database located at Almería. Wandji et al. [12] described a technique for an accurate assessment of PAR under clear-sky conditions using Kato bands [13] from libRadtran simulations. Kato bands are 32 bands of different widths which the solar spectrum can be divided into. In each of those bands, the absorption coefficient of different gases is almost constant.

The width of Kato bands depends on the distribution and structure of the absorption bands. These bands are also provided for the Satellite Application Facility on Climate Monitoring (CM-SAF), which belongs to the European Organization for the Exploitation of Meteorological Satellites (EUMETSAT); thus, PAR can be obtained, in a first approximation, using the bands included in the part of the spectrum from 400 up to 700 nm. Indeed, of the 32 Kato bands available in the entire solar spectrum, the interval of bands 7–16 includes the region corresponding to PAR (Table 1).

Table 1. Kato bands in the photosynthetically active radiation (PAR) region.

Kato Band	Wavelength Region (µm)
7	0.408–0.452
8	0.452–0.518
9	0.518–0.540
10	0.540–0.550
11	0.550–0.567
12	0.567–0.605
13	0.605–0.625
14	0.625–0.667
15	0.667–0.684
16	0.684–0.704

Therefore, GHI and PAR can be estimated in a first approximation from satellite-derived Kato bands. On the other hand, the linear relationship between both variables is usually the basis of the empirical models used to obtain the PAR value [14] and, thus, PAR can be approximated from the knowledge of the ratio between both variables and the GHI value. According to these considerations, this ratio was the variable used for the analysis performed in this work. The spatial and temporal variability of solar radiation advises a clustering analysis [15–18], which provides the groups within which the variable of interest, namely, the ratio between both radiations, is coherent. Once the regions with similar behavior in terms of the PAR/GHI ratio have been obtained by clustering analysis, a linear regression model in each of these regions is developed. The performance of this model from satellite-derived PAR and GHI estimations is the main novelty of this work. Thus, this model allows for the estimation of PAR in any part of this region from GHI measurements carried out at this location.

2. Data

The spectral-resolved irradiance [19] containing the Kato bands was the product required from CM-SAF for this work. The CM-SAF includes the following atmospheric input information to retrieve the surface incoming solar radiation: effective cloud albedo and clear sky index, aerosol, water vapor, ozone, and surface albedo.

Solar surface irradiance on a horizontal plane is supplied by the MAGICSOL (Magic Solar Irradiance) method, which only needs the satellite information from the broadband visible channel. Thus, it can be implemented in different satellite generations [20]. On the other hand, MAGICSOL's method for clouds is based on the Heliosat algorithm [21], which determines the cloud index (n) using reflection measurements given as normalized digital counts. This index measures the reflectance detected on the sensor that is normalized by the dynamic range [22,23]. The following expression provides this index:

$$n = \frac{\rho - \rho_g}{\rho_c - \rho_g} \tag{1}$$

where ρ is the instantaneous planetary albedo, which is estimated from the digital count of the satellite sensor; ρ_c is the cloud albedo estimated from the brightest pixel; and ρ_g is the ground albedo, estimated from the darkest pixel.

However, the spectral-resolved irradiance requires modifications of the original method [24,25]. Indeed, the spectral effect of clouds is treated using the spectral corrections of the broadband effective cloud albedo, which is carried out by the application of the radiative transfer model.

The CM-SAF surface solar radiation datasets, which have a native spatial resolution of $0.05° \times 0.05°$, have been already validated in previous studies [26,27]. Besides this, a spatial distribution of the errors for these datasets was found in a previous study [28], where the surface solar radiation estimates derived from SAFs were compared with gridded daily solar radiation estimates obtained from station measurements of Joint Research Centre Monitoring Agricultural ResourceS (JRC-MARS) database.

The area requested for this work covers from 44°N to 35.3°N latitude and 9.5°W to 3.5°E longitude, where the study region (mainland Spain) is included. The spatial resolution was $0.1° \times 0.1°$ and the mean daily data during 1991–2011 (21 years) were used. After this, for the clustering analysis, the data were grouped in months in order to reduce the high temporal resolution. This restricts the computational complexity and avoids fluctuations that can introduce noise into a climatological study.

On the other hand, PAR and GHI daily ground measurements were taken in order to validate the model obtained at three sites: Plataforma Solar de Almería (PSA-CIEMAT), from 24 February 2016 to 22 June 2017; Centro de Desarrollo de Energías Renovables (CEDER-CIEMAT), from 26 January 2016 to 20 August 2017; and Santiago de Compostela (Santiago-EOAS, [29]), from 1 January 2016 to 31 December 2017.

Regarding the two first stations, the PAR sensor, an Eko ML-020P model, was installed over a horizontal plane on the top of a weather house at 3 agl m of altitude. There were also other sensors installed for the characterization of solar radiation (global, direct, and diffuse). The placement of instruments was free of obstacles, such as mountains, buildings, and trees. Therefore, any radiation measurement at these sites can represent the conditions above the canopy layer. GHI was recorded using a CM21 (Kipp & Zonen) pyranometer at the first two stations. Data from those stations were monitored and collected continuously every 1 min (on average).

Regarding the Santiago-EOAS station, the PAR at ground level was measured at this meteorological site over a horizontal plane at 1.5 agl m by using a multiband GUV-2511 radiometer. This site is located at the top of a hill in a flat grassland terrain without any obstacle in the surrounding area. Therefore, any radiation measurement at this site can represent the conditions above the canopy layer. The 10 min average PAR solar radiation data were collected. On the other hand, GHI data from MeteoGalicia were measured using a pyrradiometer PH.SCHENK Type 8111.

For this work, all data were transformed into daily data.

3. Methodology

The methodology applied is based on the clustering analysis performed on the PAR/GHI ratio estimations. The role of clustering is to identify regions with different cloud patterns throughout the year and, thus, to develop specific models for each region that can be sensitive to the different cloud dynamics of each region.

The clustering technique used for this study was *k*-means, which is one of the most widely used methods. The *k*-means algorithm [30–32] is based on the minimization of the sum of squared distances between the centroid of the group and each object of this group. *k*-means is implemented as follows: (a) initial clusters are randomly selected; (b) distances between data and centroids of each cluster are determined; (c) data are assigned to clusters so that their centroids are the nearest; (d) according to the new data, new centroids are obtained for each cluster; and (e) the process is repeated until the sum of distances between the data and centroids of clusters converges.

On the other hand, the optimal number of clusters for the clustering must be determined. Indeed, the optimal number of clusters allows for identification of the most significant clusters. If a smaller number of clusters was considered, certain zones with differing behavior would not be taken into account. In contrast, considering too many clusters would make similarly behaving regions appear different. In this work, the optimal number was determined using the so-called silhouette method [33], which validates the consistency within clusters.

After this, a linear regression model to obtain PAR from GHI, trained with satellite data, was produced for each cluster. Finally, this model was validated with the available ground measurements from three stations.

Next, the steps followed in the methodology, as well as the justification and limitations of the applied method, are shown.

3.1. Steps Followed

The methodology includes the following steps:

- Step 1: Obtaining GHI and PAR Estimations

GHI and PAR estimations were obtained by summing the satellite-derived Kato bands. In the case of PAR, the interval of bands is {7–16}. In the case of GHI, all bands provided for CM-SAF were used, which had the interval {4–27}. The percentage of the radiation included in the three first bands, which are not included in CM-SAF, can be considered to be negligible [34]. Regarding the five last bands that were not included {28–32}, although the percentage included in this interval must be small, an approximation was conducted considering a triangle whose base was the width of the interval and height the radiation corresponding to Band 27. From these estimations, the PAR/GHI ratio is available for the following analysis.

- Step 2: Clustering Analysis

The *k*-means algorithm was applied to the clustering analysis. As mentioned, the silhouette method [33] was used in order to determine the optimal number of clusters. For each individual object *i* of the cluster, the silhouette width value, s(*i*), is defined as

$$s\left(i\right) = \frac{b\left(i\right) - a\left(i\right)}{\max\left(a\left(i\right), b\left(i\right)\right)} \tag{2}$$

where $a(i)$ is the average dissimilarity of the object *i* regarding all other data within the same cluster; $b(i)$ is the lowest average dissimilarity of *i* regarding any neighbor cluster; and s(*i*) is in the interval range of $[-1,1]$. If $s(i) \approx +1$, then *i* is well matched within the group, as $a(i) << b(i)$. In contrast, if $s(i) \approx -1$, then *i* is mislabeled as $a(i) >> b(i)$. When $s(i) \approx 0$, *i* is between two groups because $a(i) = b(i)$. The quality of the whole structure of the cluster is measured by the average silhouette width (ASW), which is defined as

$$ASW = \frac{1}{N} \sum_{i=1}^{n} s(i). \tag{3}$$

Thus, the silhouette plots may be used to determine the natural number of clusters for a certain dataset so the highest ASW shows the optimal number of clusters.

As indicated, the variable of interest for the analysis was the PAR/GHI ratio, which uses 12 features (12 months) at each grid point, so the distances between pairs of data were calculated considering these 12 features.

- Step 3: Obtaining Regression Models

A regression linear model trained with only the satellite data within each region was produced. The satellite-derived PAR and GHI values corresponding to the same grid point and same day were the pairs of points used for the regression. Due to the temporal variability affecting the solar radiation, a different model for each month (12 models for each region) was obtained.

- Step 4: Validation

The regression models were validated with the ground measurements obtained for two years (2016 and 2017) in the three sites previously indicated (PSA-CIEMAT, CEDER-CIEMAT, and Santiago-EOAS). As mentioned, the linear relationship between PAR and GHI is usually the basis of the empirical models used to obtain the PAR values. Thus, the coherence between both variables was analyzed by checking this linear relationship and removing the data corresponding to extreme outliers. These outliers are the points that are very far off from the regression line and the process to obtain them was the following:

(a) Obtaining the regression line between PAR measured versus GHI measured;
(b) Obtaining the distances between the PAR measured values and PAR values obtained by the regression line;
(c) Obtaining the interquartile range and the 25th and 75th percentiles of these distances; and
(d) Determining the points with a PAR measured that is either higher than the 75th percentile plus three times the interquartile range or lower than the 25th percentile minus three times this interquartile range. These points are the extreme values [35].

Once the data were filtered according to the former process, the GHI ground measurements were introduced in the models obtained in Step 3 (Obtaining Regression Models) using the satellite-derived estimates. After this, the obtained PAR values were compared with the measured PAR values at these three stations.

3.2. Justification of Method

Satellite-derived irradiances (broadband and spectral) have uncertainties as a result of several factors of different natures (systematic errors, approximations made in the model, etc.). In fact, CM-SAF provides, apart from Kato bands, the global irradiance for the overall spectrum and this value is not fully consistent with the approximation from the sum of Kato bands. Thus, it is very common to correct and improve these retrievals by identifying bias and errors with the help of ground data. The motivation to develop a PAR model from the satellite-derived data in which the more accurate GHI ground measurements are used for its application is twofold: on the one hand, the improvement of the retrieval provided by the satellite and on the other hand, the fact that Kato bands are not available in the whole dataset of the CM-SAF product (CM-SAF only provides the Spectral-Resolved Irradiance, containing the Kato bands, until 31 December 2011). Thus, this model will help to extend the usability, allowing PAR computation from the whole period of global irradiance data in the CM-SAF database.

3.3. Limitations of Method

Two limitations can be found:

- The lack of more ground measurements prevents correcting the model using such measurements. In fact, this work is supported by the Spanish Ministry of Economy, Industry and Competitiveness (Project CGL2016-79284-P AEI/FEDER/UE), which is devoted to reducing this lack of measurements via the installation of a network of stations.
- The assumption of the PAR/GHI ratio estimation provided by the satellite is accurate enough and, thus, a model based on this ratio can be used to obtain PAR from ground GHI values. This assumption is based on the fact that both satellite-derived radiation types are obtained by the same method (summing Kato bands). However, there are no simultaneous ground and satellite data that can be used to assess this accuracy.

4. Results and Discussion

4.1. Determination of the Optimal Number of Clusters According to the Silhouette Method

The silhouette width according to the number of clusters is shown in Figure 1.

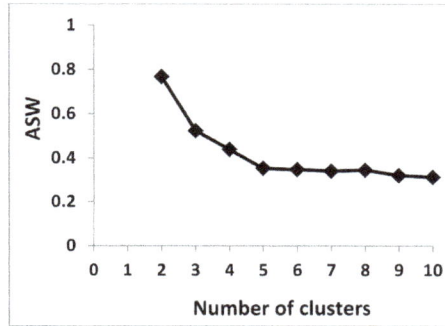

Figure 1. Average silhouette width versus number of clusters (from *k*-means).

The highest value of silhouette is obtained for two clusters and, thus, according to the (previously mentioned) silhouette method, this value is the optimum number of clusters.

4.2. Clustering Analysis

Figure 2 shows the two regions based on the *k*-means algorithm, which uses the PAR/GHI ratio as variable of interest.

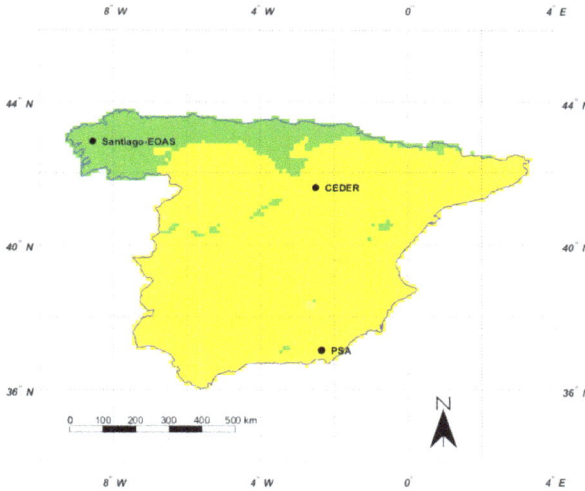

Figure 2. Clusters by the *k*-means algorithm (2 clusters).

The two regions are clearly different. One of them extends along the north of the Iberian Peninsula, which also includes other punctual zones. The other region covers most of the territory.

In order to see the difference between both regions more clearly, complementary information about the annual variability of the radiation was calculated. Table 2 shows the size and the mean PAR/GHI values during different months for each cluster.

Table 2. Number of points for the two clusters and mean PAR/global horizontal irradiance (GHI) values for the different months (N: Number of points).

Region	N	January	February	March	April	May	June	July	August	September	October	November	December
green	959	0.44	0.44	0.43	0.43	0.43	0.43	0.43	0.43	0.43	0.43	0.44	0.44
yellow	4515	0.43	0.43	0.42	0.43	0.42	0.42	0.42	0.42	0.42	0.43	0.43	0.43

The mean values of the small region (green) are slightly higher. In addition, the values corresponding to winter months are also slightly higher than the values of other months.

The division obtained has also physical meaning since it is consistent with the global climatology of Spain [27,36,37]. Indeed, in the northern edge of the territory, the climate is Oceanic, with continuous clouds and precipitation over the year. However, in the rest of the territory, the Mediterranean climate with hot summer is dominant. The climate of a region is obviously related to the solar radiation reaching Earth's surface and, thus, the clustering achieved must agree with these climatological features. The north edge in Figure 2 is clearly associated with the Oceanic climate where abundant clouds decrease the solar activity [38]. This activity increases in the rest of the territory, which is characterized by a Mediterranean climate. However, we must recall that the variable used for the study was the PAR/GHI ratio, and not the solar radiation. This ratio depends on the attenuation, which affects the different bands of the solar spectrum. Regarding the spectral attenuation caused by clouds, the scattering is nonselective [39] and, thus, there are no important differences between the behavior of this ratio in cloudy and noncloudy zones. The case of absorption is different as water has absorption mainly in the infrared region [40], which affects the attenuation of the global radiation, but not the attenuation in the PAR band. Thus, in cloudy zones, an increase in the PAR/GHI ratio can be expected.

Regarding the punctual zones of the green region, they belong to important mountain ranges where cloudiness is abundant, which is the same as in the zone associated with the Oceanic climate.

4.3. Regression Model

The model obtained according to Step 3 (Obtaining Regression Models) of the methodology from the satellite-derived PAR and GHI values is the following:

$$PAR = a\,GHI + b \tag{4}$$

where *a* and *b* take the values shown in Table 3.

Table 3. Values of slope and intercept for the model.

		January	February	March	April	May	June	July	August	September	October	November	December
Green	*a*	0.41	0.41	0.40	0.41	0.39	0.39	0.39	0.40	0.41	0.42	0.41	0.42
cluster	*b*	0.99	1.76	2.75	2.88	7.61	8.69	10.18	6.38	4.50	1.68	1.36	0.65
Yellow	*a*	0.42	0.42	0.41	0.41	0.38	0.39	0.36	0.39	0.39	0.41	0.41	0.41
cluster	*b*	0.35	0.49	1.37	2.15	10.94	9.37	18.05	8.63	7.12	2.25	1.45	1.25

4.4. Validation

The former regression model was validated with ground measurements obtained at three stations. Two of these stations are included within the larger cluster (yellow): PSA-CIEMAT and CEDER-CIEMAT. The other station, Santiago-EOAS, belongs to the small cluster. Once the lags were deducted, the numbers of points (days) for the study were: 483 from PSA-CIEMAT, 549 from CEDER-CIEMAT, and 368 from Santiago-EOAS. These stations are also shown in Figure 2. These ground measurements were filtered according to Step 4 (Validation) of the methodology. At this end, the regression lines between PAR ground measurements and GHI ground measurements were obtained (Figure 3).

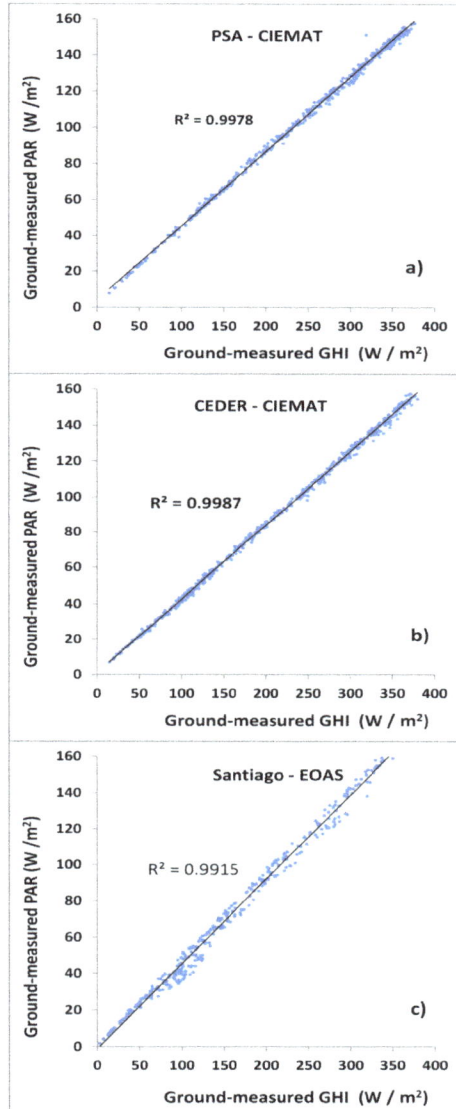

Figure 3. PAR measured versus GHI measured for validation stations: (**a**–**c**) correspond to PSA-CIEMAT, CEDER-CIEMAT, and Santiago-EOAS stations, respectively.

Figure 3 clearly shows the good linear relationship between both variables. Only one data point of PSA-CIEMAT had to be removed according to the coherence filter applied due to some punctual incidence on the ground sensors. The ground-measured and filtered GHI values at these places were introduced in Equation (4) and the obtained PAR values were compared with the ground-measured PAR values. This comparison can be appreciated in the corresponding correlograms (Figure 4).

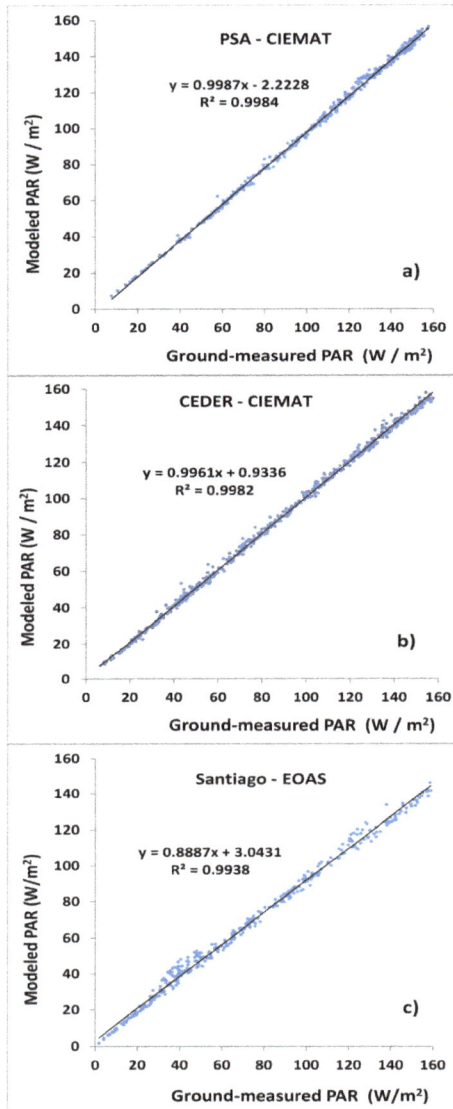

Figure 4. Correlograms for validation stations: (**a–c**) correspond to PSA-CIEMAT, CEDER-CIEMAT, and Santiago-EOAS stations, respectively.

The statistics obtained from the correlograms: determination coefficients (R^2), slopes, and intercepts, as well as the mean bias errors (MBE) and the root mean square errors (RMSE) have been included in Table 4, in order to assess the goodness degree of the model at the places with ground measurements.

Table 4. Statistics of validation.

Station	R^2	Slope	Intercept (W/m^2)	MBE (W/m^2)	MBE (%)	RMSE (W/m^2)	RMSE (%)
PSA	0.998	0.999	−2.223	−2.356	−2.3	2.827	2.8
CEDER	0.998	0.996	0.934	0.598	0.7	1.912	2.2
Santiago-EOAS	0.994	0.889	3.043	−4.741	6.8	7.247	10.4

In all cases, the correlation is very high (determination coefficients are higher than 0.99). However, the stations of the center and south zone (PSA-CIEMAT and CEDER-CIEMAT) show better behavior compared to the station of the north zone (Santiago de Compostela), especially in the case of CEDER-CIEMAT. According to the results of Santiago-EOAS (MBE = −4.741 and slope = 0.889), the model slightly underestimates the PAR in the north region. This underestimation could be due to a low estimation of the satellite-derived PAR/GHI relation. However, since there are no simultaneous ground and satellite data to assess the accuracy in the estimation of this relation, the mean value of the PAR/GHI ratio has been obtained from the ground measurements and compared with the mean values corresponding to the north zone (Table 2). The relation from the ground measurements (0.46) is slightly higher than these means, which are in the range of 0.43–0.44, and this helps to understand the underestimation observed. On the other hand, the errors shown in Table 4 can also be due to the satellite-derived GHI error itself. Indeed, according to a previous study [28], higher errors in GHI are appreciated in the northern zone of Spain, which is consistent with the findings of another study [27], in which a similar behavior to those shown in Table 4 was observed. Indeed, in that study, the satellite-derived GHI estimates were compared with the ground measurements at three stations of Spain (sited at north zone, center, and Mediterranean coast) and according to the results, the highest errors were located at the northern station and the lowest at the center station.

5. Conclusions

A model based on the known linear relationship between GHI and PAR was implemented to estimate PAR from GHI measurements in this present study. The model has been developed using satellite-derived GHI and PAR estimations, which is the main novelty of this work. These estimations were achieved using the Kato bands provided for CM-SAF. The ratio between both variables was considered as the variable of interest in order to split the study area into regions within which the relation between PAR and GHI was similar. This consideration seems suitable since the division obtained provided two regions in accordance with the climatological features of mainland Spain. Indeed, the different regimes of clouds and precipitation characteristics of each climate affect the ratio of PAR/GHI in a different way. On the one hand, the northern area, along with some small and punctual zones, is associated with the Oceanic climate. On the other hand, the rest of the territory has a dominant Mediterranean climate. In addition, a separation across different months was included in order to consider the different seasonal behavior. In fact, according to Table 2, in both zones, the values corresponding to the winter months are slightly higher than the values of other months.

The validation of the model carried out with the three stations (two included within the south in the yellow zone) show correlograms with very high determination coefficients (higher than 0.99), as well as slopes that are practically equal to 1 for the largest region. According to all statistics of validation, the behavior of the model is better in this region, as the model slightly underestimates the PAR in the north region. This underestimation could be due to a previous underestimation of the relation of PAR/GHI from the satellite. Indeed, according to Table 2, the monthly mean values of the north region (green) fluctuate between 0.43 and 0.44, while this relation measured at the station has a mean value of 0.46. On the other hand, the errors in the spatial distribution shown in this table can also be due to the satellite-derived GHI error distribution itself as the highest errors are located in the northern zone and the lowest are in the center zone of the country.

Remote Sens. **2018**, *10*, 849

On the other hand, the proposed method could be used for obtaining PAR historical values from the satellite-derived GHI estimates. The model coefficients have been derived using a long series of daily data (21 years) and, thus, they should show high temporal stability, at least, for periods with available satellite data.

Finally, according to the results of the work, there is the need for a PAR station network in order to allow for the usual correction of the satellite-derived solar radiation estimations.

Author Contributions: J.M.V. conceived the proposed model, which was designed with the rest of co-authors; J.M.V., R.X.V., A.A.N., L.F.Z. and A.P.-G. processed satellite data, and they processed and validated PSA-CIEMAT and CEDER-CIEMAT data; J.A.S., R.M.-G., D.C., and J.J.C., processed and validated Santiago–EOAS data.

Acknowledgments: This work was supported by the Spanish Ministry of Economy, Industry and Competitiveness (MINECO) [Project CGL2016-79284-P AEI/FEDER/UE]. Authors also acknowledge the data provided by EUMETSAT for the SRI product, MeteoGalicia (Consellería de Medio Ambiente e Ordenación do Territorio, Xunta de Galicia) for global irradiance data at Santiago-EOAS station, and University of Santiago de Compostela for the PAR data.

Conflicts of Interest: The authors declare no conflict of interest.

References

1. Akitsu, T.; Kume, A.; Hirose, Y.; Ijima, O.; Nasahara, K.N. On the stability of radiometric ratios of photosynthetically active radiation to global solar radiation in Tsukuba, Japan. *Agric. For. Meteorol.* **2015**, *209*, 59–68. [CrossRef]
2. Alados, I.; Foyo-Moreno, I.; Alados-Arboledas, L. Photosynthetically active radiation: Measurements and modelling. *Agric. For. Meteorol.* **1996**, *78*, 121–131. [CrossRef]
3. Escobedo, J.F.; Gomes, E.N.; Oliveira, A.P.; Soares, J. Modeling hourly and daily fractions of UV, PAR and NIR to global solar radiation under various sky conditions at Botucatu, Brazil. *Appl. Energy* **2009**, *86*, 299–309. [CrossRef]
4. Hu, B.; Wang, Y.; Liu, G. Measurements and estimations of photosynthetically active radiation in Beijing. *Atmos. Res.* **2007**, *85*, 361–371. [CrossRef]
5. Jacovides, C.P.; Tymvios, F.S.; Boland, J.; Tsitouri, M. Artificial Neural Network models for estimating daily solar global UV, PAR and broadband radiant fluxes in an eastern Mediterranean site. *Atmos. Res.* **2015**, *152*, 138–145. [CrossRef]
6. Sudhakar, K.; Srivastava, T.; Satpathy, G.; Premalatha, M. Modelling and estimation of photosynthetically active incident radiation based on global irradiance in Indian latitudes. *Int. IJEEE* **2013**, *4*, 21. [CrossRef]
7. Zhang, X.; Zhang, Y.; Zhoub, Y. Measuring and modelling photosynthetically active radiation in Tibet Plateau during April-October. *Agric. For. Meteorol.* **2000**, *102*, 207–212. [CrossRef]
8. Liang, S.; Zheng, T.; Liu, R.; Fang, H.; Tsay, S.-C.; Running, S. Estimation of incident photosynthetically active radiation from Moderate Resolution Imaging Spectrometer data. *J. Geophys. Res. Atmos.* **2006**, *111*, D15208. [CrossRef]
9. Zheng, T.; Liang, S.; Wang, K. Estimation of incident photosynthetically active radiation from GOES visible imagery. *J. Appl. Meteorol. Climatol.* **2008**, *47*, 853–868. [CrossRef]
10. Rubio, M.A.; López, G.; Tovar, J.; Pozo, D.; Batlles, F.J. The use of satellite measurements to estimate photosynthetically active radiation. *Phys. Chem. Earth Parts A/B/C* **2005**, *30*, 159–164. [CrossRef]
11. Alados-Arboledas, L.; Olmo, F.J.; Alados, I.; Perez, M. Parametric models to estimate photosynthetically active radiation in Spain. *Agric. For. Meteorol.* **2000**, *101*, 187–201. [CrossRef]
12. Wandji, W.; Espinar, B.; Blanc, P.; Wald, L. Estimating the photosynthetically active radiation under clear skies by means of a new approach. *Adv. Sci. Res.* **2015**, *12*, 5–10. [CrossRef]
13. Seiji, K.; Ackerman, Y.P.; Mather, J.H.; Clothiaux, E.E. The k-distribution method and correlated-k approximation for a shortwave radiative transfer model. *J. Quant. Spectrosc. Radiat. Transf.* **1999**, *62*, 109–121.
14. López, G.; Rubio, M.A.; Martínez, M.; Batlles, F.J. Estimation of hourly global photosynthetically active radiation using artificial neural network models. *Agric. For. Meteorol.* **2001**, *107*, 279–291. [CrossRef]
15. Gastón-Romeo, M.; Leon, T.; Mallor, F.; Ramírez-Santigosa, L. A Morphological Clustering Method for daily solar radiation curves. *Sol. Energy* **2011**, *85*, 1824–1836. [CrossRef]

16. Ghayekhloo, M.; Ghofrani, M.; Menhaj, M.B.; Azimi, R. A novel clustering approach for short-term solar radiation forecasting. *Sol. Energy* **2015**, *122*, 1371–1383. [CrossRef]

17. Jiménez-Pérez, P.F.; Mora-López, L. Modeling and forecasting hourly global solar radiation using clustering and classification techniques. *Sol. Energy* **2016**, *135*, 682–691. [CrossRef]

18. Polo, J.; Gastón, M.; Vindel, J.M.; Pagola, I. Spatial variability and clustering of global solar irradiation in Vietnam from sunshine duration measurements. *Renew. Sustain. Energy Rev.* **2015**, *42*, 1326–1334. [CrossRef]

19. Stengel, M.; Kniffka, A.; Meirink, J.F.; Lockhoff, M.; Tan, J.; Hollmann, R. Claas: The CM SAF cloud property dataset using SEVIRI. *Atmos. Chem. Phys. Discuss.* **2013**, *13*, 26451–26487. [CrossRef]

20. Posselt, R.; Mueller, R.; Stöckli, R.; Trentmann, J. Spatial and Temporal Homogeneity of Solar Surface Irradiance across Satellite Generations. *Remote Sens.* **2011**, *3*, 1029–1046. [CrossRef]

21. Müller, R.; Pfeifroth, U.; Träger-Chatterjee, C.; Cremer, R.; Trentmann, J.; Hollmann, R. *Surface Solar Radiation Data Set—Heliosat (SARAH)—Edition 1*; EUMETSAT: Darmstadt, Germany, 2015.

22. Perez, R.; Ineichen, P.; Moore, K.; Kmiecik, M.; Chain, C.; George, R.; Vignola, F. A new operational model for satellite-derived irradiances: Description and validation. *Sol. Energy* **2002**, *73*, 307–317. [CrossRef]

23. Zelenka, A.; Perez, R.; Seals, R.; Renné, D. Effective accuracy of satellite-derived hourly irradiances. *Theor. Appl. Climatol.* **1999**, *62*, 199–207. [CrossRef]

24. Mueller, R.; Behrendt, T.; Hammer, A.; Kemper, A. A New Algorithm for the Satellite-Based Retrieval of Solar Surface Irradiance in Spectral Bands. *Remote Sens.* **2012**, *4*, 622–647. [CrossRef]

25. Mueller, R.; Behrendt, T. Algorithm Theoretical Baseline Document: Spectrally Resolved Solar Surface Irradiance SRI. 2013. Available online: http://www.cmsaf.eu/EN/Documentation/Documentation/ATBD/pdf/SAF_CM_DWD_ATBD_SRI_1.pdf?__blob=publicationFile&v=4 (accessed on 4 April 2018). [CrossRef]

26. Riihelä, A.; Carlund, T.; Trentmann, J.; Müller, R.; Lindfors, A.V. Validation of CM SAF Surface Solar Radiation Datasets over Finland and Sweden. *Remote Sens.* **2015**, *7*, 6663–6682. [CrossRef]

27. Vindel, J.M.; Navarro, A.A.; Valenzuela, R.X.; Ramírez, L. Temporal scaling analysis of irradiance estimated from daily satellite data and numerical modelling. *Atmos. Res.* **2016**, *181*, 154–162. [CrossRef]

28. Bojanowski, J.S.; Vrieling, A.; Skidmore, A.K. A comparison of data sources for creating a long-term time series of daily gridded solar radiation for Europe. *Sol. Energy* **2014**, *99*, 152–171. [CrossRef]

29. Pettazzi, A.; Souto, J.A.; Salsón, S.; Pérez Muñuzuri, V. EOAS, a shared joint atmospheric observation site of MeteoGalicia. In Proceedings of the 4th International Conference on Experiences with Automatic Weather Stations, Lisboa, Portugal, 24–26 May 2006.

30. Andenberg, M.R. *Cluster Analysis for Applications*; Academic: New York, NY, USA, 1973.

31. Adam, F.; Celebi, M.E. An Accelerated Nearest Neighbor Search Method for the K-Means Clustering Algorithm. In Proceedings of the Twenty-Sixth International Florida Artificial Intelligence Research Society Conference, St. Pete Beach, FL, USA, 22–24 May 2013; AAAI Press: Palo Alto, CA, USA, 2013.

32. Macqueen, J. Some methods for classification and analysis of multivariate observations. In Proceedings of the Fifth Berkeley Symposium on Mathematical Statistics and Probability, Berkeley, CA, USA, 7 January 1966; Le, L.M., Neyman, J., Eds.; University of California Press: Berkeley, CA, USA, 1967; pp. 281–297.

33. Rousseeuw, P.J. Silhouettes: A graphical aid to the interpretation and validation of cluster analysis. *J. Comput. Appl. Math.* **1987**, *20*, 53–65. [CrossRef]

34. Wandji, W.; Espinar, B.; Blanc, P.; Wald, L. How close to detailed spectral calculations is the *k*-distribution method and correlated-*k* approximation of Kato et al. (1999) in each spectral interval? *Meteorol. Z.* **2014**, *23*, 547–556. [CrossRef]

35. He, X. Quartiles and Boxplots (Modified). Normal Quantile Plots (QQ-Plot). 2012. Available online: http://www.stat.purdue.edu/~xuanyaoh/stat350/xyJan27Lec6.pdf (accessed on 4 April 2018).

36. Peña-Angulo, D.; Trigo, R.M.; Cortesi, N.; González-Hidalgo, J.C. The influence of weather types on the monthly average maximum and minimum temperatures in the Iberian Peninsula. *Atmos. Res.* **2016**, *178–179*, 217–230.

37. Agencia Estatal de Meteorología (AEMET); Instituto de Meteorología de Portugal (IM). *Iberian Climate Atlas*; AEMET-Ministerio de Medio Ambiente y Medio Rural y Marino & IM: Madrid, Spain, 2011; ISBN 978-84-7837-079-5.

38. Mueller, R.; Trentmann, J.; Träger-Chatterjee, C.; Posselt, R.; Stöckli, R. The Role of the Effective Cloud Albedo for Climate Monitoring and Analysis. *Remote Sens.* **2011**, *3*, 2305–2320. [CrossRef]

39. Woodhous, I.H. *Introduction to Microwave Remote Sensing*; Taylor & Francis Group: Boca Raton, FL, USA, 2006.
40. Hill, C.; Jones, R.L. Absorption of solar radiation by water vapor in clear and cloudy skies: Implications for anomalous absorption. *J. Geophys. Res.* **2000**, *105*, 9421–9428. [CrossRef]

remote sensing

MDPI

Article

Nowcasting Surface Solar Irradiance with AMESIS via Motion Vector Fields of MSG-SEVIRI Data

Donatello Gallucci [1,*], **Filomena Romano** [1], **Angela Cersosimo** [2], **Domenico Cimini** [1,3], **Francesco Di Paola** [1], **Sabrina Gentile** [1,3], **Edoardo Geraldi** [1,4], **Salvatore Larosa** [1], **Saverio T. Nilo** [1], **Elisabetta Ricciardelli** [1] and **Mariassunta Viggiano** [1]

[1] Institute of Methodologies for Environmental Analysis, National Research Council (IMAA/CNR), 85100 Potenza, Italy; filomena.romano@imaa.cnr.it (F.R.); domenico.cimini@imaa.cnr.it (D.C.); francesco.dipaola@imaa.cnr.it (F.D.P.); sabrina.gentile@imaa.cnr.it (S.G.); edoardo.geraldi@imaa.cnr.it (E.G.); salvatore.larosa@imaa.cnr.it (S.L.); saverio.nilo@imaa.cnr.it (S.T.N.); elisabetta.ricciardelli@imaa.cnr.it (E.R.); mariassunta.viggiano@imaa.cnr.it (M.V.)

[2] School of Engineering, University of Basilicata, 85100 Potenza, Italy; angela.cersosimo@imaa.cnr.it

[3] Center of Excellence Telesensing of Environment and Model Prediction of Severe events (CETEMPS), University of L'Aquila, 67100 L'Aquila, Italy

[4] Institute for Archaeological and Monumental Heritage, National Research Council (IBAM/CNR), 85100 Potenza, Italy

* Correspondence: donatello.gallucci@imaa.cnr.it; Tel.: +39-0971-427500

Received: 3 May 2018; Accepted: 27 May 2018; Published: 29 May 2018

check for updates

Abstract: In this study, we compare different nowcasting techniques based upon the calculation of motion vector fields derived from spectral channels of Meteosat Second Generation—Spinning Enhanced Visible and InfraRed Imager (MSG-SEVIRI). The outputs of the nowcasting techniques are used as inputs to the Advanced Model for Estimation of Surface solar Irradiance from Satellite (AMESIS), for predicting surface solar irradiance up to 2 h in advance. In particular, the first part of the methodology consists in projecting the time evolution of each MSG-SEVIRI channel (for every pixel in the spatial domain) through extrapolation of a displacement vector field obtained by matching similar patterns within two successive MSG-SEVIRI data images. Different ways to implement the above method result in substantial differences in the predicted trajectory, leading to different performances depending on the time interval of interest. All the nowcasting techniques considered here systematically outperform the simple persistence method for all MSG-SEVIRI channels and for each case study used in this work; importantly, this occurs across the entire 2 h period of the forecast. In the second part of the algorithm, the predicted irradiance maps computed with AMESIS from the forecasted radiances, are shown to be in good agreement with irradiances derived from MSG measured radiances and improve on numerical weather model predictions, thus providing a feasible alternative for nowcasting surface solar radiation. The results show that the mean values for correlation, bias, and root mean square error vary across the time interval, ranging between 0.94, $-1\,\text{W/m}^2$, $61\,\text{W/m}^2$ after 15 min, and 0.73, $-18\,\text{W/m}^2$, $147\,\text{W/m}^2$ after 2 h, respectively.

Keywords: solar irradiance; nowcasting; AMESIS; MSG; SEVIRI; radiance; brightness temperature; motion vector field

1. Introduction

Short-term forecast of cloud cover still poses a challenge to the scientific community, due to the inherent complexity and non-linearity of cloud motion in atmosphere [1–7]. This topic is of relevance to many fields, including solar energy production [4,8–12], since the presence of clouds has a significant impact on the stability and energy production of solar plants, causing dangerous fluctuations and great reduction to power supply [5]. Hence it is crucial to accurately monitor and forecast the position and trajectory of cloudy systems on very short-time scales. In the scientific literature this is referred to as nowcasting, i.e., short-term forecasting up to a few hours ahead (typically 0–2 h).

Many approaches for nowcasting have been proposed, which can be generally classified as statistical or physical/deterministic. Statistical techniques rely on a training process based on past datasets, and learn how to infer the evolution of weather parameters through identification of historical patterns. The simplest of these approaches is the persistence method, for which the current status is projected as is to the future; other statistical techniques include, but are not limited to, multivariate regression and neural network [13–15]. Among the physical approaches, cloud tracking techniques are applied by processing images from geostationary satellites [5,16–18], total sky imagery [19] or using other ground sensors [20]. In particular, satellite-based methods allow for global coverage, and high quality images are nearly continuously available. Satellite imagery is currently exploited to derive Atmospheric Motion Vectors [21] (AMVs, i.e., wind vector) as well as Cloud Motion Vectors (CMVs) [22–24]; these are obtained by analysing successive satellite images searching for the same features and to extrapolate the future trajectory on the basis of the recent past motion. The feature matching among subsequent images is performed by maximizing a pre-determined measure of similarity, which is typically either a correlation coefficient or the inverse of a mean square error. This technique was first implemented in [5] using a cross-correlation coefficient, similarly to [16], showing an improvement over the simple persistence method. In [17] a cloud-tracking technique is applied to Meteosat images using a probabilistic prediction for the cloud cover; many improvements have been proposed since then [18,25–27]. In [28,29] a variational method is used to minimise an energy like objective function incorporating the relevant features of the images analysed, while in [30] a disparity vector field for each pixel is used to perform the forecast. A huge body of literature can also be found on similar approaches, referred to as optical flow methods (see pioneer works [31,32]), which are typically applied in computer vision techniques; in practice these methods consist in the estimation of the distribution of brightness patterns of an image. A comprehensive description and analysis of these methods can be found in [4], in which an hybrid approach combining block-matching methods [16,33,34] and variational optical flow is implemented.

In this work, we exploit observations from Spinning Enhanced Visible and InfraRed Imager (SEVIRI) aboard the Meteosat Second Generation (MSG) geostationary satellite. We perform nowcasting of MSG-SEVIRI IR/VIS channels up to 2 h, by deriving motion vector fields in analogy to optical flow methods and cloud motion vector techniques. The forecasted radiances are then used as inputs to the Advanced Model for Estimation of Surface solar Irradiance from Satellite (AMESIS) [35] to predict surface solar irradiance maps. We extrapolate the motion with three different methods, compute the forecasted irradiance with AMESIS (using the forecasted radiances as input), and compare against the observed irradiance, derived with AMESIS from MSG measured radiances. To evaluate the degree of accuracy, we also include in the comparison the irradiance obtained with benchmarking methods such as simple persistence and the Weather Research and Forecasting (WRF) model [36] (see Appendix A for further details on the implementation of the model in this context). We emphasize that in this work we directly forecast MSG-SEVIRI IR/VIS radiances, while previous implemented satellite-based methods [37–41] generally pre-process the satellite images (e.g., to derive cloud index or cloud optical thickness maps), which then undergo an advection process to forecast the future position of cloud patterns. To summarise, the rationale for this work is twofold: i) to investigate and compare the performances of three variants of cloud motion technique, applied directly to the radiances measured by MSG-SEVIRI channels and ii) to define a self-consistent methodology providing nowcasting (up to 2 h) of surface solar irradiance maps, based on the integration of advection techniques and AMESIS.

The paper proceeds as follows: in Section 2 we present the methodology adopted in this work, by firstly (Section 2.1) analysing and comparing three variants of the cloud motion technique (used here to advect the radiance values of MSG-SEVIRI channels) and secondly (Section 2.2) by describing the AMESIS model used here to nowcast irradiance. In Section 3 we compare the statistical performances of each nowcasting variant previously discussed, and use the most accurate one for predicting solar irradiance maps with AMESIS; the results are then evaluated against the AMESIS-based MSG measurements and the WRF-based irradiance forecast. In Section 4 we finally draw our conclusions.

2. Materials and Methods

The method we use to forecast solar irradiance at the surface exploits the image data produced by the SEVIRI radiometer aboard the MSG geostationary satellite. MSG-SEVIRI scans the Earth's full disk every 15 min, representing a valuable resource for cloud monitoring and tracking. SEVIRI is a multispectral imager, featuring twelve bands from the Visible (three channels) to the InfraRed (nine channels) regions of the electromagnetic spectrum. Here, we only consider the eleven channels with 3 km sub-point satellite resolution, thus neglecting the 1 km High Resolution Visible (HRV) channel. While the methodology could be easily extended to incorporate the HRV data, we decided not to for a fair comparison with the WRF model, that is run on a 3 km × 3 km grid.

The variables adopted in this work to seed the nowcasting process are the radiance L_e from the first three (VIS/NIR) channels and the brightness temperature T_b from the eight (IR) channels; in this work we shall henceforth refer to both L_e and T_b as 'radiances', unless specified otherwise. Every pixel within the spatial domain under analysis is therefore associated with eleven values, corresponding to three L_e and eight T_b. Based on these data at time t_0 and previous time $t_0 - \Delta t$ (with time interval $\Delta t = 15$ min), the aim is to perform a forecast of L_e and T_b for each pixel, at subsequent times $t_0 + n\Delta t$ (with integer values $1 \leq n \leq 8$), i.e., up to 2 h ahead. The predicted values for L_e and T_b are then used as inputs into AMESIS, to produce surface solar irradiance forecast every 15 min, with 2 h time horizon. We analysed day-time scenarios, selecting the following 24 days in 2017: 01-02-03-11-14-15-24-25-26-27-28 April; 01-03-10-11-12-22-31; June 04-10-25-27-28-30 May. For each of these days we provide nowcasting in two temporal intervals, i.e., during the morning (between 08:15 and 10:00 UTC) and in the afternoon (between 13:15 and 15:00 UTC), using SEVIRI data at 07:45 (This time corresponds to $t_0 - \Delta t$ in the nowcasting scheme described in Section 2.1) and 08:00 (This time corresponds to t_0 in the nowcasting scheme described in Section 2.1) UTC for the morning, and 12:45 (This time corresponds to $t_0 - \Delta t$ in the nowcasting scheme described in Section 2.1.) and 13:00 (This time corresponds to t_0 in the nowcasting scheme described in Section 2.1) UTC for the afternoon. The nowcasting of solar irradiance is therefore provided every 15 min and up to 2 hour time horizon, for the whole set of 48 case studies. We tested our methodology on a geographic area including most of the Italian peninsula, within the ranges of North latitude [37.5°, 46.1°], and East longitude [7.5°, 18.2°]. The method is applied on the entire domain consisting of 79616 pixel (311 × 256), with approximately 4 km × 5 km typical resolution in the considered latitude range. The statistical analysis however focusses on a central inner grid (257 × 202 pixel) for reasons due to boundary conditions, as explained in Section 2.1.

In this section, we describe the algorithm used throughout this work, which essentially rests upon (i) a nowcasting scheme to predict the radiances on each channel and for every pixel of the spatial domain of interest, and (ii) the AMESIS product to calculate the solar irradiance for the corresponding pixels, based on the input forecasted values of L_e and T_b. It is worth mentioning that the two parts are self-consistent: the first one (the nowcasting scheme, discussed in Section 2.1) yields forecasted values of L_e and T_b, by means of advection techniques applied to the MSG-SEVIRI channels, while the second one (the AMESIS package described in Section 2.2) is a stand-alone model designed to monitor and calculate solar irradiance maps at the surface. We emphasize that the novelty feature in this work is twofold. On one hand we describe and compare different nowcasting techniques to directly advect L_e and T_b for every MSG-SEVIRI channel, whereas previous works focus on the advection of a posteriori derived features, such as the cloud index. On the other hand, the AMESIS software is proved to work in nowcasting mode, rather than real-time monitoring (as already demonstrated in [35]). We should further stress that while nowcasting of irradiance is usually performed by projecting the irradiance maps forward in time, in this work the irradiance forecast is calculated on the basis of the predicted values of L_e and T_b.

2.1. Part I: The Nowcasting Process

The general approach to extrapolate the future motion of clouds, based upon matching the same feature between two (or more) consecutive data images, can be implemented in several ways, depending upon the required level of optimisation and the selected correlation criterion among the features. To derive displacement vectors to forward the trajectory of clouds in time and space, we adopt a technique that is similar in spirit to an optical flow method, however implementing a few refinements, which are discussed below. The core of the proposed nowcasting technique consists in extrapolating the time and spatial evolution of radiance maps (for each of the eleven channels taken into account) by projecting the motion observed between the actual time (t_0) and the previous time step ($t_0 - \Delta t$).

Let us first describe the method to find the displacement between times $t_0 - \Delta t$ and t_0. For each channel we initially select a squared 'target' area in the data image at time $t_0 - \Delta t$, and a wider squared 'search' area (surrounding symmetrically the target area region) in the image at time t_0; the sizes of target and search areas are defined from the outset. To find the displacement of the pixel centered within the target area, between times $t_0 - \Delta t$ and t_0, we follow a variational procedure. This consists in finding—inside the search area at time t_0—the target area that minimises the Mean Square Error (MSE) against the target area at time $t_0 - \Delta t$. This process then yields the displacement vector from the pixel located in the center of the target area at time $t_0 - \Delta t$, to the central pixel of the matching target area at time t_0 (see Figure 1). Finally, by scanning over the whole spatial domain, and iterating this process, we derive a displacement field for all pixels. Importantly, the size of the target and search areas are dependent upon each other; we found the optimum values to be 5 and 19 pixels for each side of the square-shaped target and search areas, respectively. The chosen target size is a good compromise for capturing large structures motion as well as small details dynamics; besides, the search area size allows for those features to travel at a reasonable maximum speed [42]. Please note that in our approach there is no distinction between clear and cloudy pixels (i.e., no cloud mask is used in the process); the algorithm relies solely on the radiance maps measured by SEVIRI, unlike previous works that deal with specific a posteriori derived parameters (e.g., cloud optical thickness or cloud index).

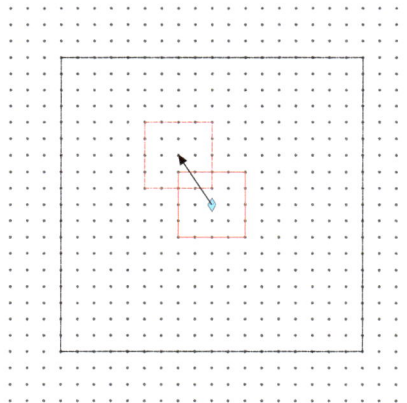

Figure 1. Graphical example of the iterative minimisation process to find the displacement vector field between time t_0 and the previous time step $t_0 - \Delta t$. The Mean Square Error is computed between the radiances within the central target area at time $t_0 - \Delta t$ (solid line, 5 × 5 pix) and each similar-shaped target area at time t_0 within the surrounding search area (dot-dashed line, 19 × 19 pix). In this example the matching target area at time t_0 (dot-dashed line, 5 × 5 pix) is found to minimise the MSE; the displacement vector then connects the central pixels within the corresponding target areas.

A few technical issues may arise when applying the technique as above. Firstly, it is unlikely that the variational procedure will yield a one-to-one correspondence, as it may occur that the same target area at time t_0 matches several target areas at time $t_0 - \Delta t$. As a consequence, some pixels may be allocated with several radiance values whereas others remain 'empty'; this non-injectivity issue is circumvented by using the mean of the several allocated radiance values in the first case, while reassigning the corresponding value from the previous time step and averaging over the neighbour pixels in the second case, respectively. While this may result in a somewhat crude approximation, we typically find that only a low percentage (less than 5%) of the pixels are affected by the above degeneracy. Secondly, cloudy systems may enter the considered spatial domain from outside the grid borders; to overcome this issue we define and focus on a smaller inner grid (which is the area of interest) within the analysed spatial domain, so that cloudy systems along the outer edges are usually buffered by the borders surrounding the inner area of interest.

The first step described so far strongly depends on the choice of the size of target and research areas, as well as the type of similarity measurement criterion for computing the correlation (here defined as the MSE). The second step features different ways to extrapolate the subsequent motion for every pixel, each of these leading to a different outcome. We now describe how these differences arise. At the end of the variational process discussed above, each pixel may be actually associated with a *in* and a *out* displacement vector, referring to the incoming direction of the radiance value entering the pixel and the outgoing direction of the radiance value leaving the pixel, respectively (Figure 2a). In case these two displacements differ (as in the example shown in Figure 2a), the forwarded trajectory becomes dependent on the type of displacement one decides to carry on in order to extrapolate the future motion. To take this subtlety into account, we have investigated and compared three types of forecast, here referred to as *out*, *in* and *hybrid* (Figure 2b).

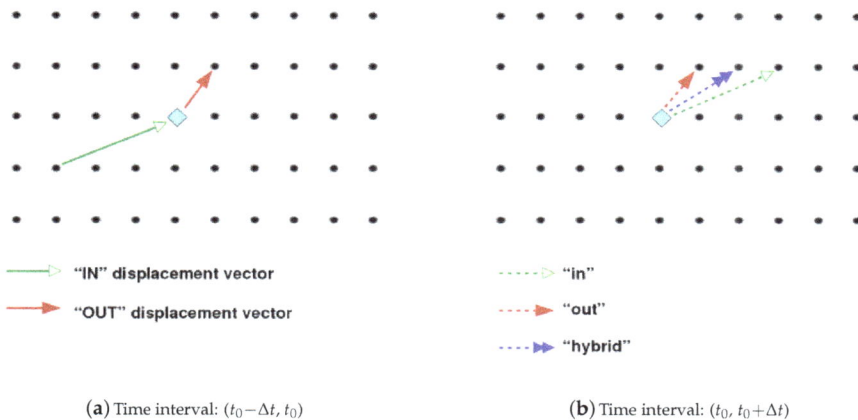

⟶▷ "IN" displacement vector	⋯▷ "in"
⟶ "OUT" displacement vector	⋯→ "out"
	⋯▶▶ "hybrid"

(**a**) Time interval: $(t_0 - \Delta t, t_0)$ (**b**) Time interval: $(t_0, t_0 + \Delta t)$

Figure 2. Left panel (**a**) Schematic of a possible outcome of the variational procedure, between time $t_0 - \Delta t$ and t_0; in this example the radiance value relative to the diamond shaped pixel at time $t_0 - \Delta t$ is assigned to the pixel pointed by the *out* displacement vector (filled red arrow) at time t_0, and replaced by the incoming radiance along the *in* displacement vector (hollow green arrow). Right Panel (**b**) the radiance value allocated to the diamond shaped pixel in the previous time step (**a**) can be displaced in the following time step $(t_0 + \Delta t)$ in three possible ways, i.e., along the *out* (filled red arrow), *in* (hollow dashed arrow) or *hybrid* (filled two-headed arrow) direction. See main text for details.

In the *out* type, we can think of the grid as a fixed frame of reference and for each pixel consider only the outgoing (*out*) displacement. In this framework, which is closely related to the Atmospheric Motion Vector approach, each location can be thought as associated with a definite intensity and direction of a local wind, and consequently each radiance value is displaced accordingly. This approach is implemented by iterating the following equation, where $\Delta y_{(i,j)}^{(out)}$ and $\Delta x_{(i,j)}^{(out)}$ are the components (over the latitude and longitude directions, respectively) of the outgoing (*out*) displacements of the radiance from the pixel location (i, j):

$$T_b(i + \Delta y_{(i,j)}^{(out)}, j + \Delta x_{(i,j)}^{(out)}, t_0 + n\Delta t) = T_b(i, j, t_0 + (n-1)\Delta t) \tag{1}$$

Please note that the indices span the ranges $1 \leq i \leq 256$ and $1 \leq j \leq 311$ within the spatial domain of interest. In the *in* type, we deal with a moving frame of reference, as we closely follow the path of each radiance value starting from the initial incoming (*in*) displacement. In this framework, each radiance value can be thought of as a particle featuring its own motion, and the iteration describing its evolution is as follows (where $\Delta y_{(i,j)}^{(in)}$ and $\Delta x_{(i,j)}^{(in)}$ are obtained by averaging over the several *in* displacements found in the variational procedure, and refer to the incoming (*in*) displacements):

$$T_b(i + n\Delta y_{(i,j)}^{(in)}, j + n\Delta x_{(i,j)}^{(in)}, t_0 + n\Delta t) = T_b(i + (n-1)\Delta y_{(i,j)}^{(in)}, j + (n-1)\Delta x_{(i,j)}^{(in)}, t_0 + (n-1)\Delta t) \tag{2}$$

The third type, *hybrid*, is a linear combination of the *out* and *in* ones, based on a displacement for each pixel obtained as the mean of the incoming and outgoing displacements (i.e., $\Delta y_{(i,j)}^{hybr} = [\Delta y_{(i,j)}^{(in)} + \Delta y_{(i,j)}^{(out)}]/2$ and $\Delta x_{(i,j)}^{hybr} = [\Delta x_{(i,j)}^{(in)} + \Delta x_{(i,j)}^{(out)}]/2$), yielding the following equation:

$$T_b(i + \Delta y_{(i,j)}^{(hybr)}, j + \Delta x_{(i,j)}^{(hybr)}, t_0 + n\Delta t) = T_b(i, j, t_0 + (n-1)\Delta t) \tag{3}$$

It should be noticed that in all former equations the brightness temperature T_b is to be replaced by the radiance L_e when considering the VIS/NIR bands. For each variant presented above, we also consider its smoothed counterpart: this is derived by averaging (every time step) the radiance value in each pixel with its surrounding eight neighbours. In Section 3 we demonstrate that the above three variants (together with their smoothed counterparts) lead to an overall good agreement with SEVIRI measurements, within the first two hours of forecast horizon.

2.2. Part II: AMESIS

The second part of the algorithm discussed in this work is based on AMESIS; in this section we therefore provide a brief overview of this model (for further details see [35]). As discussed in Section 2.1, the first part of the solar irradiance nowcasting algorithm yields the predicted values for radiance on each of the eleven SEVIRI spectral channels and for every pixel within the spatial domain under analysis; the second part, entirely based on AMESIS, takes as inputs the predicted L_e and T_b, and computes estimates for surface solar irradiance values at each pixel location. AMESIS was proved to work for surface solar irradiance monitoring purposes in the region 33–60 degrees North and 11–30 degrees East, for every season, different sun zenith angles, radiance ranges and albedo. AMESIS exploits MSG-SEVIRI data, ingesting near-real-time 15 min resolution SEVIRI spectral images. In this work, the use of AMESIS has been extended to nowcasting, i.e., by ingesting predicted values of SEVIRI radiance to provide surface solar irradiance forecast within 2 h time horizon. AMESIS starts with the pixel classification as clear, cloudy, partially cloudy, or affected by aerosol presence, exploiting Cloud Mask Coupling of Statistical and Physical methods (C-MACSP) algorithm [43]. Except for pixels detected as clear, the model retrieves the microphysical optical parameters for clouds or aerosol. Subsequently the model retrieves the surface solar irradiance on the basis of look-up tables that are periodically updated. Therefore AMESIS incorporates the effects due to aerosol, overcast and partially cloudy coverage; the retrieval of surface temperature, total integrated water vapour, cloud and aerosol microphysical parameters is fulfilled by using VIS and IR SEVIRI channels, whereas surface solar irradiance can be retrieved either with the low-resolution VIS channels or through the HRV, depending on the desired resolution. As anticipated earlier, the SEVIRI-HRV channel is not used here, and thus

AMESIS is run in low-resolution mode, allowing direct comparison with the WRF model output, also implemented on a 3 km × 3 km grid.

3. Results and Discussion

The treatment discussed in Section 2.1 is based solely on simple advection of the MSG-SEVIRI channels, thus neglecting convective processes as well as dissipative mechanisms. For this reason the quality of the algorithm has been evaluated by selecting a dataset mainly featuring broken clouds, avoiding unstable atmospheric conditions possibly causing convective initiation. We emphasize that the selected conditions (partially cloudy) heavily affect the stability of photovoltaic systems, due to rapid changes in the solar irradiance intensity.

The output of the first part of the algorithm is analysed by comparing the six variants discussed in Section 2.1, against the MSG-SEVIRI measured values; we also include in the comparison the simple persistence model (here implemented by keeping the t_0 MSG-SEVIRI channels radiance values across the entire time interval). Figure 3 shows the statistical analysis by means of the Root Mean Square Error (RMSE), Mean Bias Error (MBE) and Correlation for the two visible channels (Ch. 1, 2), the near-infrared (Ch. 4) and two infrared channels (Ch. 6, 10). In this work the BIAS is computed as the mean of the differences between the forecast and MSG measured radiances for every pixel; the MBE is obtained as the average of the BIAS over the whole sample of 48 case studies. It is worth noticing that these channels are the most relevant ones for the calculation of the solar irradiance through AMESIS. The simple persistence model is outperformed (in terms of both RMSE and Correlation) by all the other methods, at all times within the forecast interval. The improvement relative to the persistence model is negligible at the first nowcasting time step (i.e., after 15 min) and increases with time. Figure 3 shows that the *hybrid smooth* is the most performing among the variants in terms of RMSE and Correlation; for all channels the correlation ranges between the maximum value of 0.97 and the minimum value of 0.7 within the 2 h of prediction. However the *hybrid smooth* reveals a relatively higher bias compared to the other variants, while the *in* variant minimises the BIAS for the visible bands (Ch. 1, 2). We point out that the analysis is based on the average over the whole dataset (48 case studies) considered in this work, and concerns the estimate of radiances in each pixel and for every SEVIRI channel. Finally, the first stage of the algorithm (i.e., the nowcasting process) revealed the following: (i) all of the variants used here show very good agreement with the MSG-SEVIRI measured values of radiances for each channel (e.g., correlation is greater than 0.7 even after 2 h of nowcasting); (ii) all the methods clearly show very similar trends, improving on the predictions provided by the simple persistence model; (iii) relative to the visible bands (Ch. 1–2) the *hybrid smooth* and the *in* approaches are the most performing variants in terms of RMSE/Correlation and Bias, respectively. For the reasons above, in the second part of the algorithm we use both *in* and *hybrid smooth* based outputs, as inputs in AMESIS for the estimation of surface solar irradiance.

To characterise the second part of the algorithm (described in Section 2.2) and evaluate its performance, we have selected three particular case studies: these feature the lowest (10 June 2017, afternoon), intermediate (25 June 2017, morning) and highest (28 June 2017, morning) RMSE value for the forecasted radiances of the SEVIRI visible channels, according to the *hybrid smooth* performance in the nowcasting scheme analysis. We therefore expect a direct relation with the AMESIS solar irradiance output, such that the aforementioned case studies will prove to be the best, intermediate, and worst irradiance forecast within the dataset, respectively. In Figure 4 we report the statistical analysis for the above case studies, comparing the irradiance predicted by means of the *hybrid smooth* and *in* nowcasting approaches against the benchmark models of simple persistence and WRF. The statistical measures are evaluated with respect to the irradiance calculated through AMESIS based on the MSG-SEVIRI measured radiances. We should also point out that solar irradiance predictions based on *hybrid smooth*, *in* and simple persistence approaches are obtained through AMESIS by using the corresponding forecasted radiances values as inputs, whereas the WRF solar irradiance is computed within the code using the model atmospheric variables to solve a parametrized radiative transfer (see Appendix A). We include the WRF prediction in the comparison analysis in order to quantify the potential improvement by using the nowcasting approach within the first two hours of forecast, instead of the WRF prediction. The WRF irradiance predictions are available on a hourly basis, at specific times, i.e., at 14:00 UTC and 15:00 UTC for the afternoon case on the 10 June 2017 and 09:00 UTC and 10:00 UTC for the morning cases on the 25 June 2017 and 28 June 2017. We compared the above against the time collocated nowcasting-based AMESIS predictions; the times selected for the comparison correspond to the 3rd (08:45 or 13:45 UTC) and 7th (09:45 or 14:45 UTC) temporal step of the *hybrid smooth/in* nowcasting process, i.e., after 45 and 105 min, respectively. The results shown in Figure 4 reveal the following: (i) the predictions based on the *hybrid smooth* method are more accurate then the predictions based on the *in* method; (ii) the irradiance forecasted with the benchmarks models feature a systematically higher RMSE and lower Correlation, at all times and in each case study; (iii) the correlation relative to the nowcasting methods proposed in this work is relatively stronger across the entire time interval; in the worst case scenario (28 June 2017) it reaches the lowest value of 0.52 only after two hours of forecast, whereas it ranges between 0.85–1 for the two other cases, hence demonstrating the robustness of the procedure. By combining the statistical scores of the *hybrid smooth* method, obtained for the three cases analysed in Figure 4, we find that the mean value ranges between 0.94 and 0.73 for correlation, between $-1 \, \text{W/m}^2$ and $-18 \, \text{W/m}^2$ for bias and between $61 \, \text{W/m}^2$ and $147 \, \text{W/m}^2$ for the rmse (values are calculated after 15 min and 2 h, respectively). In order to give a qualitative overview of the results, the corresponding irradiance maps are reported in Figure 5, where we compare the AMESIS-based *hybrid smooth* predictions (proved to be the best nowcasting method in this work) and the WRF calculated output, against the AMESIS-based SEVIRI measurements, for the three case studies analysed (10 June 2017; 25 June 2017; 28 June 2017). The results show that irradiance maps obtained through nowcasting (*hybrid smooth*) largely reproduce the SEVIRI observed results, however featuring smooth low-irradiance regions, as expected. The WRF results tend to underestimate the cloudy regions instead, and this occurs for all the case studies examined. We should further stress that the purpose of Figure 5 is only to provide a qualitative comparison of the irradiance maps obtained with nowcast, WRF and SEVIRI measurements, for the case studies analysed. One can clearly infer the distribution of cloudy systems (regions of relatively low irradiance) in the maps, and qualitatively evaluate whether the simulated patterns match the SEVIRI observations.

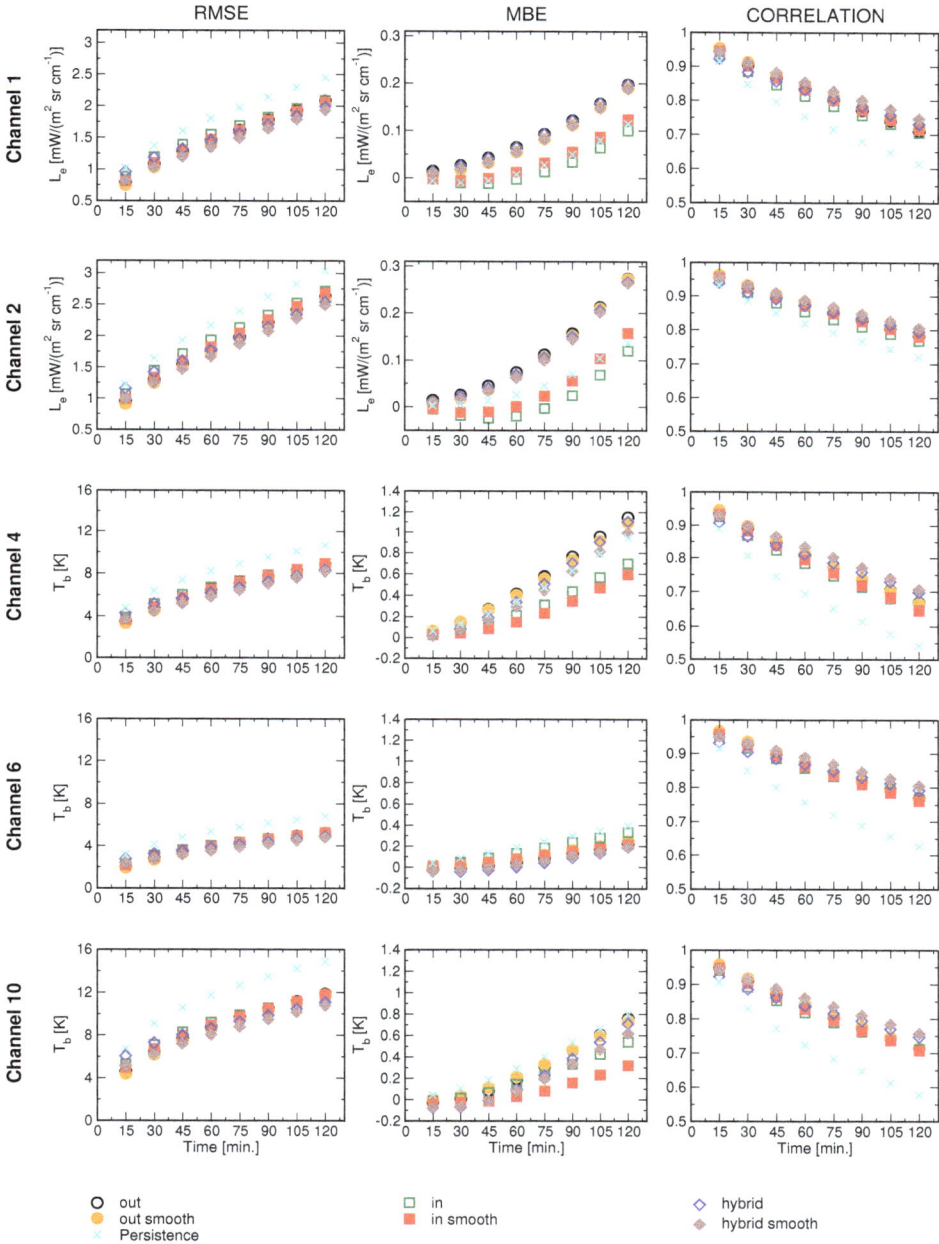

Figure 3. Statistical scores (2 h time horizon, with 15 min time step) for the SEVIRI channels 1, 2, 4, 6, 10 (from top to bottom) of the *out*, *in* and *hybrid* approaches (hollow symbols: black circles, green squares and brown diamonds respectively) and their smoothed counterparts (filled symbols), and the simple Persistence method (turquoise crosses). Root Mean Square Error (left column), Mean Bias Error (centre column) and Correlation (right column) are shown. The statistical scores represent the average over the whole sample of 48 case studies.

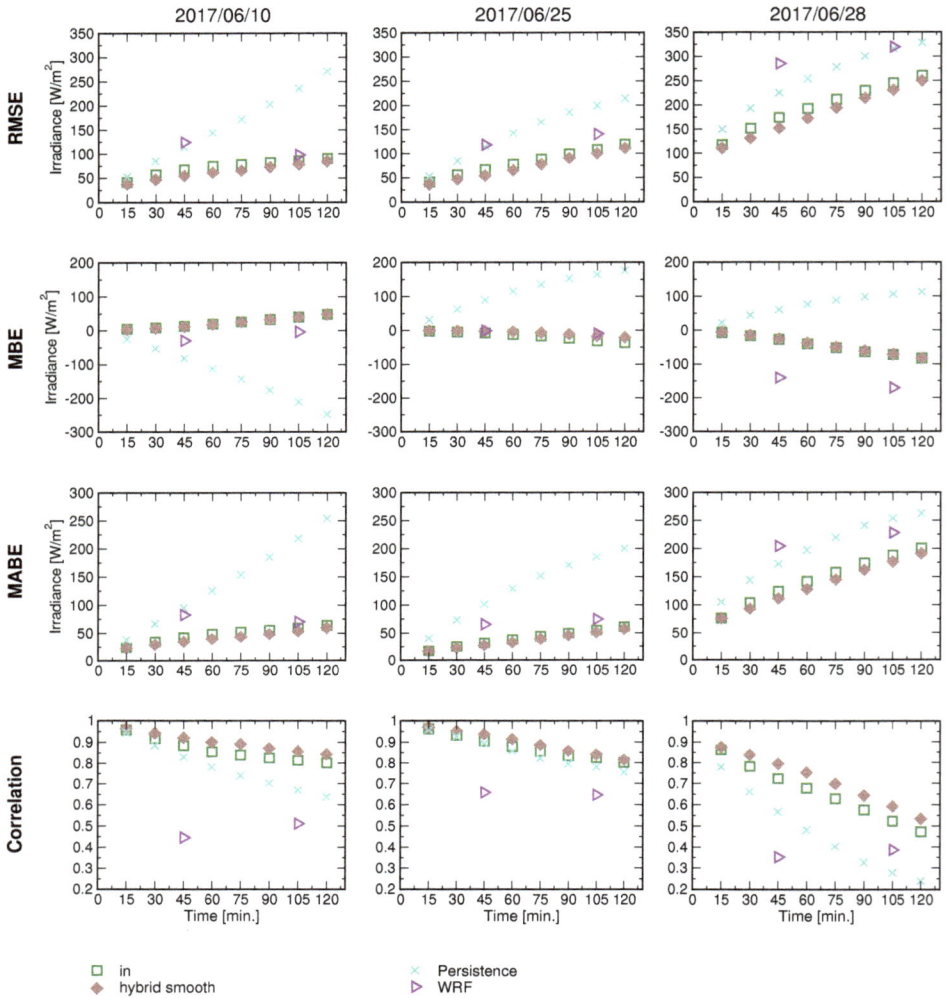

Figure 4. Statistical scores for irradiance obtained with the *hybrid smooth* (brown diamonds) and *in* (green squares) methods, the simple persistence (turquoise crosses) and the WRF (violet triangles) models, relative to three case studies (10 June 2017 left coloumn; 25 June 2017 central coloumn; 28 June 2017 right coloumn). The time origin refer to 13:00 UTC on the 10 June 2017, and to 08:00 UTC on the 25 June 2017 and 28 June 2017. The scores are based on the computation of the Root Mean Square Error, Mean Bias Error, Mean Absolute Bias Error and Correlation (first, second, third and fourth row respectively) for the *hybrid smooth* and *in* nowcasting methods, the Persistence and WRF output irradiances, each of these evaluated with respect to the irradiance calculated through AMESIS based on the MSG-SEVIRI measured radiances. Corresponding irradiance maps, relative to the 3rd (after 45 min) and 7th (after 105 min) time step, are reported in Figure 5.

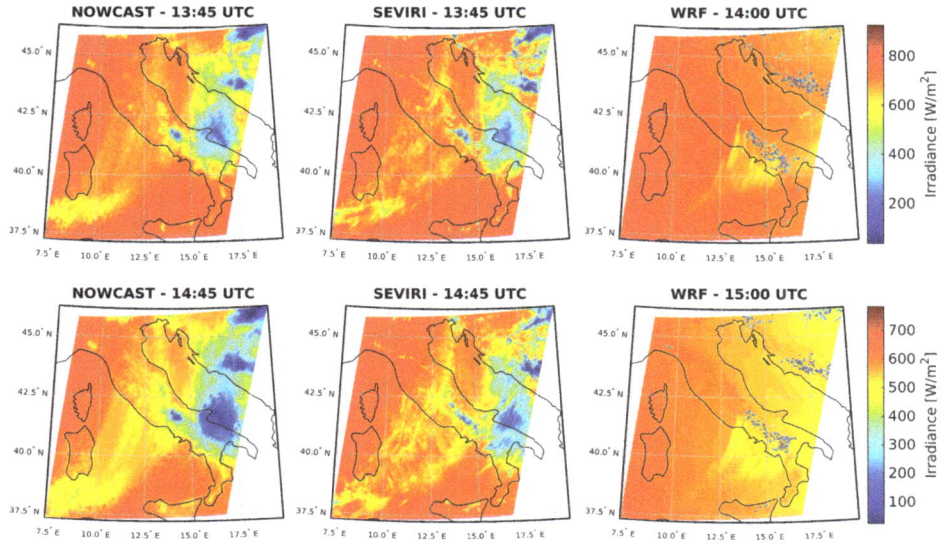

(a) 10 June 2017

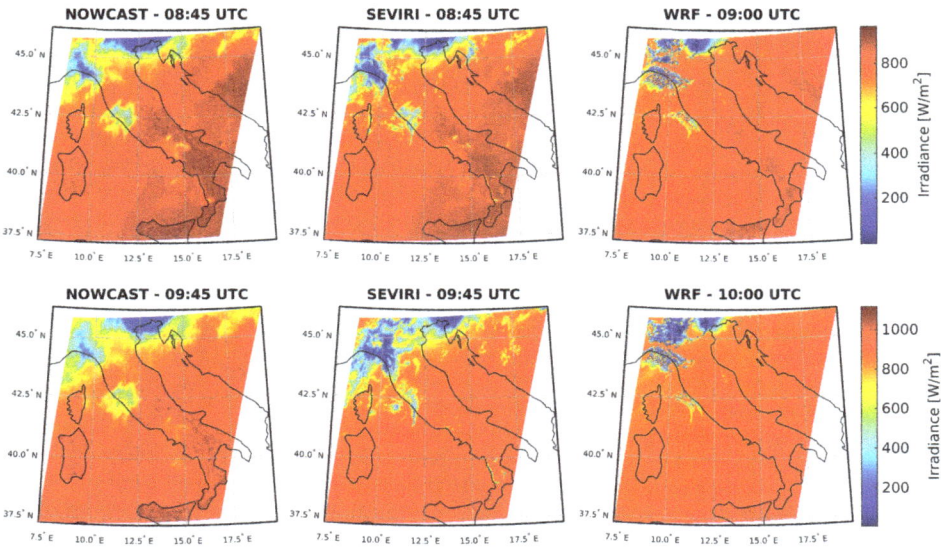

(b) 25 June 2017

Figure 5. *Cont.*

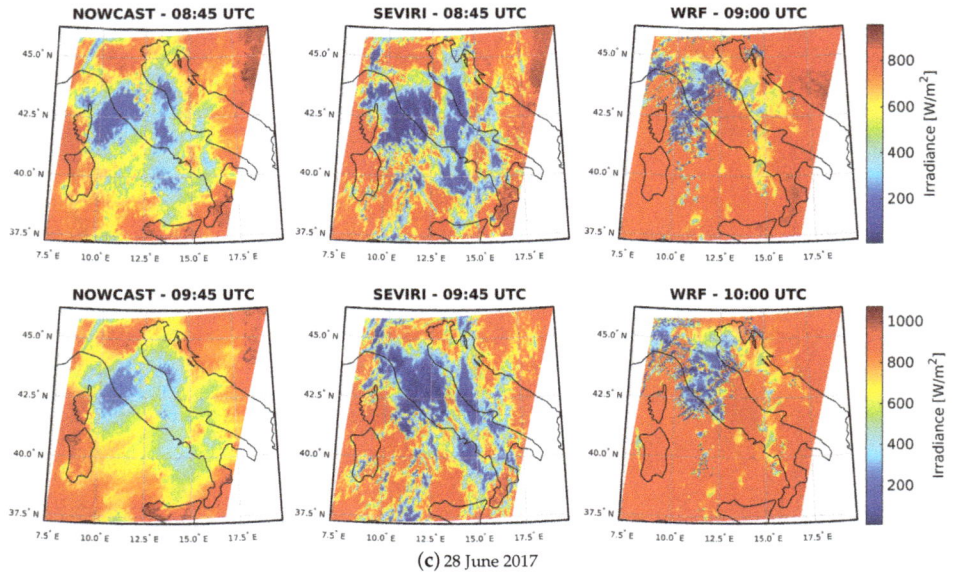

Figure 5. Comparison of irradiance maps based on (i) the estimated radiances via the *hybrid smooth* method (left column), (ii) the measured MSG-SEVIRI radiances (intermediate column) and (iii) WRF (right column), for the case studies on 10 June 2017 at 14:00 UTC ((**a**), first row) and 15:00 UTC ((**a**), second row), on 25 June 2017 at 09:00 UTC ((**b**), first row) and 10:00 UTC ((**b**), second row) and finally on 28 June 2017 at 09:00 UTC ((**c**), first row) and 10:00 UTC ((**c**), second row). While irradiance based on the *hybrid smooth* estimated radiances and MSG-SEVIRI measured radiances is calculated through AMESIS (see main text for details), the WRF output is generated by solving a simplified radiative transfer [44]. A cylindrical projection (Plate Carrée) has been used in this figure.

We now further investigate the statistics of the irradiance output derived with the most performing nowcasting method, namely the *hybrid smooth*; this is compared against the AMESIS-based irradiance derived with MSG-SEVIRI measured radiances, by evaluating scatter plots and cumulative frequency curves. In Figure 6 we show the scatter plots for the irradiances obtained with the nowcasting methodology, for the same case studies and times shown in Figure 5. We notice that nowcast predictions follow a regular distribution and feature a relatively low spread compared to the SEVIRI output, at the 3rd time step (i.e., after 45 min) of the nowcasting process. This is further supported by the cumulative frequency curves, also shown in Figure 6, which demonstrate a very good agreement between predictions and MSG measured values (except for the worst case scenario on 28 June 2017). As expected, the agreement and correlation between measured and simulated data tend to reduce after 105 min, at the 7th time step, as also shown in Figure 6. Based on the above outcome, the nowcasting approach described here may be used to complement the WRF solar irradiance prediction, especially within the first two hours forecast time. In such a way, a self-consistent approach providing an efficient monitoring, nowcasting up to 2 h and forecasting beyond 2 h of solar irradiance maps, is derived through the integration of AMESIS, cloud motion techniques, and WRF model.

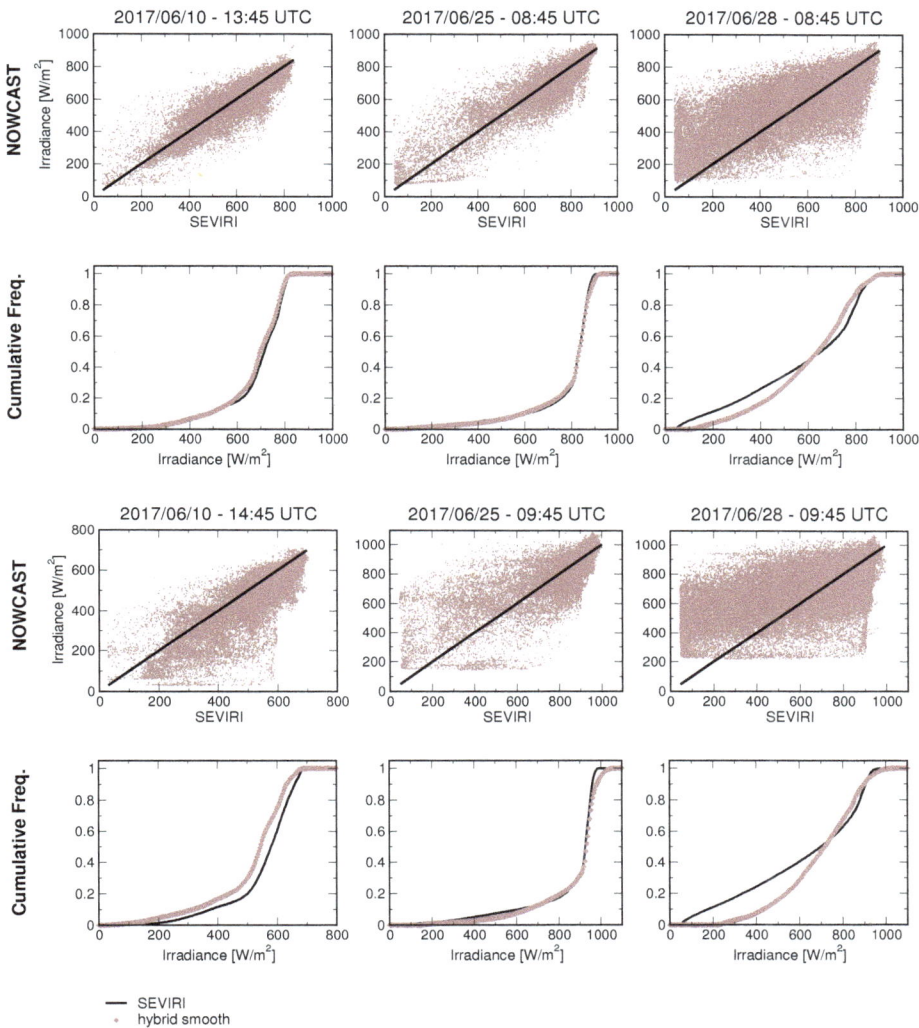

Figure 6. Scatter plots (first and third row) and cumulative frequency curves (second and fourth row) for nowcast irradiance predictions (brown diamonds) against AMESIS-based irradiance using MSG-SEVIRI measured radiances (black solid line). Prediction times refer to the 3rd time step (two top rows) and the 7th time step (two bottom rows) of the nowcasting process (as also shown in Figure 5), relative to each case study analysed (i.e., 10 June 2017, left coloumn; 25 June 2017, central coloumn; 28 June 2017, right coloumn).

4. Conclusions

In this work we present an algorithm for providing accurate short term forecast (0–2 h) of surface solar irradiance, based solely on the (i) advection of MSG-SEVIRI channels and (ii) Advanced Model for Estimation of Surface solar Irradiance from Satellite (AMESIS). The first part of the algorithm consists in an advecting technique (in analogy to optical flow methods) applied directly to the radiances and brightness temperatures of MSG-SEVIRI channels; we investigate different types of extrapolation (here referred to as *out, in, hybrid*), and select the best approach to feed AMESIS in the second part of the algorithm. For evaluation, we selected a set of 48 case studies, divided equally between day and afternoon, and featuring mainly broken clouds scenarios. The nowcasting of surface solar irradiance is provided every 15 min across the time intervals [08:15, 10:00] UTC for the morning cases and [13:15, 15:00] UTC for the afternoon ones.

The statistical analysis is based on the calculation of the RMSE, BIAS, MABE, Correlation, as well as scatter plots and cumulative frequency curves; each of these is evaluated with respect to MSG measured values and averaged over all samples. We find similar overall trend across the entire forecast time range for the nowcasting methods proposed here. However the *hybrid smooth* approach stands out for relatively higher (lower) values of Correlation (RMSE). Nonetheless all variants implemented show very good agreement with the outcome based on MSG-SEVIRI measurements, and improve on the results derived from benchmark models such as simple persistence and WRF models. This has been verified on the basis of the performance exhibited in the best (10 June 2017), intermediate (25 June 2017) and worst (28 June 2017) case scenarios relative to the RMSE. The correlation for the above cases varies on average between 0.94 after 15 min and 0.73 after 2 h, for the *hybrid smooth* nowcasting method. The corresponding range for the simple persistence method is [0.89, 0.54], while for the WRF model is on average 0.5 at all times examined.

The nowcasting approach described here is therefore proved to be a valid approach for short term (0–2 h) forecasting of solar irradiance. The treatment discussed however is based solely on advection and extrapolation of MSG-SEVIRI channels, thus neglecting convective processes and dissipative mechanisms. Further work foresees to incorporate the modeling of cloud growth and dissipation mechanisms, as well as of convection phenomena, useful for the forecasting of storm events.

Author Contributions: D.G., F.R., F.D.P., D.C., S.G. and E.G. designed the research, wrote the paper and contributed to evaluation process. A.C., S.L., S.T.N., E.R. and M.V. contributed to data processing, analysis and evaluation process. All the co-authors helped to revise the manuscript.

Funding: This work has been financed by the Italian Ministry of Economic Development (MISE) in the framework of the SolarCloud project, contract No. B01/0771/04/X24.

Conflicts of Interest: The authors declare no conflict of interest. The founding sponsors had no role in the design of the study; in the collection, analyses, or interpretation of data; in the writing of the manuscript, and in the decision to publish the results.

Appendix A. The Weather Research and Forecast Model

The Advanced Research WRF (ARW) modeling system is a numerical weather prediction model developed by the joint effort of different research institutes coordinated by the National Center for Atmospheric Research (NCAR, http://www.wrf-model.org). It has been designed for flexible purposes from research to operational issues. It is suitable over a wide range of scales for the horizontal resolution from climate studies (thousands of kilometers) to Large Eddy Simulations (a few meters). WRF is composed of several initialization programs for idealized and real-data simulations; it solves the fully compressible non-hydrostatic equations and uses a mass-based terrain-following coordinate with a Vertical grid-spacing varying with height. For the time integration it uses a 2nd- or 3rd-order Runge-Kutta scheme with a time-split small step for acoustic and gravity-wave modes. The horizontal grid is staggered Arakawa-C. The model can perform simulations with a one way or two-way nesting with multiple domains. WRF provided full physics options for land-surface, planetary boundary layer, atmospheric and surface radiation, microphysics and cumulus convection [36]. The WRF 3.8.1 version SOLAR has been used for this study. The outputs are from the operational chain run at CNR-IMAA since November 2016, developed in the framework of the Solar Cloud project (funded by the Italian Ministry of Economic Development) to provide detailed forecasts of solar irradiance variables to solar energy industry operators. The adopted configuration is characterized by two-way nested domains: the mother domain covers the whole Mediterranean basin with 9 km spatial resolution, whereas the inner domain covers the whole Italian peninsula with 3 km spatial resolution. The simulation is initialized using GFS forecast at 0.25 degree, upgraded every 6 hours with 35 vertical levels. The hydrometeors are calculated using (i) the Aerosol–aware Thompson Scheme [45], (ii) the Yonsei non-local-K scheme with explicit entrainment layer [46] for the Planetary Boundary Layer (PBL) parameterization, and (iii) the RRTMG [47] for the longwave and the shortwave radiation scheme. The Kain-Fritsch [48] cumulus parameterization is used only for the coarser grid and the precipitation is explicitly computed (no cumulus scheme) for the inner domain. The Rapid Radiative Transfer Model for climate and weather models (RRTMG) performs the radiative calculations given clear or cloudy sky conditions. This parameterization neglects partial cloudiness as nearly every grid box is considered completely cloudy if the microphysics parameterization contains cloud hydrometeors or is considered cloud-free if no hydrometeors are predicted. This scheme uses look-up tables to compute the absorbed, emitted, reflected, and transmitted components of broadband solar and longwave radiation within specific intervals (bins) of wavelength.

References

1. Menzel, W.P. Cloud Tracking with Satellite Imagery: From the Pioneering Work of Ted Fujita to the Present. *Bull. Am. Meteorol. Soc.* **2001**, *82*, 33–47. [CrossRef]
2. Batlles, F.J.; Alonso, J.; López, G. Cloud Cover Forecasting from METEOSAT Data. *Energy Procedia* **2014**, *57*, 1317–1326. [CrossRef]
3. Arbizu-Barrena, C.; Ruiz-Arias, J.A.; Rodríguez-Benítez, F.J.; Pozo-Vázquez, D.; Tovar-Pescador, J. Short-term solar radiation forecasting by advecting and diffusing MSG cloud index. *Sol. Energy* **2017**, *155*, 1092–1103. [CrossRef]
4. Peng, Z.; Yu, D.; Huang, D.; Heiser, J.; Kalb, P. A hybrid approach to estimate the complex motions of clouds in sky images. *Sol. Energy* **2016**, *138*, 10–25. [CrossRef]
5. Leese, J.A.; Novak, C.S.; Clark, B.B. An Automated Technique for Obtaining Cloud Motion from Geosynchronous Satellite Data Using Cross Correlation. *J. Appl. Meteorol.* **1971**, *10*, 118–132. [CrossRef]
6. Bedka, K.M.; Velden, C.S.; Petersen, R.A.; Feltz, W.F.; Mecikalski, J.R. Comparisons of Satellite-Derived Atmospheric Motion Vectors, Rawinsondes, and NOAA Wind Profiler Observations. *J. Appl. Meteorol. Climatol.* **2009**, *48*, 1542–1561.
7. Bedka, K.M.; Mecikalski, J.R. Application of Satellite-Derived Atmospheric Motion Vectors for Estimating Mesoscale Flows. *J. Appl. Meteorol.* **2005**, *44*, 1761–1772, doi:10.1175/JAM2264.1. [CrossRef]
8. Heinemann, D.; Lorenz, E.; Girodo, M. Forecasting of Solar Radiation. 2006. Available online: http://citeseerx.ist.psu.edu/viewdoc/download?doi=10.1.1.526.2530&rep=rep1&type=pdf (accessed on 25 May 2018).
9. Cebecauer, T.; Suri, M.; Perez, R. High Performance MSG Satellite Model for Operational Solar Energy Applications. *ASES Ann. Conf.* **2010**. Available online: http://proceedings.ases.org/wp-content/uploads/2014/02/2010-086small.pdf (accessed on 25 May 2018).
10. Perez, R.; Kivalov, S.; Schlemmer, J.; Hemker, K.; Zelenka, A. Improving the Performance of Satellite-To-Irradiance Models Using the Satellite's Infrared Sensors. In Proceedings of the 39th ASES National Solar Conference 2010, SOLAR 2010; Volume 1. Available online: http://proceedings.ases.org/wp-content/uploads/2014/02/2010-038small.pdf (accessed on 25 May 2018).
11. Perez, R.; Kankiewicz, A.; Schlemmer, J.; Hemker, K.; Kivalov, S. A New Operational Solar Resource Forecast Model Service for PV Fleet Simulation. 2014. pp. 0069–0074. Available online: http://www.asrc.albany.edu/people/faculty/perez/2014/fcst.pdf (accessed on 25 May 2018).
12. Pelland, S.; Remund, J.; Kleissl, J.; Oozeki, T.; De Brabandere, K. Photovoltaic and Solar Forecasting: State of the Art. 2013. Available online: https://www.nachhaltigwirtschaften.at/resources/iea_pdf/reports/iea_pvps_task14_report_2013_photovoltaic_and_solar_forecasting_state_of_the_art.pdf (accessed on 25 May 2018).
13. Mellit, A. Artificial Intelligence technique for modelling and forecasting of solar radiation data: A review. *Int. J. Artif. Intell. Soft Comput.* **2008**, *1*, 52–76. [CrossRef]
14. Pedro, H.T.; Coimbra, C.F. Assessment of forecasting techniques for solar power production with no exogenous inputs. *Sol. Energy* **2012**, *86*, 2017–2028. [CrossRef]
15. Marquez R, Gueorguiev VG, C.C. Forecasting of Global Horizontal Irradiance Using Sky Cover Indices. *ASME. J. Sol. Energy Eng.* **2012**, *135*. Available online: https://pdfs.semanticscholar.org/12b0/b506a1b3d5eedcc9a749512b71026669ac29.pdf (accessed on 25 May 2018). [CrossRef]

16. Hamill, T.M.; Nehrkorn, T. A Short-Term Cloud Forecast Scheme Using Cross Correlations. *Weather Forecast.* **1993**, *8*, 401–411. Available online: https://www.esrl.noaa.gov/psd/people/tom.hamill/crosscorr_cloud.pdf (accessed on 25 May 2018). [CrossRef]

17. Hammer, A.; Heinemann, D.; Lorenz, E.; Lückehe, B. Short-term forecasting of solar radiation: A statistical approach using satellite data. *Sol. Energy* **1999**, *67*, 139–150. [CrossRef]

18. Lorenz, E.; Hammer, A.; Heinemann, D. Short term forecasting of solar radiation based on satellite data. In Proceedings of the EUROSUN2004 (ISES Europe Solar Congress), Freiburg, Germany, 20–23 June 2004.

19. Cheng, H.Y. Cloud tracking using clusters of feature points for accurate solar irradiance nowcasting. *Renew. Energy* **2017**, *104*, 281–289. [CrossRef]

20. Bosch, J.L.; Zheng, Y.; Kleissl, J. Deriving Cloud Velocity From an Array of Solar Radiation Measurements. *Energy Sustain.* **2012**, 1059–1065, doi:10.1115/ES2012-91369. [CrossRef]

21. Velden, C.S.; Olander, T.L.; Wanzong, S. The Impact of Multispectral GOES-8 Wind Information on Atlantic Tropical Cyclone Track Forecasts in 1995. Part I: Dataset Methodology, Description, and Case Analysis. *Mon. Weather Rev.* **1998**, *126*, 1202–1218. [CrossRef]

22. Guillot, E.M.; Haar, T.H.V.; Forsythe, J.M.; Fletcher, S.J. Evaluating Satellite-Based Cloud Persistence and Displacement Nowcasting Techniques over Complex Terrain. *Weather Forecast.* **2012**, *27*, 502–514. [CrossRef]

23. Nonnenmacher, L.; Coimbra, C.F. Streamline-based method for intra-day solar forecasting through remote sensing. *Sol. Energy* **2014**, *108*, 447–459. [CrossRef]

24. Schroedter-Homscheidt, M.; Gesell, G. Verification of sectoral cloud motion based direct normal irradiance nowcasting from satellite imagery. *AIP Conf. Proc.* **2016**, *1734*, 150007. [CrossRef]

25. Hammer, A.; Heinemann, D.; Hoyer, C.; Kuhlemann, R.; Lorenz, E.; Müller, R.; Beyer, H.G. Solar Energy Assessment Using Remote Sensing Technologies. *Remote Sens. Environ.* **2003**, *86*, 423–432. Available online: http://www.sciencedirect.com/science/article/pii/S003442570300083X (accessed on 25 May 2018). [CrossRef]

26. Perez, R.; Kivalov, S.; Schlemmer, J.; Hemker, K.; Renné, D.; Hoff, T.E. Validation of Short and Medium Term Operational Solar Radiation Forecasts in the US. *Sol. Energy* **2010**, *84*, 2161–2172. Available online: http://www.sciencedirect.com/science/article/pii/S0038092X10002823 (accessed on 25 May 2018). [CrossRef]

27. Escrig, H.; Batlles, F.; Alonso, J.; Baena, F.; Bosch, J.; Salbidegoitia, I.; Burgaleta, J. Cloud detection, classification and motion estimation using geostationary satellite imagery for cloud cover forecast. *Energy* **2013**, *55*, 853–859. [CrossRef]

28. Heas, P.; Memin, E.; Papadakis, N.; Szantai, A. Layered Estimation of Atmospheric Mesoscale Dynamics From Satellite Imagery. *IEEE Trans. Geosci. Remote Sens.* **2007**, *45*, 4087–4104. [CrossRef]

29. Heas, P.; Memin, E. Three-Dimensional Motion Estimation of Atmospheric Layers From Image Sequences. *IEEE Trans. Geosci. Remote Sens.* **2008**, *46*, 2385–2396. [CrossRef]

30. Sirch, T.; Bugliaro, L.; Zinner, T.; Möhrlein, M.; Vazquez-Navarro, M. Cloud and DNI nowcasting with MSG/SEVIRI for the optimized operation of concentrating solar power plants. *Atmos. Meas. Tech.* **2017**, *10*, 409–429. [CrossRef]

31. Berthold K.P. Horn, B.G.S. Determining Optical Flow. *Proc. SPIE* **1981**, *0281*. [CrossRef]

32. Lucas, B.D.; Kanade, T. An Iterative Image Registration Technique with an Application to Stereo Vision. 1981; pp. 674–679. Available online: https://cecas.clemson.edu/~stb/klt/lucas_bruce_d_1981_1.pdf (accessed on 25 May 2018).

33. Leese, J.A.; Novak, C.S.; Taylor, V.R. The Determination of Cloud Pattern Motions from Geosynchronous Satellite Image Data. *Pattern Recognit.* **1970**, *2*, 279–292. Available online: http://www.sciencedirect.com/science/article/pii/003132037090018X (accessed on 25 May 2018). [CrossRef]

34. Evans, A.N. Cloud motion analysis using multichannel correlation-relaxation labeling. *IEEE Geosci. Remote Sens. Lett.* **2006**, *3*, 392–396. [CrossRef]

35. Geraldi, E.; Romano, F.; Ricciardelli, E. An Advanced Model for the Estimation of the Surface Solar Irradiance Under All Atmospheric Conditions Using MSG/SEVIRI Data. *IEEE Trans. Geosci. Remote Sens.* **2012**, *50*, 2934–2953. [CrossRef]

36. Skamarock, W.C.; Klemp, J.B.; Dudhia, J.; Gill, D.O.; Barker, D.M.; Wang, W.; Powers, J.G. A description of the Advanced Research WRF Version 3; NCAR Technical Note, NCAR/TN-475+STR; 2008; 125p. Available online: http://opensky.ucar.edu/islandora/object/technotes:500 (accessed on 25 May 2018).

37. Cano, D.; Monget, J.; Albuisson, M.; Guillard, H.; Regas, N.; Wald, L. A Method for the Determination of the Global Solar Radiation from Meteorological Satellite Data. *Sol. Energy* **1986**, *37*, 31–39. Available online: http://www.sciencedirect.com/science/article/pii/0038092X86901040 (accessed on 25 May 2018). [CrossRef]

38. Perez, R.; Ineichen, P.; Moore, K.; kmiecik, M.; Chain, C.; Georges, R.; Vignola, F. A New Operational Model for Satellite-Derived Irradiances: Description and Validation. *Sol. Energy* **2002**, *73*, 307–317. Available online: https://archive-ouverte.unige.ch/unige:17201 (accessed on 25 May 2018). [CrossRef]

39. Blanc, P.; Gschwind, B.; Lefèvre, M.; Wald, L. The HelioClim Project: Surface Solar Irradiance Data for Climate Applications. *Remote Sens.* **2011**, *3*, 343–361. Available online: http://www.mdpi.com/2072-4292/3/2/343 (accessed on 25 May 2018). [CrossRef]

40. Wenhui Jiang, Fei Su, J.Z. Short-term forecasting of cloud images using local features. *Proc. SPIE* **2014**, *9069*. [CrossRef]

41. Hammer, A.; Kühnert, J.; Weinreich, K.; Lorenz, E. Correction: Hammer, J., et al. Short-Term Forecasting of Surface Solar Irradiance Based on Meteosat-SEVIRI Data Using a Nighttime Cloud Index. *Remote Sens.* **2015**, *7*, 9070–9090; reprinted in *Remote Sens.* **2015**, *7*, 13842. [CrossRef]

42. Forsythe, M. Atmospheric Motion Vectors: Past, Present and Future. ECMWF Seminar on Recent Development in the Use of Satellite Observations in NWP. 2007. Available online: https://www.ecmwf.int/sites/default/files/elibrary/2008/9445-atmospheric-motion-vectors-past-present-and-future.pdf (accessed on 25 May 2018).

43. Ricciardelli, E.; Romano, F.; Cuomo, V. Physical and statistical approaches for cloud identification using Meteosat Second Generation-Spinning Enhanced Visible and Infrared Imager Data. *Remote Sens. Environ.* **2008**, *112*, 2741–2760. [CrossRef]

44. Jimenez, P.A.; Hacker, J.P.; Dudhia, J.; Haupt, S.E.; Ruiz-Arias, J.A.; Gueymard, C.A.; Thompson, G.; Eidhammer, T.; Deng, A. WRF-Solar: Description and Clear-Sky Assessment of an Augmented NWP Model for Solar Power Prediction. *Bull. Am. Meteorol. Soc.* **2016**, *97*, 1249–1264. [CrossRef]

45. Thompson, G.; Eidhammer, T. A Study of Aerosol Impacts on Clouds and Precipitation Development in a Large Winter Cyclone. *J. Atmos. Sci.* **2014**, *71*, 3636–3658. [CrossRef]

46. Hong, S.Y.; Noh, Y.; Dudhia, J. A New Vertical Diffusion Package with an Explicit Treatment of Entrainment Processes. *Mon. Weather Rev.* **2006**, *134*, 2318–2341. [CrossRef]

47. Iacono, M.J.; Delamere, J.S.; Mlawer, E.J.; Shephard, M.W.; Clough, S.A.; Collins, W.D. Radiative forcing by long-lived greenhouse gases: Calculations with the AER radiative transfer models. *J. Geophys. Res. Atmos.* **2008**, *113*. Available online: http://adsabs.harvard.edu/abs/2008JGRD..11313103I (accessed on 25 May 2018). [CrossRef]

48. Kain, J.S. The Kain–Fritsch Convective Parameterization: An Update. *J. Appl. Meteorol.* **2004**, *43*, 170–181. [CrossRef]

remote sensing

MDPI

Article

Earth-Observation-Based Estimation and Forecasting of Particulate Matter Impact on Solar Energy in Egypt

Panagiotis G. Kosmopoulos [1,*], Stelios Kazadzis [1,2], Hesham El-Askary [3,4,5], Michael Taylor [6], Antonis Gkikas [7], Emmanouil Proestakis [7], Charalampos Kontoes [7] and Mohamed Mostafa El-Khayat [8]

[1]   Institute for Environmental Research and Sustainable Development, National Observatory of Athens (IERSD/NOA), 15236 Athens, Greece; stelios.kazadzis@pmodwrc.ch
[2]   Physikalisch-Meteorologisches Observatorium Davos, World Radiation Center (PMOD/WRC), CH-7260 Davos, Switzerland
[3]   Center of Excellence in Earth Systems Modeling & Observations, Chapman University, Orange, CA 92866, USA; elaskary@chapman.edu
[4]   Schmid College of Science and Technology, Chapman University, Orange, CA 92866, USA
[5]   Department of Environmental Sciences, Faculty of Science, Alexandria University, Moharem Bek, Alexandria 21522, Egypt
[6]   Department of Meteorology, University of Reading, Reading RG6 6BB, UK; Michael Taylor michael.taylor@reading.ac.uk
[7]   Institute for Astronomy, Astrophysics, Space Applications and Remote Sensing, National Observatory of Athens (IAASARS/NOA), 15236 Athens, Greece; agkikas@noa.gr (A.G.); proestakis@noa.gr (E.P.); kontoes@noa.gr (C.K.)
[8]   New and Renewable Energy Authority (NREA), Cairo 4544, Egypt; mohamed.elkhayat@yahoo.com
*   Correspondence: pkosmo@meteo.noa.gr

Received: 24 October 2018; Accepted: 21 November 2018; Published: 23 November 2018

Abstract: This study estimates the impact of dust aerosols on surface solar radiation and solar energy in Egypt based on Earth Observation (EO) related techniques. For this purpose, we exploited the synergy of monthly mean and daily post processed satellite remote sensing observations from the MODerate resolution Imaging Spectroradiometer (MODIS), radiative transfer model (RTM) simulations utilizing machine learning, in conjunction with 1-day forecasts from the Copernicus Atmosphere Monitoring Service (CAMS). As cloudy conditions in this region are rare, aerosols in particular dust, are the most common sources of solar irradiance attenuation, causing performance issues in the photovoltaic (PV) and concentrated solar power (CSP) plant installations. The proposed EO-based methodology is based on the solar energy nowcasting system (SENSE) that quantifies the impact of aerosol and dust on solar energy potential by using the aerosol optical depth (AOD) in terms of climatological values and day-to-day monitoring and forecasting variability from MODIS and CAMS, respectively. The forecast accuracy was evaluated at various locations in Egypt with substantial PV and CSP capacity installed and found to be within 5–12% of that obtained from the satellite observations, highlighting the ability to use such modelling approaches for solar energy management and planning (M&P). Particulate matter resulted in attenuation by up to 64–107 kWh/m^2 for global horizontal irradiance (GHI) and 192–329 kWh/m^2 for direct normal irradiance (DNI) annually. This energy reduction is climatologically distributed between 0.7% and 12.9% in GHI and 2.9% to 41% in DNI with the maximum values observed in spring following the frequent dust activity of Khamaseen. Under extreme dust conditions the AOD is able to exceed 3.5 resulting in daily energy losses of more than 4 kWh/m^2 for a 10 MW system. Such reductions are able to cause financial losses that exceed the daily revenue values. This work aims to show EO capabilities and techniques to be incorporated and utilized in solar energy studies and applications in sun-privileged locations with permanent aerosol sources such as Egypt.

Keywords: solar energy; aerosol impact; earth observation

---

1. Introduction

The various observed and predicted changes in global climate stem from the short-sighted use of fossil fuels as an energy source. To mitigate climate change while permitting continued industrial development, it is necessary to make greater use of renewable energy sources as soon as possible [1]. To this direction, renewables currently account for more than 22% of total global electricity generation, of which last year more than 400 GW (32.4%) were produced from solar energy [2]. Over the last 5 years (2013–2017), an estimated 15 Gt CO_2eq of emissions was avoided through renewables, compared to the emissions that would otherwise have occurred from fossil fuels-based power [3]. As a result, the exploitation of renewables is becoming a main requirement in order to meet the sustainable development goals (SDG), which were institutionalized by the United Nations [4], without decelerating economic growth and reducing welfare.

The above situation is most relevant in developing countries like Egypt, which have historically been reliant on fossil fuels for electricity and the market for renewable energy was underdeveloped, without clear business models and practices to make energy more reliable and more affordable for citizens. Nevertheless, given its geographical location, the most important potential source of renewable energy for Egypt is the Sun. A country with high average solar energy potential [5] and a massive land mass, well positioned to benefit from the continued growth in solar power generation [6]. In order to succeed in providing 22% of its energy supply by renewables before 2030 [7,8]. By the end of 2015, 70 MW of solar power was already operational in Egypt, with 1.8 GW in project development [9], while almost 7,600 km^2 of desert were allocated in 2014 for future renewable energy projects, with permits for land allocation already obtained by the New and Renewable Energy Authority (NREA). As cloudy conditions in Egypt are rare, aerosols, mainly dust aerosols, are the most common source of solar irradiance attenuation [10,11], causing performance problems in the photovoltaic (PV) and concentrated solar power (CSP) plants. In various cases aerosol and dust are able to cause solar energy losses of the order of 80% and 50%, respectively [12–16]. In particular, the main source of aerosols in Egypt is Saharan dust and more specifically the Khamassen dust storms, which is a fifty days phenomenon (Khamaseen in Arabic means "fifty"), frequent from mid/March through April [17–22]. Aerosols are also responsible for changes in the radiative forcing (RF) of the Earth-Atmosphere system through their interaction with solar radiation [23]. As defined by the recent World Meteorological Organization report on "Aerosol Measurements, Procedures and recommendations" [24], "Aerosol Optical Depth (AOD) is the most important aerosol parameter, in terms of climate sensitivity along with well mixed greenhouse gases, for determining the direct radiative effect and forcing".

At the same time, continuous monitoring of the impact of particulate matter on solar energy has become an important activity at many research and operational weather centres [12,25] due to the growing interest from the solar energy industry. In brief, aerosols, reduce the energy generation potential of solar panels by absorbing and scattering light, reducing the strength of the direct beam (from which energy generation is most efficient). Electricity supplied to the grid must balance demand such that unexpected fluctuations in the power generated by the solar facility are costly since they require the use of rapid-response generators (e.g., natural gas). Being able to predict solar generation allows cheaper energy sources to be used. As a result, the need for improved EO-based estimation and forecasting services of AOD and solar energy potential is substantial in order to fulfil the increasing integration of solar systems into the electricity grid and load exchanges with direct impacts for the transmission and distribution system operators (TSO and DSO, respectively) and their coordination [26]. The lack of forecasts or inaccurate forecasts results in an inefficient operation of the electricity system and can even endanger the security of supply. The prediction of AOD in Numerical Weather Prediction (NWP) models faces a number of challenges owing to the complexity of atmospheric aerosol processes and their sensitivity to the underlying meteorological conditions [27–29]. At the moment, there are numerous aerosol monitoring satellite sensors and operational forecasting services which provide the AOD at a high spatial and temporal resolution, while accurate predictions of the irradiance received at individual PV (where GHI is needed) or CSP installations (where DNI applies) are able to be estimated by various solar energy nowcasting and forecasting methods and systems [30–35], which are able to use as input the aerosol information from the aforementioned sources.

Many recent papers have studied the impact of aerosols and dust on the GHI and DNI [15,36–38] and particularly in Egypt [39–42]. In this study we investigate the particulate matter impact on solar radiation and energy in the region of Egypt by analysing long-term and forecast data sets of AOD in conjunction with a state-of-the-art real-time RTM technique. This technique [34] was developed, used and applied within several EU-funded projects (e.g., Geo-Cradle; http://geocradle.eu/en/) as the so-called Solar Energy Nowcasting SystEm (SENSE). SENSE is based on the synergy of RTM simulations, machine learning and real-time atmospheric inputs from satellites and models. In order to estimate and forecast the aerosol and dust impact on the solar irradiances, we integrated MODIS observations in a daily and climatological basis or CAMS 1 day forecasts, which is a combination of NWP modelling and measuring approach [43,44], to the SENSE. The analysis was performed by: (i) calculating a 16-year AOD climatology from MODIS and quantifying the corresponding impact on GHI and DNI, (ii) using the MODIS daily AOD observation values for the last 3 years (2015–2017) to evaluate the CAMS forecasts for the whole Egyptian domain and for specific locations with high solar energy exploitation potential and (iii) proposing and testing three energy M&P techniques; the CAMS as an holistic approach, the MODIS persistence (PERS) based on the previous day values and the MODIS climatology (CLIM) by using the 16-year average values. Finally, we made a brief financial analysis for a hypothetical scenario of a 10 MW system in order to quantify the impact of aerosol and dust presence on the energy production from PV and CSP systems and on the annual, monthly and daily revenues under climatological and extreme dust event conditions. Section 2 presents data, methods and techniques used. Section 3 describes the solar power and energy results including the financial analysis for the aforementioned scenarios and in Section 4 we present our conclusions on the proposed EO solutions.

2. Data and Methodology

2.1. Data

2.1.1. Model Forecasts

For aerosols, dust estimation and forecasting we used the CAMS 1-day total AOD and dust AOD forecasts at 550 nm, which are based on the Monitoring Atmospheric Composition and Climate (MACC) reanalysis tool and its aerosol type classification identifier [45]. The CAMS data set includes modelling of aerosols and satellite AOD data assimilation from MODIS and other data sources for consistent bias correction purposes [43,46]. The modelling part uses the ECMWF physical parameterizations for aerosol interaction processes and follows the corresponding particulate matter treatment in the LOA/LMD-Z model [47,48]. As presented in Reference [34] the main uncertainty of SENSE is linked with the uncertainty of the model inputs. In this case with the aerosol related ones, most importantly AOD. Reported CAMS AOD forecast uncertainty ranges from −0.1 to 0.2 in terms of mean bias against Aeronet sun photometer data [49] in winter and summer months respectively. In addition to the results of [49] we have used the El Farafra Aeronet site in Egypt and we have compared 180 existing Level 2 days of AOD data for the 2015–2017 period, with the CAMS AOD data used in this study. We found a mean AOD bias of 0.107 showing a CAMS overestimation, with a correlation coefficient of 0.74 which is in a relative agreement with [46,49]. [46] report also that the spatial agreement of CAMS AOD compared to MODIS, is very good confirming the capture of dust outbreaks and their spatiotemporal evolution.

Aerosol classification, in brief, is based on annual or monthly climatology derived from the emission database for global atmospheric research and the speciated particulate emission wizard as described by [50], especially for dust particles, is a combination of source functions [51,52], 10 m wind fields, land coverage, soil moisture and albedo in the ultraviolet—visible spectral region [53].

The CAMS AOD and dust AOD 1 day forecasts were obtained for the period from January 2015 to December 2017, at 3 hour time steps and 0.4 degree spatial resolution for the regions of Alexandria, Cairo, Suez, Hurghada, Aswan, Luxor, Marsamatrouh and Asyut as described in Table 1. NREA proposed these specific locations because of their appropriateness for the installation of solar farms that are able to cover the energy requirements of nearby residential areas and support the Egyptian electricity grid. Subsequently, this aerosol forecast information was the main input parameter, together with solar elevation, to the proposed RTM methodology for the determination of impacts on solar irradiances.

Table 1. Coordinates (degrees), population and average height (meters above sea level) of the specific locations in Egypt.

Location	Population	Code	Latitude	Longitude	Height (m.a.s.l.)
Alexandria	5,172,000	ALE	31.2001	29.9187	12
Cairo	9,153,000	CAI	30.0444	31.2357	75
Suez	744,000	SUE	29.9668	32.5498	5
Hurghada	288,000	HUR	27.2579	33.8116	14
Aswan	290,000	ASW	24.0889	32.8998	194
Luxor	507,000	LUX	25.6872	32.6396	76
Marsamatrouh	448,000	MAR	31.3543	27.2373	30
Asyut	4,123,000	ASY	27.1783	31.1859	70

2.1.2. Satellite Observations

MODIS, onboard the polar orbiting Aqua satellite, has provided cloud-free multi-wavelength aerosol retrievals, among other EO, since 2002. The primary aerosol product is the AOD, reported at 550 nm, which is retrieved via the implementation of three individual algorithms operating separately over dark continental [54,55] and maritime targets [56,57] while thanks to the deployment of the enhanced Deep Blue (DB) algorithm [58], aerosol observations are possible over land areas characterized either by limited vegetation coverage, depending on the season, or by high surface albedo (i.e., deserts), excluding snow/ice covered regions [59]. These AOD retrievals, are merged providing almost full spatial coverage of the planet [60]. In the present study, MODIS-Aqua observations acquired from different collections (i.e., versions of the retrieval algorithm) and at different spatiotemporal resolutions (i.e., levels) have been utilized. More specifically, the Level 2 Collection 6 MODIS-Aqua AODs, over the period 2002–2017, as well as the corresponding L3 C061 datasets, over the period 2015–2017, have been processed. The former data are provided in 5 min intervals (i.e., swaths) and their spatial resolution is 10 km × 10 km (nadir view) while the latter ones, aggregated to $1° × 1°$ lat-lon grid, are available on a daily basis. The climatology of AOD in Egypt and its impact on solar energy potential was calculated in order to identify in monthly basis the attributes of the local and regional climatological conditions that favour the presence of aerosols from local emissions and/or long-range transport. Regarding the fine resolution of MODIS data, these have been regridded at an equal $0.1° × 0.1°$ projection and then the monthly values, used as inputs to the RTM, have been calculated for the entire domain of Egypt. From the raw L2 and L3 files, both accessible at the Level-1 and Atmosphere Archive & Distribution System Distributed Active Archive Centre (https://ladsweb.modaps.eosdis. nasa.gov/), the scientific data sets named as "AOD_550_Dark_Target_Deep_Blue_Combined" and "AOD_550_Dark_Target_Deep_Blue_ Combined_Mean", respectively, have been extracted and analysed. There are four Quality Assurance (QA) flags (0: No Confidence, 1: Marginal Confidence, 2: Good Confidence and 3: Very Good Confidence) assigned to each MODIS L2 AOD retrieval indicating its "reliability." In the "merged" L2 AODs, the QA flags vary among the Dark Target (DT) (QA≥1 over ocean and QA=3 over land) and DB (QA≥2) algorithms while the derivation of the L3 AODs is relied on spatial averages of L2 pixels weighted by their QA confidence level [61]. Finally, MODIS AOD uncertainty and validation are presented in References [59,60,62] showing an agreement with Aeronet measurements in the regions of North Africa and Middle East. In particular, the bias of DB AOD product is −0.036 and the corresponding of DT is reduced to −0.013 [60].

2.2. Methodology

2.2.1. Radiative Transfer Modelling Technique

We used an existing technique, which is based on RTM simulations produced by libRadtran [63,64], machine learning in the form of a continuous function-approximating model, or a Neural Network (NN) model and a variety of atmospheric inputs covering clear-sky and all-sky conditions. This technique is the so-called SENSE system and its technical background and validation were described in detail in Reference [34]. In brief, we first developed a large scale look-up-table (LUT) with more than 2.5 million RTM simulations by using the pseudo-Spherical Discrete Ordinate Radiative Transfer solver [65] and with input parameters the solar zenith angle (SZA), the AOD, the ice and water cloud optical thicknesses, the Angstrom exponent (AE), the single-scattering albedo (SSA), the total ozone column (TOC) and the columnar water vapor (WV) (Abbreviations presents the complete list of abbreviations). All the technical and structural information about the RTM simulations, the LUT construction and specific features are presented in Reference [66]. Then, a series of NNs were trained on solar irradiances spectra and on integrated irradiances to produce instantaneous results covering the wavelength region between 285 and 2700 nm. For multivariate input-output data, feed-forward NNs with a minimum of one layer of "hidden" neurons have been shown to be a universal function approximation [67]. For our approach we connected the input-output vectors via two network layers—one containing the hidden neurons with tanh activation functions and another containing output neurons with linear activation functions [66,68] as depicted in Figure 1. This configuration allows the continuous and nonlinear functional approximation that relates the output vector (e.g., surface solar radiation; SSR) with the input vector (e.g., combination of atmospheric parameters).

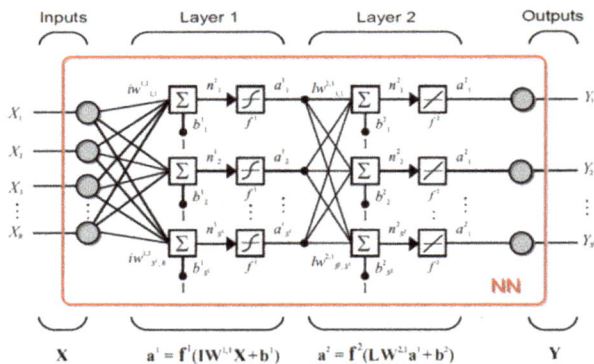

Figure 1. Schematic showing the NN architecture connecting the input and outputs parameters [66].

The above computing architectures, in conjunction with operational inputs from EO data sources like satellites and models, brought the "birth" of the SENSE system, which is capable of producing solar power and energy results in terms of GHI and DNI of the order of 1 million simulations in less than 1 minute in high spectral resolution (1 nm) depending spatially and temporally on the input parameter resolution (e.g., MODIS 0.1 degree and 1 day, CAMS 0.4 degree and 3 hours). SENSE was applied to various solar energy related applications through the EU-funded coordination and support action Geo-Cradle project by developing targeted and subsequent solar energy applications (http://solea.gr/). In this study, since cloudy conditions in Egypt are rare, we focused on the aerosols' quantification of impact on SSR by retrieving and exploiting the AOD from MODIS observations and the CAMS forecasts through the SENSE. In our RTM simulations we used the default aerosol model of [69] for ordinary particulate matter conditions (e.g., rural type aerosol in the boundary layer, background aerosol above 2 km, spring-summer conditions and a visibility of 50 km), which is the simplest way to include aerosols in libRadtran [63,64]. Since we specified the AOD impact on SSR, we simulated this parameterization by overwriting the default parameters with the integrated AOD using the aerosol_set_tau command [63]. More recent aerosol modelling efforts (e.g., the software package of Optical Properties of Aerosols and Clouds; OPAC [70]) are based on the aerosol component descriptions of [69], demonstrating the reliability and durability of the original aerosol model approaches. The parameterization was band-based [71] (correlated K-approximation) and for the gas absorption the molecular bands provided by Low-resolution atmospheric Transmittance and Radiance (LOWTRAN), while the code for spectral irradiance for the extra-terrestrial solar source spectrum was implemented.

Concerning other than AOD aerosol optical properties, water vapor and traces gases, that affect solar irradiance, input parameters were set to constant monthly climatological values from relevant data sources. In particular, the WV climatology was retrieved by the medium resolution imaging spectrometer onboard the European Space Agency's environmental satellite, the TOC from Ozone Monitoring Instrument (2008–2017) and the parameters SSA and AE were retrieved by the AeroCom database [72]. The impact of TOC on SSR is of the order of 0.5% for 100 Dobson unit differences, while for WV columns ranging between 0.5 and 2 cm the SSR difference is almost 3–5% under low SZA (< 15 degrees) [34]. Regarding SSA and AE, we have used for both the constant value of 0.9 for the Egypt region. In order to assess the impact of the day-to-day variability of these properties on the SSR-related outputs and analysis, we have used their mean and standard deviation from the El Farafra Aeronet site for the period 2015–2017, which was found to be 0.91 ± 0.02 for SSA_{440nm} and 0.96 ± 0.51 for $AE_{440-870nm}$, with the mean AOD_{440nm} at 0.2. Since the impact of these parameters in the total solar irradiance is a function of AOD (and solar elevation) we have calculated the k = 1 uncertainties on the solar output based on the statistical standard deviations of the measured SSA and AE. Under climatological AOD levels, that is, 0.2, the SSA uncertainty is ± 0.22 for GHI and it is not affecting the DNI. The corresponding AE uncertainties were found to be ± 0.5 and ± 2.9 for GHI and DNI respectively. For a case of a dust episode (e.g., $AOD_{440nm}=0.8$) the uncertainties of SSA_{440nm} increase to ± 0.8 for GHI and for AE they are ± 2.8 and ± 11.4 respectively, pointing out that they become significant only for DNI and AE variability, while for all other cases is less than 4%. These results are comparable with similar sensitivity analysis of radiative transfer studies [66,73] considering overall these differences as a scale of error introduced by this approach.

The solar irradiance outputs were produced in terms of solar power (in W/m^2) and based on the time dimension provided by the CAMS and MODIS temporal resolution we calculated the corresponding daily, monthly and annual sum of solar energy (in kWh/m^2). The reliability of the RTM techniques of SENSE were tested against ground-based measurements from southern Africa to northern Europe [34] by comparing the simulated outputs with selected stations from the baseline solar radiation network as well as under high aerosol loads [15], while CAMS outputs are continuously validated through analytical reports [74].

2.2.2. Energy Management and Planning (M&P)

Accurate solar energy forecasts are crucial in the energy exchange marketplace, where on-the-spot energy prices are defined by supply and demand equilibriums [2]. Simultaneously, knowledge of the solar energy potential of each location is a basic condition for solar farm investment for effective energy planning and determination of the break-even point, at which the total cost and total revenue are equal. Therefore, the energy M&P requires interactive decision making solutions in order to forecast the input and output loads of the solar facilities. Based on the combination of SENSE with the CAMS aerosol input, we propose this synergy as a robust approach that provides operationally aerosol and dust impact on solar energy. Additionally, we test other EO—based solutions which use as aerosol input information to SENSE the MODIS observations by the following ways: (i) by exploiting the previous day of observations and apply them to the current day as persistent (PERS) aerosol conditions, forming a forecasting technique that from hereafter we will call MODIS PERS and (ii) by calculating the MODIS AOD climatology (CLIM) and use monthly averages, an approximation method that we will call MODIS CLIM. The utility behind these two approaches has to do with the ease of application but with consequential uncertainties, especial under unusually high aerosol loads (e.g., dust events). Figure 2 depicts the procedural flows, starting from the two aerosol data sources (CAMS and MODIS), converting the AOD values to solar irradiances (GHI and DNI) through the SENSE and the corresponding solar energy forecasting outputs. SENSE-CAMS solution is operational-ready with continuous provision of modelled AOD inputs, while SENSE-MODIS solution is the observational AOD solutions but under deferred and homogenized aerosol actual conditions. This modelling scheme is able to act as an holistic approach for energy M&P in sun-privileged locations like Egypt with dominant particulate matter sources.

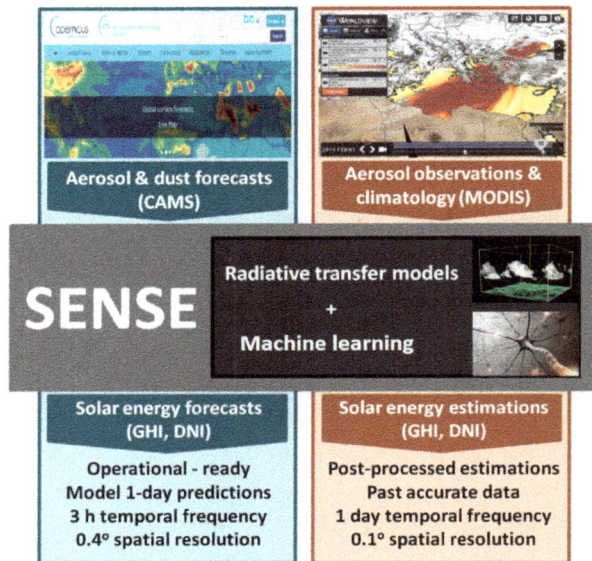

Figure 2. Flowchart of the SENSE scheme. The initial data sources followed by the observational or forecasted aerosol inputs to the SENSE and the analogue solar energy related outputs.

2.2.3. Financial Analysis

For the financial analysis we simulated a hypothetical scenario of a PV and a CSP system with nominal power of 10 MW assumed to be installed in Cairo, Asyut and Aswan. Figure 3 presents all the studied locations, including CAI, ASY and ASW, located along the river Nile and we will describe the financial analysis in Section 3.4. These locations were selected because of their different latitudes to represent conditions with various aerosol sources and solar energy potential levels. The system specifications were classified into the exploitation of GHI from PV technologies and DNI from CSP plants. The annual energy production results were cross validated with existing solar farms in Morocco, California and South Africa. Particularly, in Morocco the Noor 1 CSP (160 MW) produces almost 370 GWh on an annual basis, while in California, a CSP of nominal power 392 MW gives back annually 1,079 GWh. In South Africa, from a 100 MW CSP they exploit 480 GWh and from 96 MW and 75 MW PVs they take back energy output of about 180 and 150 GWh respectively.

Bringing the above solar farm projects into the solar energy potential levels of the selected locations in Egypt we found the corresponding energy outputs [8], which reflect the local latitudinal conditions. As a result, a 10 MW system in the region of Egypt is able to produce annually almost 25,687 MWh by using a required area of 130,000–150,000 m^2 for PV (depends on the material used, for example, crystalline silicon, cadmium telluride) and 280,000–360,000 m^2 for CSP installation (depends on the technology used, for example parabolic trough, solar tower).

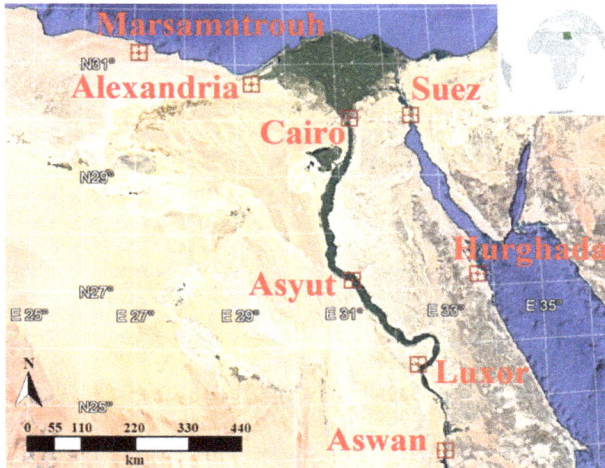

Figure 3. Study region and the specific locations of ALE, CAI, SUE, HUR, ASW, LUX, MAR and ASY. In CAI, ASY and ASW a financial analysis was additionally performed.

Concerning the system and calculation assumptions, for the PV calculations, a realistic efficiency value of 12% has been used alongside a spatial coverage of 80% and material combined losses of 29% for the most common material used, which is the crystalline silicon [75]. For the CSP, the energy storage facilities have been considered, for a required capacity of 14 hours which ensures full self-sufficiency, as well as its heat losses, the losses by shading, incidence angle modifier, the end losses and the peak optical efficiency [76,77]. For the hypothetical system scenario and its nominal power, the actual system power performance in MW has been used instead of MWp, where the peak power rating on a solar system represents the most power that it would produce under ideal conditions for solar production. For the calculation of the provided financial analysis results, the associated revenue is straightforward; one needs to multiply the produced energy by the price (USD/kWh) applicable to this hypothetical 10 MW system scenario. NREA proposed a realistic selling-electricity-to-the-grid price value (feed-in tariff) of 0.0382 USD/kWh [6]. As real new projects are procured or launched globally and the "feed-in" prices are reduced in correlation to reduced investment and operation costs (for both CSP and PV projects), prices like the used in this study might be also reduced by the time a project kicks-off. As a result, this price should be seen as an assumption that should be further substantiated in direct contact with the Egyptian Authorities. Another assumption is the maintenance of the PV and CSP plants including the solar panels cleaning after for example dust deposition [13]. Finally, the presented financial analysis was expressed in terms of Energy Production (EP), Daily Revenue (DR) and Financial Losses (FL). EP is the sum of the generated energy in kWh/m^2. DR is the EP multiplied by the feed-in tariff price. FL are described by the equation FL = (EP$_{possible}$ − EP$_{actual}$) * price, where EP$_{possible}$ is the possible EP under aerosol-free conditions and EP$_{actual}$ is the actual EP taking into account the particulate matter impact.

3. Results and Discussion

3.1. Climatological Impact

Figure 4a shows the 16-year climatology of AOD from MODIS for the greater Egypt region. It is a combination of the MODIS algorithms DT and DB Level 2 which provides reliable aerosol optical properties for arid regions like Egypt at high spatial resolution (0.1 degree). The AOD at 550 nm was found to range from 0.034 to 0.966 indicating the strong particulate matter background of the region especially in spring and summer months. Summer was found to present high aerosol loads [10,16,22] mainly because particle accumulation is favoured in this season by the absence of precipitations and by atmospheric stability [78]. On the other hand, the highest values (>0.8) are in April when particles produced by natural processes like the wind-erosion of desert surfaces and in particular the Khamaseen dust storms [17–22]. The Nile Delta was depicted also from February to October with large AOD values caused by burning activities of local agricultural wastes [19,78], while other highlighted locations like the Red Sea and the central parts of Egypt are a combination of aerosol sources as other studies found [22].

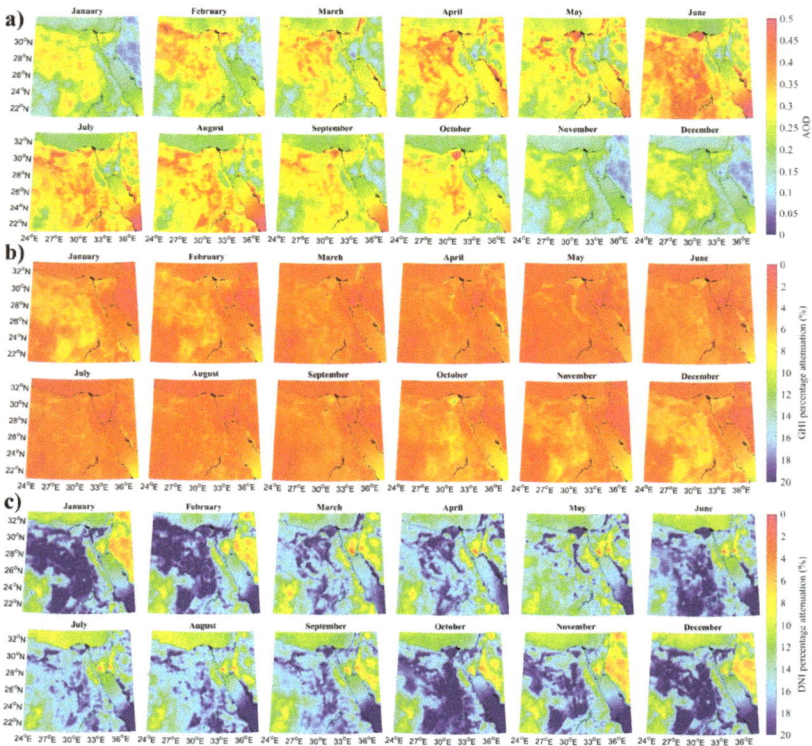

Figure 4. Monthly averages of (**a**) AOD at 550 nm in Egypt using the DT and DB Combined Level 2 product of MODIS for the period 2002–2017, (**b**) GHI and (**c**) DNI solar energy percentage attenuations relative to the aerosol-free simulations under MODIS-based AODs.

These aerosol patterns denatured into GHI and DNI percentage attenuations in Figure 4b,c, respectively. The percentage attenuations were calculated by the RTM calculations using as aerosol input the AOD and were compared to clean and clear sky conditions with the aerosols set to zero value, while the simulated time for both MODIS aerosol and aerosol-free conditions was at local noon.

For the RTM calculations the SENSE was used which produced almost 1.5 million simulations for the implementation of these results. The range of GHI attenuation was found to be 0.7 to 12.9% while for the DNI component the corresponding attenuation values range from 2.9% to 41.0% highlighting the months and regions with the highest AOD climatological conditions as well as the fact that the majority of Egypt presents attenuation values larger than 15–20%. These results are comparable with similar approaches [13] and indicate that the most important irradiance attenuation in the region is the particulate matter [10,11]. The corresponding impact from clouds is minimal reaching values of 2.8% under the annual period [79].

3.2. Performance of CAMS

Figure 5 describes the correlation of the CAMS forecasted AOD with the MODIS AOD observations (a), as well as the corresponding surface solar radiation levels (b) by using as inputs to SENSE the CAMS 1 day forecasts and the MODIS daily AOD observation values. We note that the comparison was performed for the locations of ALE, CAI, SUE, HUR, ASW, LUX, MAR and ASY for the past 3 years (2015–2017) during the MODIS overpass time positions. The coefficient of determination (R) for the AOD data sources is 0.521, while for the corresponding irradiance levels this correlation measure is significantly improved reaching R = 0.998. This means that the observed AOD differences between CAMS and MODIS present minor affectability on SSR, with the spread increasing at higher SSR levels [12]. Such data behaviour shows that the AOD absolute differences in Egypt with standard deviation (SD) of 0.137 results absolute differences in solar radiation less than 1% under low radiation levels (SD = 11.72 W/m^2) as discussed in various similar comparison approaches [15,46,74].

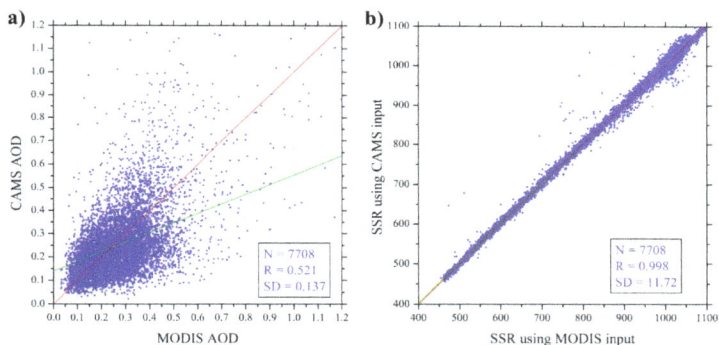

Figure 5. Scatterplots of (**a**) the CAMS forecasted AOD as compared to the MODIS observed values and (**b**) the SENSE simulated surface solar radiation (SSR) using as input the CAMS forecasted AOD as compared to the SENSE SSR using as input the MODIS AOD in Egypt for the period 2015–2017.

A more analytical description of the CAMS performance against MODIS observations is depicted in Figure 6. The CAMS AOD was plotted for a week with the MODIS daily and climatological values for the region of Aswan (a) as well as the simulated GHI (b) and DNI (c). The higher frequency of data from CAMS (1 per 3 hours) provides a more detailed monitoring of AOD and solar irradiances, while the differences from using MODIS AOD and CAMS inputs to SENSE are lower than 50 W/m^2 for GHI but are able to reach 150 W/m^2 for DNI. We note that the use of MODIS daily and climatological AOD encompasses the assumption of persistent aerosol conditions for the daily (MODIS daily value) and monthly (MODIS climatological value) time period performed. Furthermore, the comparison of the CAMS forecasts against the MODIS's 'truth' representation (real satellite observations), shows that the AOD differences presented in (a), have a minor impact on GHI, while on DNI highlight a known underestimation of CAMS under higher aerosol loads [15,46,74].

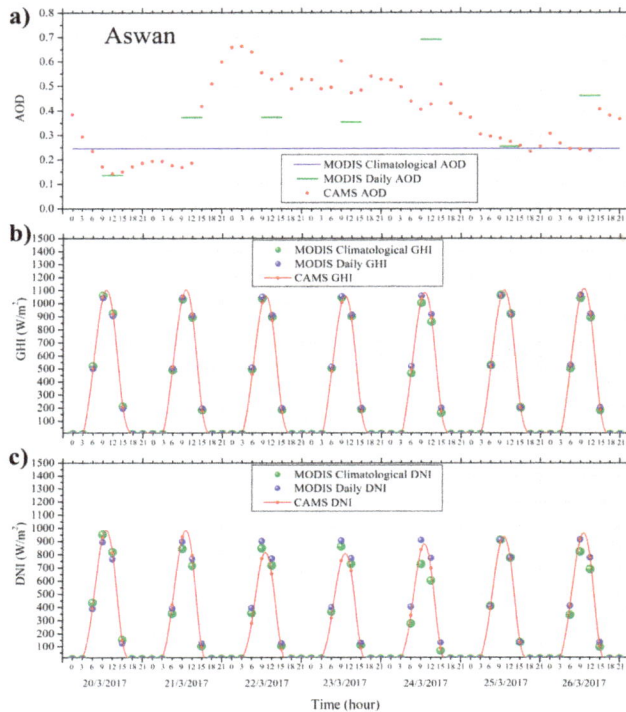

Figure 6. The AOD from MODIS climatology, MODIS daily observations and CAMS forecasts for a week period (20–26 March 2017) in Aswan (**a**) and the simulated by SENSE GHI (**b**) and DNI (**c**) using as inputs the SZA and the aforementioned AOD sources collocated to the CAMS temporal resolution of 3 hours.

3.3. Performance of M&P Techniques

The main scope for this comparison of M&P techniques performance is to highlight the impact of the different temporal resolutions between CAMS forecasts and MODIS observations (1 per day for MODIS PERS and 1 per month for MODIS CLIM). The MODIS CLIM is able to provide information about the aerosol background for each location but is unable to monitor the intraday aerosol variability. The MODIS PERS makes the assumption that the AOD is persistent from the previous day's observation. This is useful for accurate aerosol levels but does not account for upcoming dust events or other spontaneous aerosol events. The CAMS 1-day forecasts provides information about the total aerosol and the dust particle levels based on the MACC classification as described in the previous Section 2.1.1. In any case this comparison between the forecasting techniques is able to provide useful information to the energy managing authorities and investors about the current potential EO solutions and to consider the opportunity cost from each aerosol, dust and energy forecasting approach.

Figure 7 depicts the monthly 3-year average (2015–2017) forecasting behaviour of CAMS, MODIS PERS and MODIS CLIM approaches as absolute energy losses for GHI and DNI (in kWh/m^2) for the eight locations in Egypt. In general, the greatest losses are during the spring and summer months, while the average losses of GHI range from 5 to 24 kWh/m^2 and for DNI reaches 72 kWh/m^2. By comparing the CLIM technique with the CAMS we found an overestimation in summer months at ALE and MAR locations and an underestimation at LUX [80]. ASW and ASY are being affected by continuous periods of dust transport and we highlight the high mean losses in ASW and the remarkably good agreement for the CAMS and PERS techniques in ASY in April, where Khamaseen dust storms are frequent in this month, showing that under such persistently high particulate matter levels the variability is lower, marking prediction easier. Overall, good agreement of CAMS and PERS is presented in all locations except CAI because of the complexity of the multi-source aerosol conditions [79].

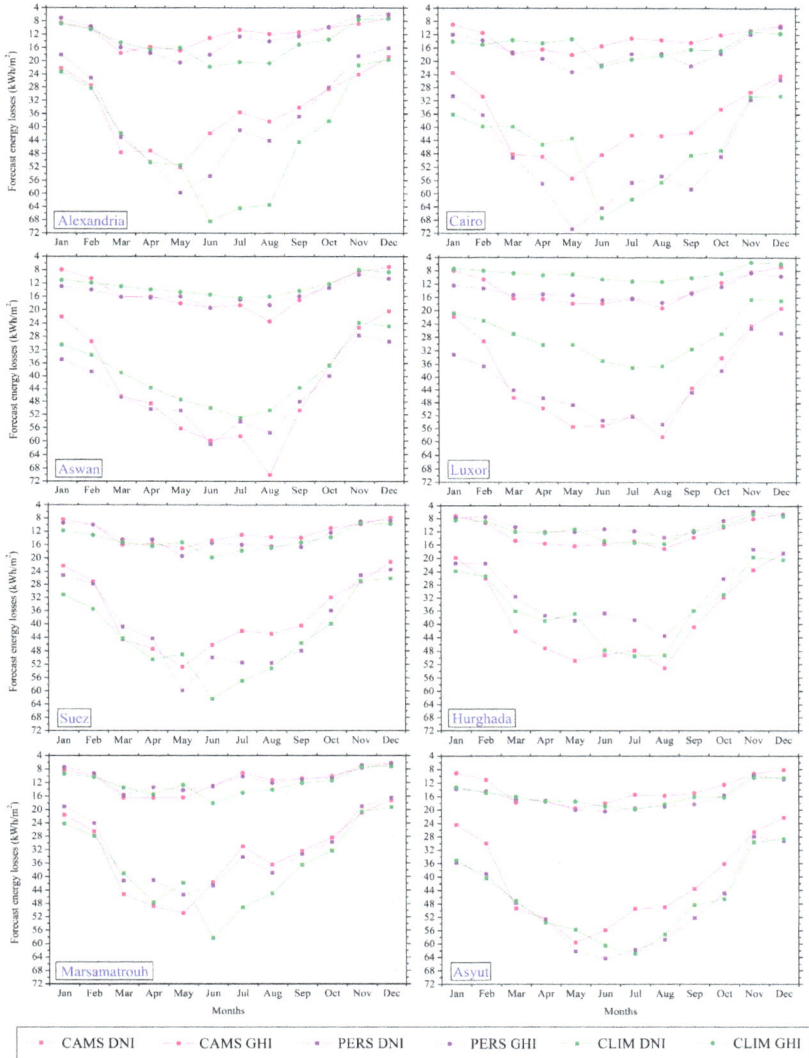

Figure 7. Monthly mean forecast solar energy losses in kWh/m² for the regions of ALE, CAI, SUE, HUR, ASW, LUX, MAR and ASY. The AOD forecasting techniques of CAMS, MODIS PERS and MODIS CLIM were applied as inputs to the SENSE producing the solar energy potential in terms of GHI (circles) and DNI (squares). The CAMS produces 1-day forecasts with 3 hour temporal resolution, the PERS uses the MODIS AOD values of the previous day for the 1-day forecast as persistent aerosol conditions and the CLIM uses the monthly mean MODIS AOD values as steady aerosol conditions for every single time step of the whole month.

For more detail of the above differences, in Figure 8 we present the daily forecast solar energy losses in March to observe the analytical GHI and DNI losses of the 3 M&P approaches at three representative locations in terms of latitude and aerosol sources (CAI, ASY and ASW) [19–22]. At this temporal resolution, the basic assumption of PERS forecasting technique emerges. We observe that after the appearance of high aerosol loads, this method is unable to detect sudden changes and hence losses the actual particulate matter impact on solar energy (e.g., ASY at 18 March). On the other hand, the CLIM approach gives a fairly constant result, ignoring all fluctuations of aerosol load. Therefore, CAMS varies more than CLIM and monitors the AOD and its impact on solar energy having to deal with its modelling nature and the subsequently indeterminacies [74].

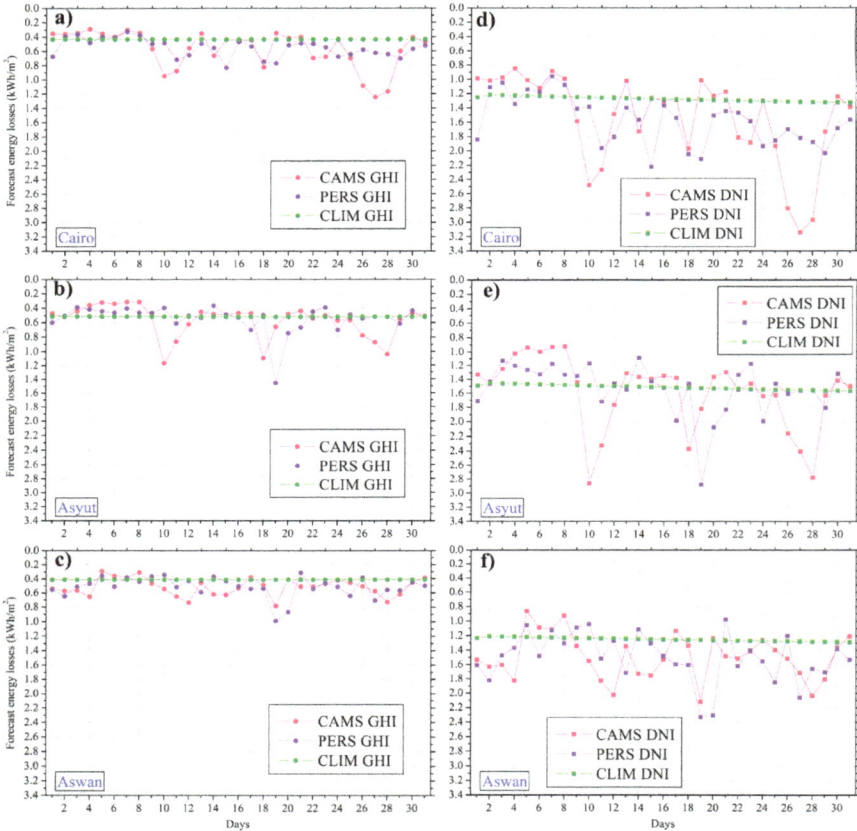

Figure 8. Daily mean forecast solar energy losses in kWh/m^2 for the regions of CAI (**a,d**), ASY (**b,e**) and ASW (**c,f**). The AOD forecasting techniques of CAMS, MODIS PERS and MODIS CLIM were applied as inputs to the SENSE producing the solar energy potential in terms of GHI (**a–c**) and DNI (**d–f**).

Finally, Figure 9 presents time series of simulated GHI in Aswan and DNI in Asyut, using CAMS AOD inputs to SENSE, in the form of contour plots for the past 3 years as well as direct comparison against the MODIS PERS and MODIS CLIM approaches. The percentage differences are larger near sunset and sunrise and during winter months (i.e., smaller absolute values) reaching 8–10% for GHI and exceeding 20% for DNI, conditions that in both cases followed by low solar energy potential. The assessment of such differences can be a useful tool for future scientific or solar sector oriented business plan studies, as it will directly contribute to the particulate matter related uncertainties introduced to solar radiation and energy calculations and/or forecasts [81,82].

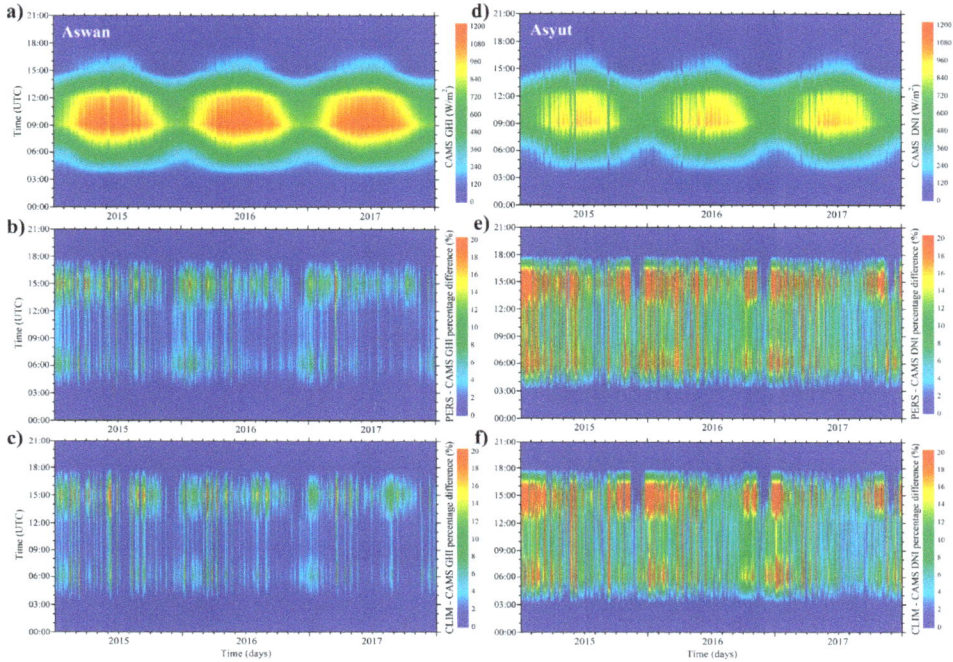

Figure 9. Contour plots of the GHI in Aswan (**a–c**) and DNI in Asyut (**d–f**) as simulated by SENSE using as AOD input the CAMS 1-day forecasts (**a,f**) and the percentage differences for GHI (**b,c**) and DNI (**e,f**) respectively as compared to the MODIS PERS and MODIS CLIM forecasting approaches for the period 2015–2017.

3.4. Economic Impact

Figure 10 presents the financial analysis results for the three specific locations of Figure 3 focusing on the hypothetical 10 MW system. The economic and energy impact was quantified in terms of monthly means, total FL and solar energy potential, by using the CAMS forecasts, the SENSE and the information of Section 2.2.3. As we move to lower latitudes the PV energy potential as well as the annual revenue increase both, starting from 2620 kWh/m^2 in CAI and reaching almost 2746 kWh/m^2 in ASW with the revenue difference reaching almost 50,000 USD. The energy losses because of the total AOD are 161 kWh/m^2 in CAI, 169 in ASY and 175 in ASW and the corresponding energy losses due to dust AOD are 64, 88 and 107 kWh/m^2 for CAI, ASY and ASW, respectively. In CSP systems, the annual aerosol and dust impact on the produced solar energy is much larger, since DNI is more affected than GHI [15]. Indicatively, in CAI the losses are 469 kWh/m^2 under total AOD and 192 under dust presence, in ASY are 499 and 269 kWh/m^2 and in ASW 524 and 329 kWh/m^2. The economic impact for PV and CSP indicates that the annual FL in ASW are almost 70,000 and 200,000 USD respectively, because of the total AOD. We note that the annual revenues are of the order of 1,098,449 and 831,697 USD for PV and CSP plants respectively. In order to understand the relevance of these FL, the corresponding annual operating and maintenance costs of such a 10 MW system in ASW are 221,000 and 340,000 USD for PV and CSP plants [83]. These costs include general site inspections, cleaning of the systems (mechanical maintenance and mirror cleaning), checking of various components (e.g., inverters, mounting/tracking, storage), local taxes, site security and administration costs [84]. Assessing also the effect of the CAMS

AOD uncertainty mentioned in Section 2.1.1 on the SENSE solar energy output, we found in ASW a range of −3.6 to 1.8% (the minus symbol corresponds to an underestimation) for GHI and −12.3 to 5.7% for DNI. The magnitude of these percentages translates into annual uncertainty on the financial calculations of −39,132 to 18,816 USD for the 10 MW PV and −101,052 to 46,668 USD for the CSP.

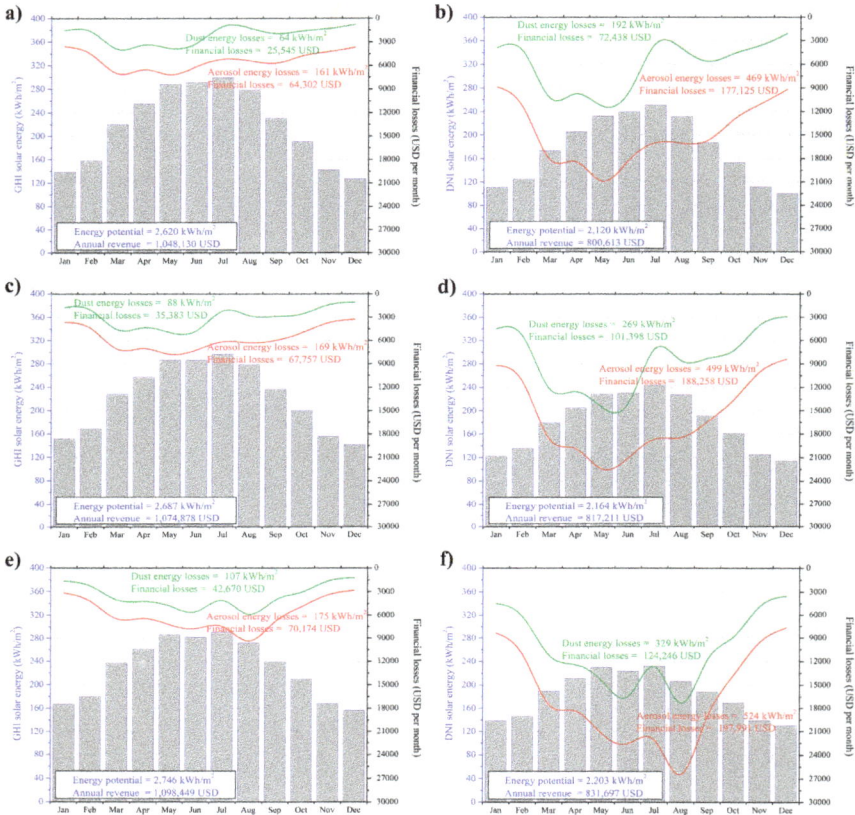

Figure 10. Financial analysis of the aerosol and dust impacts on the produced solar energy from PV (**a,c,e**) and CSP (**b,d,f**) installations with nominal power of 10MW in the regions of CAI (**a,b**), ASY (**c,d**) and ASW (**e,f**). The impact was quantified in terms of monthly mean and total financial losses and solar energy potential.

Finally, Figure 11 represents the economic impact of forecasting solar energy under extreme dust event conditions. We studied the dust event of the 18th of March 2017 at the region of Asyut (inset map shown for the same date at 10:45 UTC from MODIS true colour imaging) as well as the previous and next day in order to identify the differences and the overall energy and financial impact. The AOD (a) exceeds 3 at the peak of the event as modelled by CAMS and reaches almost 3.5 on MODIS observations. Figure 11 b and c present the forecasted solar power and financial impact for the supposed 10 MW PV (b) and CSP (c) plants. The impact was quantified in terms of EP, DR and total FL as described in Section 2.2.3. The blue and red insets show the corresponding solar power and financial losses respectively, by using as input the MODIS observations, for reference purposes. The previous and next day of the dust event were also presented in order to quantify the magnitude of the extreme dust case on solar radiation and energy.

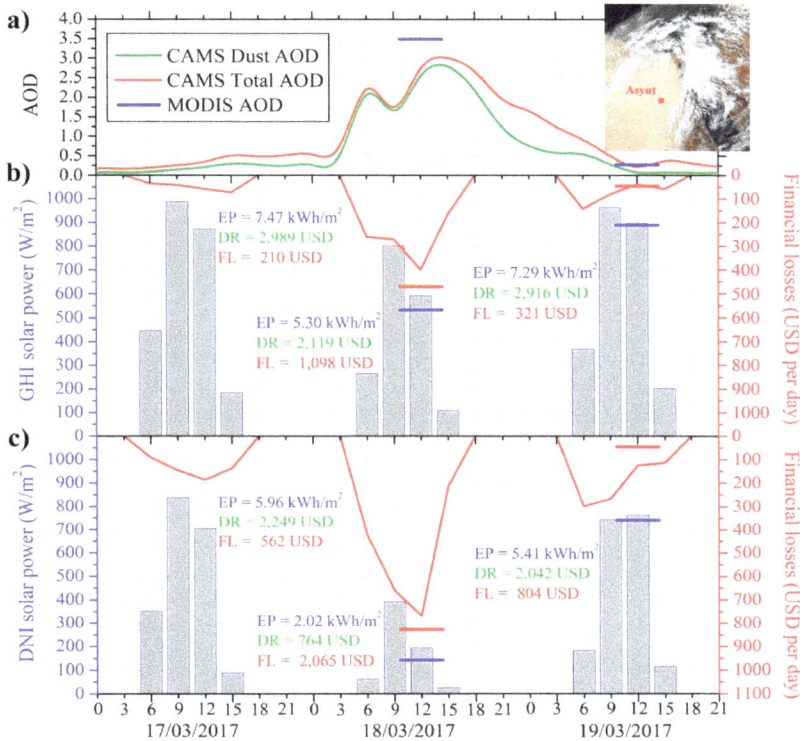

Figure 11. Temporal evolution and financial analysis of an extreme dust event impact (18 March 2017) on the CAMS AOD forecasted values (**a**) and on the produced solar energy from PV (**b**) and CSP (**c**) installations with nominal power of 10 MW in the region of ASY. The impact was quantified in terms of EP, DR and total FL. The blue and red insets show the corresponding solar power and financial losses respectively, using as input the MODIS observations.

For the previous and next day, the daily energy production for PV systems was 7.47 kWh/m² and 7.29 respectively, while at the peak of the dust event the EP was 5.3 kWh/m². For the CSP case, the daily energy losses as compared to the previous and next day were almost 4 kWh/m², meaning that for a 10 MW system the daily FL are able to exceed the DR values [85]. This fact highlights also the impact of not having energy forecasts, since the FL represent the economic difference of the actual DR from the possible DR under aerosol-free conditions. So, the lack of forecasts means that the hypothetical aerosol-free DR during the 18th of March 2017 would have been estimated close to 2,829 USD (DR+FL) and therefore indicates the usefulness of the studied approach using the SENSE. The FL under such aerosol conditions for the 10 MW CSP system are 2,065 USD with the actual DR not exceeding the 764 USD. We note that all the above energy calculations and results were cross validated against real production data from the Egyptian energy market. Indicatively, a 10 MW solar plant project in Aswan, is able to produce in a daily basis in March almost 7.43 kWh/m² or 80 MWh in total [86]. The corresponding values we simulated are 80.4, 57.1 and 78.5 MWh during 17, 18 and 19 March 2017, respectively. The lower EP during dust events is able to impact the local economy, since the required daily energy will be covered by the grid in higher prices (22 and 33 USD/MWh at off-peak and peak-time tariff) [87]. Therefore, the daily EP deficit on the 18th of March 2018 translates into an additional M&P cost of 484–726 USD, as to purchase from the Egyptian grid the remaining ≈23 MWh. As a result, the ability to forecast the forthcoming dust events or in general the irregular high AOD differences, is translated to more efficient energy M&P, minimizing the energy inadequacy by exploiting alternative energy sources or storing from the previous day energy into batteries for the case of PV systems or into melted salt as thermal energy for the case of CSP plants [76].

4. Summary and Conclusions

This study presented the estimation and forecasting techniques for the impact of particulate matter on solar energy in Egypt by exploiting the synergy of EO-based aerosol observations and forecasts from MODIS and CAMS with a state-of-the-art solar irradiance simulation system (SENSE). AOD was used as the main input parameter to SENSE and the aerosol effects on solar radiation revealed that the accuracy of estimated solar energy potential depends on the aerosol input data sources dealing with a trade-off between temporal frequency, availability of data and the overall M&P usefulness and reliability.

We firstly described the modelling scheme and we proposed three different forecasting approaches (CAMS, PERS and CLIM) to investigate potential solutions for the quantification of the aerosol effects on solar energy production. The study was performed for the whole Egypt region as well as for eight specific locations with highly installed and planned solar energy capacity. The climatological analysis showed a dependence on seasonal variability with the highest attenuation to occur during spring and summer, reaching values of 8% for GHI and exceeding 20% in DNI for the majority of the Egyptian region. The evaluation of the PERS and CLIM forecasting solutions indicated alternative roadmaps for the M&P, revealing a "shifted-reality" for the PERS, which is manageable and useful under steady atmospheric conditions but fails to predict upcoming large AOD values and differences. On the other hand, the CLIM is applicable only for large time-horizon averages (e.g., monthly means) keeping the background particulate matter information but missing the continuous variability which is a major M&P requirement. Both solutions presented higher differences as compared to CAMS in winter months and at large SZA, conditions followed by minimum solar energy potential impact.

Overall, the combination of CAMS 1 day forecasts with the SENSE is a promising tool for the solar energy management community. We simulated a hypothetical energy financial scenario of a 10 MW system under various latitudes, time-horizons and atmospheric conditions. In a climatological basis, such a system incurs the largest energy losses in the CSP form and in spring and summer months reaching almost 20% as compared to the annual energy production and the followed annual revenue. In the day-to-day and intra-day time-horizon monitoring scenario, we found that the FL are able to reach the 50% of the DR for the PV cases and overcome the DR by almost 270% for the CSP plants, which translates into daily energy losses of 4 kWh/m^2, highlighting the holistic usefulness for M&P market operations.

The findings show the potential of such EO-based techniques for solar energy applications and electricity grid TSO and DSO support services. As a result, the exploitation of EO data and solar energy management systems like the SENSE, are able to provide advanced solar energy related services, in support of large scale solar farm projects, grid operators, national and private electrical transmission and handling entities, so as to guarantee the uninterrupted energy flow and the power grid stability.

Author Contributions: P.G.K. and S.K. conceived and designed the research. P.G.K. analysed the data and wrote the manuscript. S.K. and H.E.-A revised the manuscript and adjusted the approach. M.T. and P.G.K. developed the initial NN. S.K. is the principal investigator and P.G.K. is the developer of the SENSE. C.K. is the principal investigator of the GEO-Cradle project within which the SENSE was designed, developed and applied. A.G. and E.P. downloaded and processed the MODIS data. M.M.E.-K. provided the Egyptian energy market details and the determinant financial sizes.

Funding: This research was partly funded by the COST Action "InDust" under grant agreement CA16202, supported by COST (European Cooperation in Science and Technology) and more specifically the Short Term Scientific Mission project "Finding".

Acknowledgments: P.G.K., S.K., H.E.-A. and C.K. acknowledge the GEO-Cradle project which has received funding from the EU's Horizon 2020 research and innovation programme under grant agreement No 690133. A.G. acknowledges the Dust-Glass project funded form the EU's H2020 research and innovation programme under the Marie Sklodowska-Curie grant agreement No 749461. E.P. acknowledges support through the Stavros Niarchos

Foundation. We acknowledge Mossad El-Metwally and Stefane Alfaro for establishing and maintaining the El Farafra Aeronet site used in this study.

Conflicts of Interest: The authors declare no conflict of interest.

Abbreviations

AE	Angstrom Exponent
AeroCom	Aerosol Compositions between Observations and Models
AOD	Aerosol Optical Depth
CAMS	Copernicus Atmosphere Monitoring Service
CLIM	Climatology
COT	Cloud Optical Thickness
CSP	Concentrated Solar Power
DB	Deep Blue
DNI	Direct Normal Irradiance
DR	Daily Revenue
DSO	Distribution System Operator
DT	Dark Target
ECMWF	European Centre for Medium-Range Weather Forecasts
EO	Earth Observation
EP	Energy Production
EU	European Union
FL	Financial Losses
GHI	Global Horizontal Irradiance
LUT	Look Up Table
M&P	Management and Planning
MACC	Monitoring Atmospheric Composition and Climate
MODIS	Moderate resolution Imaging Spectroradiometer
NN	Neural Network
NREA	New and Renewable Energy Authority
NWP	Numerical Weather Prediction
PERS	Persistence
PV	Photovoltaic
QA	Quality Assurance
R	Coefficient of Determination
RTM	Radiative Transfer Model
SD	Standard Deviation
SENSE	Solar Energy Nowcasting SystEm
SSA	Single Scattering Albedo
SSR	Surface Solar Radiation
SZA	Solar Zenith Angle
TOC	Total Ozone Column
TSO	Transmission System Operator
WV	Columnar Water Vapor

References

1. Solangi, K.H.; Islam, M.R.; Saidur, R.; Rahim, N.A.; Fayaz, H. A review on Global Solar Energy Policy. *Renew. Sustain. Energy Rev.* **2011**, *15*, 2149–2163. [CrossRef]
2. REN21. Renewables Global Futures Report: Great Debates towards 100% Renewable Energy 2017. Available online: http://www.ren21.net/wp-content/uploads/2017/03/GFR-Full-Report-2017.pdf (accessed on 27 September 2018).
3. International Energy Agency (IEA). *Renewables: Analysis and Forecasts to 2022*; International Energy Agency: Paris, France, 2017.

4.   UN. Progress towards the Sustainable Development Goals, Report of the Secretary-General 2017. Available online: http://www.un.org/ga/search/view_doc.asp?symbol=E/2017/66&Lang=E (accessed on 12 October 2018).

5.   Omran, M.A. Analysis of solar radiation over Egypt. *Theor. Appl. Clim.* **2000**, *67*, 225–240. [CrossRef]

6.   El-Sobki, M.S. Electrical sector in Egypt between challenges and opportunities—Full scale program for renewable energy in Egypt. Presented at the World Future Energy Summit, Abu Dhabi, UAE, 27 January 2015; Available online: http://www.mesia.com/wp-content/uploads/Sobki%20-%20NREA%20-%20AbuDhabi-January%2020-2015.pdf (accessed on 12 October 2018).

7.   Khalil, A.; Mubarak, A.; Kaseb, S. Road map for renewable energy research and development in Egypt. *J. Adv. Res.* **2010**, *1*, 29–38. [CrossRef]

8.   International Renewable Energy Agency (IRENA). *Renewable Energy Outlook: Egypt*; International Renewable Energy Agency: Abu Dhabi, UAE, 2018; ISBN 978-92-9260-069-3.

9.   Middle East Solar Industry Association (MESIA). Solar Outlook Report 2018. Available online: https://www.mesia.com/wp-content/uploads/2018/03/MESIA-OUTLOOK-2018-Report-7March2018.pdf (accessed on 8 November 2018).

10.   Nabat, P.; Somot, S.; Mallet, M.; Sevault, F.; Driouech, F.; Meloni, D.; Di Sarra, A.; Di Biagio, C.; Formenti, P.; Sicard, M.; et al. Dust aerosol radiative effects during summer 2012 simulated with a coupled regional aerosol-atmosphere-ocean model over the Mediterranean. *Atmos. Chem. Phys.* **2015**, *15*, 3303–3326. [CrossRef]

11.   Khalil, S.A.; Shaffie, A.M. Evaluation of transposition models of solar irradiance over Egypt. *Renew. Sustain. Energy Rev.* **2016**, *66*, 105–119. [CrossRef]

12.   Schroedter-Homscheidt, M.; Oumbe, A.; Benedetti, A.; Moncrette, J.J. Aerosols for concentrated solar electricity production forecasts: Requirement quantification and ECMWF/MACC aerosol forecast assessment. *Bull. Am. Meteorol. Soc.* **2013**, *94*, 903–914. [CrossRef]

13.   Maghami, M.R.; Hizam, H.; Gomes, C.; Radzi, M.A.; Rezadad, I.; Hajighorbani, S. Power loss due to soiling on solar panel: A review. *Renew. Sustain. Energy Rev.* **2016**, *59*, 1307–1316. [CrossRef]

14.   Rieger, D.; Steiner, A.; Bachmann, V.; Gasch, P.; Forstner, J.; Deetz, K.; Vogel, B.; Vogel, H. Impact of the 4 April 2014 Saharan dust outbreak on the photovoltaic power generation in Germany. *Atmos. Chem. Phys.* **2017**, *17*, 13391–13415. [CrossRef]

15.   Kosmopoulos, P.G.; Kazadzis, S.; Taylor, M.; Athanasopoulou, E.; Speyer, O.; Raptis, P.I.; Marinou, E.; Proestakis, E.; Solomos, S.; Gerasopoulos, E.; et al. Dust impact on surface solar irradiance assessed with model simulations, satellite observations and ground-based measurements. *Atmos. Meas. Tech.* **2017**, *10*, 2435–2453. [CrossRef]

16.   Neher, I.; Buchmann, T.; Crewell, S.; Evers-Dietze, B.; Pfeilsticker, K.; Pospichal, B.; Schirrmeister, C.; Meilinger, S. Impact of atmospheric aerosols on photovoltaic energy production Scenario for the Sahel zone. *Energy Procedia* **2017**, *125*, 170–179. [CrossRef]

17.   El-Askary, H.; Sarkar, S.; Kafatos, M.; El-Gahzawi, T. A multi-sensor approach to dust storm monitoring over the Nile Delta. *IEEE Trans. Geosc. Remote Sens.* **2003**, *41*, 2386–2391. [CrossRef]

18.   El-Askary, H.; Farouk, R.; Ichoku, C.; Kafatos, M. Transport of dust and anthropogenic aerosol across Alexandria, Egypt. *Ann. Geophys.* **2009**, *27*, 2869–2879. [CrossRef]

19.   Prasad, A.K.; El-Askary, H.; Kafatos, M. High altitude dust transport over Nile Delta during biomass burning season. *Environ. Pollut.* **2010**, *158*, 3385–3391. [CrossRef] [PubMed]

20.   Marey, H.S.; Gille, J.C.; El-Askary, H.; Shalaby, E.A.; El-Raey, M.E. Aerosol climatology over Nile Delta based on MODIS, MISR and OMI satellite data. *Atmos. Chem. Phys.* **2011**, *11*, 10637–10648. [CrossRef]

21.   Cowie, S.M.; Knippertz, P.; Marsham, H. A climatology of dust emission events from northern Africa using long-term surface observations. *Atmos. Chem. Phys.* **2014**, *14*, 8579–8597. [CrossRef]

22.   Shokr, M.; El-Tahan, M.; Ibrahim, A.; Steiner, A.; Gad, N. Long-term, high-resolution survey of atmospheric aerosols over Egypt with NASA's MODIS data. *Remote. Sens.* **2017**, *9*, 1027. [CrossRef]

23.   IPCC. *Climate Change: The Physical Science Basis*; Contribution of Working Group I to the Fifth Assessment Report of the Intergovernmental Panel on Climate Change; Stocker, T.F., Qin, D., Plattner, G.K., Tignor, M., Allen, S.K., Boschung, J., Nauels, A., Xia, Y., Bex, V., Midgley, P.M., Eds.; Cambridge University Press: Cambridge, UK; New York, NY, USA, 2013. [CrossRef]

24. World Meteorological Organization (WMO). *Global Atmospheric Watch (GAW) Aerosol Measurement Procedures, Guidelines and Recommendations*, 2nd ed.; GAW Report No. 227; World Meteorological Organization: Geneva, Switzerland, 2016.

25. Li, X.; Wagner, F.; Peng, W.; Yang, J.; Mauzerall, D.L. Reduction of solar photovoltaic resources due to air pollution in China. *Proc. Natl. Acad. Sci. USA* **2017**, *2017*, 11462. [CrossRef] [PubMed]

26. Gerard, H.; Puente, E.I.R.; Six, D. Coordination between transmission and distribution system operators in the electricity sector: A conceptual framework. *Util. Policy* **2018**, *50*, 40–48. [CrossRef]

27. Haywood, J.M.; Allan, R.P.; Culverwell, I.; Slingo, T.; Milton, S.; Edwards, J.; Clerbaux, N. Can desert dust explain the outgoing longwave radiation anomaly over the Sahara during July 2003? *J. Geoph. Res.* **2005**, *110*, D05105. [CrossRef]

28. Rodwell, M.J.; Jung, T. Understanding the local and global impacts of model physics changes: An aerosol example. *Q. J. R. Meteorol. Soc.* **2008**, *134*, 1479–1497. [CrossRef]

29. Benedetti, A.; Reid, J.S.; Knippertz, P.; Marsham, J.H.; Di Giuseppe, F.; Remy, S.; Basart, S.; Boucher, O.; Brooks, I.M.; Menut, L.; et al. Status and future of numerical atmospheric aerosol prediction with a focus on data requirements. *Atmos. Chem. Phys.* **2018**, *18*, 10615–10643. [CrossRef]

30. Takenaka, H.; Nakajima, T.Y. Estimation of solar radiation using a neural network based on radiative transfer. *J. Geophys. Res.* **2011**, *116*, D08215. [CrossRef]

31. Inman, R.H.; Pedro, H.T.C.; Coimbra, C.F.M. Solar forecasting methods for renewable energy integration. *Prog. Energy Combust. Sci.* **2013**, *39*, 535–576. [CrossRef]

32. Lefevre, M.; Oumbe, A.; Blanc, P.; Espinar, B.; Gschwind, B.; Qu, Z.; Wald, L.; Schroedter-Homscheidt, M.; Hoyer-Klick, C.; Arola, A.; et al. McClear: A new model estimating downwelling solar radiation at ground level in clear-sky conditions. *Atmos. Meas. Tech.* **2013**, *6*, 2403–2418. [CrossRef]

33. Kosmopoulos, P.G.; Kazadzis, S.; Lagouvardos, K.; Kotroni, V.; Bais, A. Solar energy prediction and verification using operational model forecasts and ground-based solar measurements. *Energy* **2015**, *93*, 1918–1930. [CrossRef]

34. Kosmopoulos, P.G.; Kazadzis, S.; Taylor, M.; Raptis, P.I.; Keramitsoglou, I.; Kiranoudis, C.; Bais, A.F. Assessment of the surface solar irradiance derived from real-time modelling techniques and verification with ground-based measurements. *Atmos. Meas. Tech.* **2018**, *11*, 907–924. [CrossRef]

35. Qu, Z.; Oumbe, A.; Blanc, P.; Espinar, B.; Gesell, G.; Gschwind, B.; Gschwind, B.; Kluser, L.; Lenevre, M.; Sabonet, L.; et al. Fast radiative transfer parameterisation for assessing the surface solar irradiance: The Heliosat-4 method. *Energy Meteorol.* **2017**, *26*, 33–57. [CrossRef]

36. Allen, R.J.; Norris, J.R.; Wild, M. Evaluation of multidecadal variability in CMIP5 surface solar radiation and inferred underestimation of aerosol direct effects over Europe, China, Japan, and India. *J. Geophys. Res. Atmos.* **2013**, *118*, 6311–6336. [CrossRef]

37. Ishii, T.; Otani, K.; Takashima, T.; Xue, Y. Solar spectral influence on the performance of photovoltaic (PV) modules under fine weather and cloudy weather conditions. *Prog. Photovolt. Res. Appl.* **2013**, *21*, 481–489. [CrossRef]

38. Dirnberger, D.; Blackburn, G.; Müller, B.; Reise, C. On the impact of solar spectral irradiance on the yield of different PV technologies. *Sol. Energy Mater. Sol. Cells* **2015**, *132*, 431–442. [CrossRef]

39. El-Metwally, M. Simple new methods to estimate global solar radiation based on meteorological data in Egypt. *Atmos. Res.* **2004**, *69*, 217–239. [CrossRef]

40. El-Metwally, M.; Alfaro, S.C.; Abdel Wahab, M.M.; Favez, O.; Mohamed, Z.; Chatenet, B. Aerosol properties and associated radiative effects over Cairo (Egypt). *Atmos. Res.* **2011**, *99*, 263–276. [CrossRef]

41. El-Metwally, M.; Alfaro, S.C. Correlation between meteorological conditions and aerosol characteristics at an East-Mediterranean coastal site. *Atmos. Res.* **2013**, *132–133*, 76–90. [CrossRef]

42. Eissa, Y.; Korany, M.; Aoun, Y.; Boraiy, M.; Abdel-Wahab, M.M.; Alfaro, S.C.; Blanc, P.; El-Metwally, M.; Ghedira, H.; Hungershoefer, K.; et al. Validation of the surface downwelling solar irradiance estimates of the HelioClim-3 database in Egypt. *Remote. Sens.* **2015**, *7*, 9269–9291. [CrossRef]

43. Dee, D.P.; Uppala, S. Variational bias correction of satellite radiance data in the ERA-Interim reanalysis. *Q. J. R. Meteorol. Soc.* **2009**, *135*, 1830–1841. [CrossRef]

44. Inness, A.; Baier, F.; Benedetti, A.; Bouarar, I.; Chabrillat, S.; Clark, H.; Clerbaux, C.; Coheur, P.; Engelen, R.J.; Errera, Q.; et al. The MACC reanalysis: An 8 yr data set of atmospheric composition. *Atmos. Chem. Phys.* **2013**, *13*, 4073–4109. [CrossRef]

45. Penning de Vries, M.J.M.; Beirle, S.; Hormann, C.; Kaiser, J.W.; Stammes, P.; Tilstra, L.G.; Tuinder, O.N.E.; Wagner, T. A global aerosol classification algorithm incorporating multiple satellite data sets of aerosol and trace gas abundances. *Atmos. Chem. Phys.* **2015**, *15*, 10597–10618. [CrossRef]

46. Eskes, H.; Huijnen, V.; Arola, A.; Benedictow, A.; Blechschmidt, A.M.; Botek, E.; Boucher, O.; Bouarar, I.; Chabrillat, S.; Cuevas, E.; et al. Validation of reactive gases and aerosols in the MACC global analysis and forecast system. *Geosci. Model Dev.* **2015**, *8*, 3523–3543. [CrossRef]

47. Boucher, O.; Pham, M.; Venkataraman, C. *Simulation of the Atmospheric Sulfur Cycle in the LMD GCM: Model Description, Model Evaluation, and Global and European Budgets*; Note 23; Inst. Pierre-Simon Laplace: Paris, France, 2002.

48. Reddy, M.S.; Boucher, O.; Bellouin, N.; Schulz, M.; Balkanski, Y.; Dufresne, J.L.; Pham, M. Estimates of global multi-component aerosol optical depth and direct radiative perturbation in the Laboratoire de Météorologie Dynamique general circulation model. *J. Geophys. Res.* **2005**, *110*, D10S16. [CrossRef]

49. Huijnen, V.; Eskes, H.J.; Basart, S.; Benedictow, A.; Blechschmidt, A.M.; Chabrillat, S.; Christophe, Y.; Cuevas, E.; Flentje, H.; Jones, L.; et al. Validation Report of the CAMS Near-Real-Time Global Atmospheric Composition Service. System Evolution and Performance Statistics. Copernicus Atmosphere Monitoring Service (CAMS) Report 2015, CAMS84_1_D1.1_201512. Available online: https://atmosphere.copernicus. eu/sites/default/files/repository/CAMS84_1_D1.1_201512_0.pdf (accessed on 8 November 2018).

50. Dentener, F.; Kinne, S.; Bond, T.; Boucher, O.; Cofala, J.; Generoso, S.; Ginoux, P.; Gong, S.; Hoelzemann, J.J.; Ito, A.; et al. Emissions of primary aerosol and precursor gases in the years 2000 and 1750 prescribed data-sets for AeroCom. *Atmos. Chem. Phys.* **2006**, *6*, 4321–4344. [CrossRef]

51. Guelle, W.; Schulz, M.; Balkanski, Y.; Dentener, F. Influence of the source formulation on modeling the atmospheric global distribution of the sea salt aerosol. *J. Geophys. Res.* **2001**, *106*, 27509–27524. [CrossRef]

52. Schulz, M.; de Leeuw, G.; Balkanski, Y. Sea-salt aerosol source functions and emissions. In *Emission of Atmospheric Trace Compounds*; Granier, C., Artaxo, P., Reeves, C.E., Eds.; Kluwer Acad.: Norwell, MA, USA, 2004; pp. 333–354.

53. Morcrette, J.J.; Beljaars, A.; Benedetti, A.; Jones, L.; Boucher, O. Sea-salt and dust aerosols in the ECMWF IFS model. *Geophys. Res. Lett.* **2008**, *35*, L24813. [CrossRef]

54. Levy, R.C.; Remer, L.A.; Dubovik, O. Global aerosol optical properties and application to Moderate Resolution Imaging Spectroradiometer aerosol retrieval over land. *J. Geophys. Res. Atmos.* **2007**, *112*, D13210. [CrossRef]

55. Levy, R.C.; Remer, L.A.; Kleidman, R.G.; Mattoo, S.; Ichoku, C.; Kahn, R.; Eck, T.F. Global evaluation of the Collection 5 MODIS dark-target aerosol products over land. *Atmos. Chem. Phys.* **2010**, *10*, 10399–10420. [CrossRef]

56. Remer, L.A.; Kaufman, Y.J.; Tanre, D.; Mattoo, S.; Chu, D.A.; Martins, J.V.; Li, R.R.; Ichoku, C.; Levy, R.C.; Kleidman, R.G.; et al. The MODIS aerosol algorithm, products, and validation. *J. Atmos. Sci.* **2005**, *62*, 947–973. [CrossRef]

57. Remer, L.A.; Kleidman, R.G.; Levy, R.C.; Kaufman, Y.J.; Tanré, D.; Mattoo, S.; Martins, J.V.; Ichoku, C.; Koren, I.; Yu, H.; et al. Global aerosol climatology from the MODIS satellite sensors. *J. Geophys. Res. Atmos.* **2008**, *113*, D14S07. [CrossRef]

58. Hsu, N.C.; Jeong, M.J.; Bettenhausen, C.; Sayer, A.M.; Hansell, R.; Seftor, C.S.; Huang, J.; Tsay, S.C. Enhanced Deep Blue aerosol retrieval algorithm: The second generation. *J. Geophys. Res.* **2013**, *118*, 9296–9315. [CrossRef]

59. Levy, R.C.; Mattoo, S.; Munchak, L.A.; Remer, L.A.; Sayer, A.M.; Patadia, F.; Hsu, N.C. The Collection 6 MODIS aerosol products over land and ocean. *Atmos. Meas. Tech.* **2013**, *6*, 2989–3034. [CrossRef]

60. Sayer, A.M.; Munchak, L.A.; Hsu, N.C.; Levy, R.C.; Bettenhausen, C.; Jeong, M.J. MODIS Collection 6 aerosol products: Comparison between Aqua's e-Deep Blue, Dark Target, and "merged" data sets, and usage recommendations. *J. Geophys. Res. Atmos.* **2014**, *119*, 13965–13989. [CrossRef]

61. Hubanks, P.A.; Platnick, S.; King, M.D.; Ridgway, B. *MODIS Atmosphere L3 Gridded Product Algorithm Theoretical Basis Document & Users Guide*; ATBD Reference Number: ATBD-MOD-30; MODIS Atmosphere Documents 2016; Available online: https://modis-images.gsfc.nasa.gov/index.html or https://modis-images.gsfc.nasa.gov/_docs/L3_ATBD_C6.pdf (accessed on 22 October 2018).

62. Sayer, A.M.; Hsu, N.C.; Bettenhausen, C.; Jeong, M.J. Validation and uncertainty estimates for MODIS Collection 6 "Depp Blue" aerosol data. *J. Geophys. Res.* **2013**, *118*, 7864–7873. [CrossRef]

63. Mayer, B.; Kylling, A. Technical note: The libRadtran software package for radiative transfer calculations—Description and examples of use. *Atmos. Chem. Phys.* **2005**, *5*, 1855–1877. [CrossRef]

64. Emde, C.; Buras-Schnell, R.; Kylling, A.; Mayer, B.; Gasteiger, J.; Hamann, U.; Kylling, J.; Richter, B.; Pause, C.; Dowling, T.; et al. The libRadtran software package for radiative transfer calculations (version 2.0.1). *Geosci. Model Dev.* **2016**, *9*, 1647–1672. [CrossRef]

65. Dahlback, A.; Stamnes, K. A new spherical model for computing the radiation field available for photolysis and heating at twilight. *Planet Space Sci.* **1991**, *39*, 671–683. [CrossRef]

66. Taylor, M.; Kosmopoulos, P.G.; Kazadzis, S.; Keramitsoglou, I.; Kiranoudis, C.T. Neural network radiative transfer solvers for the generation of high resolution solar irradiance spectra parameterized by cloud and aerosol parameters. *J. Quant. Spectr. Radiat. Transf.* **2015**, *168*, 176–192. [CrossRef]

67. Hornik, K.; Stinchcombe, M.; White, H. Multilayer Feedforward Networks Are Universal Approximators. *Neural Netw.* **1989**, *2*, 359–366. [CrossRef]

68. Taylor, M.; Kazadzis, S.; Tsekeri, A.; Gkikas, A.; Amiridis, V. Satellite retrieval of aerosol microphysical and optical parameters using neural networks: A new methodology applied to the Sahara desert dust peak. *Atmos. Meas. Tech.* **2014**, *7*, 3151–3175. [CrossRef]

69. Shettle, E.P. Models of aerosols, clouds and precipitation for atmospheric propagation studies. In Proceedings of the AGARD Conference 454 on Atmospheric Propagation in the UV, Visible, IR and MM-Region and Related System Aspects, Copenhagen, Denmark, 9–13 October 1989; Available online: http://www.dtic.mil/dtic/tr/fulltext/u2/a221594.pdf (accessed on 8 November 2018).

70. Hess, M.; Koepke, P.; Schult, I. Optical Properties of Aerosols and Clouds: The Software Package OPAC. *Bull. Am. Meteor. Soc.* **1998**, *79*, 831–844. [CrossRef]

71. Kato, S.; Ackerman, T.; Mather, J.; Clothiaux, E. The k-distribution method and correlated-k approxiamation for shortwave radiative transfer model. *J. Quant. Spectrosc. Radiat. Transf.* **1999**, *62*, 109–121. [CrossRef]

72. Kinne, S.; Schulz, M.; Textor, C.; Guibert, S.; Balkanski, Y.; Bauer, S.E.; Berntsen, T.; Berglen, T.F.; Boucher, O.; Chin, M.; et al. An AeroCom initial assessment—Optical properties in aerosol component modules of global models. *Atmos. Chem. Phys.* **2006**, *6*, 1815–1834. [CrossRef]

73. Kumar, S.; Devara, P.C.S. Along-term study of aerosol modulation of atmospheric and surface solar heating over Pune, India. *Tellus B Chem. Phys. Meteorol.* **2012**, *64*, 18420. [CrossRef]

74. Eskes, H.J.; Douros, J.; Akritidis, D.; Antonakaki, T.; Blechschmidt, A.M.; Clark, H.; Gielen, C.; Hendrick, F.; Kapsomenakis, J.; Kartsios, S.; et al. Validation of CAMS Regional Services: Concentrations Above the Surface, Status Update for March–May 2017. Copernicus Atmosphere Monitoring Service (CAMS) Report 2017. Available online: https://atmosphere.copernicus.eu/sites/default/files/2018-08/CAMS84_2015SC2_D84.5.1.8_D84.6.1.3_2017MAM_v1.pdf (accessed on 12 October 2018).

75. PVGIS. Photovoltaic Geographical Information System. Available online: http://re.jrc.ec.europa.eu/pvgis/ (accessed on 12 October 2018).

76. Eck, M.; Hirsch, T.; Feldhoff, J.F.; Kretschmann, D.; Dersch, J.; Gavilan Morales, A.; Gonzales-Martinez, L.; Bachelier, C.; Platzer, W.; Riffelmann, K.J.; et al. Guidelines for CSP yield analysis—Optical losses of line focusing systems; definitions, sensitivity analysis and modeling approaches. *Energy Procedia* **2014**, *49*, 1318–1327. [CrossRef]

77. Ouali, H.A.L.; Merrouni, A.A.; Moussaoui, M.A.; Mezrhab, A. Electricity yield analysis of a 50 MW solar power plant under Moroccan climate. In Proceedings of the International Conference on Electrical and Information Technologies (ICEIT) 2015, Marrakech, Morocco, 25–27 March 2015. [CrossRef]

78. El-Metwally, M.; Alfaro, S.C.; Abdel Wahab, M.M.; Zakey, A.S.; Chatenet, B. Seasonal and inter-annual variability of the aerosol content in Cairo (Egypt) as deduced from the comparison of MODIS aerosol retrievals with direct AERONET measurements. *Atmos. Res.* **2010**, *97*, 14–25. [CrossRef]

79. Robaa, S.M. A study of solar radiation climate at Cairo urban area, Egypt and its environs. *Int. J. Clim.* **2006**, *26*, 1913–1928. [CrossRef]

80. Kohil, E.E.; Saleh, I.H.; Ghatass, Z.F. A study of atmospheric aerosol optical properties over Alexandria city—Egypt. *J. Phys. Conf. Ser.* **2017**, *810*, 012033. [CrossRef]

81. Bellouin, N.; Boucher, O.; Haywood, J.; Reddy, M.S. Global estimate of aerosol direct radiative forcing from satellite measurements. *Nature* **2005**, *7071*, 1138–1141. [CrossRef] [PubMed]

82. Lee, L.A.; Reddington, C.L.; Carslaw, K.S. On the relationship between aerosol model uncertainty and radiative forcing uncertainty. *Proc. Natl. Acad. Sci. USA* **2016**, *113*, 5820–5827. [CrossRef] [PubMed]

83. IRENA. Renewable Power Generation Costs in 2014, Report 2015. Available online: http://www.irena.org/-/media/Files/IRENA/Agency/Publication/2015/IRENA_RE_Power_Costs_2014_report.pdf (accessed on 9 November 2018).

84. Haney, J.; Burstein, A. PV System Operations and Maintenance Fundamentals. Solar America Board for Codes and Standards Report 2013. Available online: http://www.solarabcs.org/about/publications/reports/operations-maintenance/pdfs/SolarABCs-35-2013.pdf (accessed on 12 November 2018).

85. IRENA. Renewable Energy Integration in Power Grids 2015, IEA-ETSAP and IRENA Technology Brief E15. Available online: http://www.irena.org/DocumentDownloads/Publications/IRENA-ETSAP_Tech_Brief_Power_Grid_Integration_2015.pdf (accessed on 7 August 2018).

86. Evenflow SPRL. Business Plan for the Establishment, Operation and Exploitation of a Solar Farm: Aswan's Solar Plant Project, Report 2017. Available online: http://solea.gr/wp-content/uploads/2018/03/Aswan-Solar-Plant-Business-Plan.pdf (accessed on 11 November 2018).

87. MOEE. Ministry of Electricity and Renewable Energy of Egypt. Electricity Pricelist 2017–2018. Available online: http://www.moee.gov.eg/english_new/home.aspx (accessed on 10 November 2018).

remote sensing

MDPI

Article

Characterizing Variability of Solar Irradiance in San Antonio, Texas Using Satellite Observations of Cloudiness

Shuang Xia [1,2], Alberto M. Mestas-Nuñez [2,*], Hongjie Xie [2], Jiakui Tang [2,3] and Rolando Vega [4]

[1] Texas Sustainable Energy Research Institute, University of Texas at San Antonio, San Antonio, TX 78249, USA; shuang.xia@my.utsa.edu

[2] Laboratory for Remote Sensing and Geoinformatics, Department of Geological Sciences, University of Texas at San Antonio, San Antonio, TX 78249, USA; hongjie.xie@utsa.edu (H.X.); jktang@ucas.ac.cn (J.T.)

[3] University of Chinese Academy of Sciences, Beijing 100049, China

[4] CPS Energy, San Antonio, TX 78205, USA; RVega-Avila@cpsenergy.com

* Correspondence: Alberto.Mestas@utsa.edu; Tel.: +1-210-458-6690

Received: 29 October 2018; Accepted: 9 December 2018; Published: 12 December 2018

Abstract: Since the main attenuation of solar irradiance reaching the earth's surface is due to clouds, it has been hypothesized that global horizontal irradiance attenuation and its temporal variability at a given location could be characterized simply by cloud properties at that location. This hypothesis is tested using global horizontal irradiance measurements at two stations in San Antonio, Texas, and satellite estimates of cloud types and cloud layers from the Geostationary Operational Environmental Satellite (GOES) Surface and Insolation Product. A modified version of an existing solar attenuation variability index, albeit having a better physical foundation, is used. The analysis is conducted for different cloud conditions and solar elevations. It is found that under cloudy-sky conditions, there is less attenuation under water clouds than those under opaque ice clouds (optically thick ice clouds) and multilayered clouds. For cloud layers, less attenuation was found for the low/mid layers than for the high layer. Cloud enhancement occurs more frequently for water clouds and less frequently for mixed phase and cirrus clouds and it occurs with similar frequency at all three levels. The temporal variability of solar attenuation is found to decrease with an increasing temporal sampling interval and to be largest for water clouds and smallest for multilayered and partly cloudy conditions. This work presents a first step towards estimating solar energy potential in the San Antonio area indirectly using available estimates of cloudiness from GOES satellites.

Keywords: clear sky index; solar irradiance; downward shortwave radiation; global horizontal irradiance; solar variability; cloud categories; GOES satellites

1. Introduction

Radiation from the sun is the primary energy source for the Earth [1,2]. Accurate measurements of broadband shortwave irradiance are crucial for renewable energy resource assessments and climate change research [3]. There is also a growing demand for integrating solar energy into the electricity grid with accurate characterization of solar irradiance variability in order to provide a better quality of service. The temporal variability of solar radiation is largely due to rapid changes in atmosphere conditions, especially clouds, which create significant fluctuations in voltage [4–9]. The insolation, cloud amount, cloud type, cloud height, and surface properties determine whether clouds cause a radiative excess or deficit in a given region. Considering the importance of the stability of an energy distribution network and considering also that the reduction of solar radiation in the atmosphere due to clouds is typically larger (~30%) than that due to aerosols under clear sky (~10%) [10], it would be useful to understand how solar energy varies under different cloud conditions.

To quantify the effects of clouds on radiation at a specific location, a clear sky index (CSI) can be used [11–13]. The CSI is defined as the ratio of the actual ground irradiance to the irradiance under a cloud-free sky [14]. Thus, when calculating CSI, the selection of a good clear sky model is important. Reno et al. [15] presented an overview of clear sky models of global horizontal irradiance including very simple models that are solely dependent upon solar zenith angle and quite complex models that add dependencies on various atmospheric parameters. They found that the Ineichen model [16,17], which accounts for solar zenith angle, air pressure, temperature, relative humidity, aerosol content, Rayleigh and site elevation, compared well (error of ~5.0%) with the REST2 (Reference Evaluation of Solar Transmittance, 2-bands) model developed by Gueymard [18,19].

Cloud properties including cover, transmittance, moving velocity, type, and height, all influence solar irradiance and its variability on the ground [8,20–24]. Udelhofen and Cess [25] applied spectral analysis to cloud cover anomalies over the United States for the period 1900–1987 and found that the coherence between cloud cover and sun spots numbers (which is a proxy for solar variability) at a period of 11 years was significant (~0.7). Reno and Stein [23] hypothesized that the variability of cloud properties at a given location and time could be used to model the variability of ground solar irradiance at that location. Using global horizontal irradiance observations at two locations in Las Vegas, Nevada and cloud types from Geostationary Operation Environmental Satellites (GOES), they found that the temporal variability of ground irradiance is higher with water clouds and generally lower with opaque ice clouds. Nguyen et al. [26] validated the sky-imager-based global horizontal irradiance variability with ground observations using the temporal variability index introduced by Stein et al. [22] and found a high correlation (0.91).

Previously, we compared satellite-derived global solar irradiance from GOES with ground observations at two stations in San Antonio, Texas and found overall a good agreement on hourly and daily timescales [27]. The irradiance data from these two ground stations are used here to study the temporal variability of global horizontal irradiance on sub-hourly time scales. The specific objective of this study is to use the global horizontal irradiance measured at these two San Antonio sites and simultaneous GOES-derived cloud properties (i.e., cloud type and height) to test the hypothesis of Reno and Stein [23], namely that the global horizontal irradiance variability at these stations could be characterized by satellite-derived cloud properties at those locations. In addition, a modified version of the global horizontal irradiance variability index of Reno and Stein [23], but bearing a more physical foundation, is used here. In this paper, the datasets used are presented in Section 2, the analysis methods in Section 3, and the results in Section 4. The paper ends with a discussion in Section 5 and conclusions in Section 6.

2. Data

2.1. Ground Observations of Solar Irradiance

In situ measurements of global horizontal irradiance (G_h) come from the two ground stations shown in Figure 1, namely the main campus of the University of Texas at San Antonio (UTSA) and Alamo 1 Solar Farm (ASF). At the UTSA station (29.5833° N, 98.6199° W, 305 m elevation above sea level), instantaneous values of G_h were recorded every 5 min from 1 May to 25 October 2015 by a LI-200R pyranometer (0.4–1.1 µm). At the ASF site (29.7010° N, 98.4432° W, 164 m elevation above sea level), instantaneous values of G_h were recorded at irregular time intervals which is against best practices. The distribution of the sampling has a mean of 0.16 min and standard deviation of 0.23 min from July to September 2014 by a CMP11_L pyranometer (0.285–2.8 µm). A second ASF dataset based on temporal averages is used for the period September 2015–October 2016 in which the raw data was averaged every 15 min. The uncertainties of the instruments, LI-200R and CMP11_L, are respectively 3% and <2%, according to the manufacturer specifications. The bandwidth of the LI-200R pyranometer is much narrower than that of the CMP11_L pyranometer. Slight sensor differences in offset and/or gain and the pyranometer calibration at the UTSA station might have induced some errors, resulting in slightly smaller G_h than that recorded at the ASF station [27]. Since at large zenith angles the accuracies of radiative transfer models and pyranometers degrade rapidly [27], and slight changes in actual radiance can cause large changes in the clear sky index, observations with solar zenith angles >75° were, therefore, not included in this study.

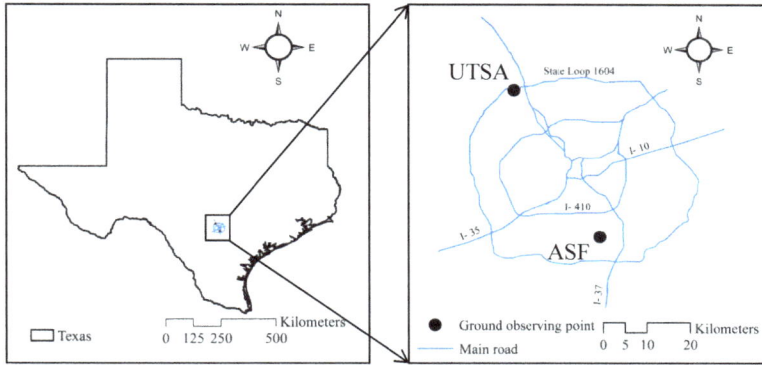

Figure 1. Map of the state of Texas showing the City of San Antonio (**left panel**). The expanded region (**right panel**) shows the sites of the ground observing points (University of Texas at San Antonio [UTSA] and Alamo 1 Solar Farm [ASF]) for solar irradiance (black solid dots) relative to the city's main roads (reproduced from Reference [27]).

2.2. Satellite Observations of Clouds

The satellite cloud data used in this paper comes from the GOES Surface and Insolation Products (GSIP) available from the National Oceanic Atmospheric Administration (NOAA). The GSIP dataset is based on remote sensing measurements obtained using the visible and infrared channels of the GOES satellites. Both GOES-East and GOES-West satellites are used, which respectively provide hourly instantaneous snapshots 45 min past the hour and on the hour since January 1996 [28]. The approach of GSIP for computing solar radiation is to first retrieve cloud properties and then use these properties as inputs to the Satellite Algorithm for Shortwave Radiation Budget (SASRAB) model [29]. The algorithm used to determine the dominant cloud types is the AVHRR Pathfinder Atmospheres-Extended (PATMOS-X) model [29–31]. In this study, hourly cloud property data (classified by cloud types and cloud layers, as defined in Table 1) with a resolution of 2.3 km (longitude) × 4.9 km (latitude) for San Antonio, Texas are obtained from GSIP.

Table 1. Cloud categories for National Oceanic Atmospheric Administration (NOAA) cloud-types (same categories used in Reference [23]) and cloud-layer classifications from Geostationary Operational Environmental Satellite (GOES) surface and insolation products.

Classification	Categories	Description
Cloud type	0	clear
	1	partly (partly cloudy/fog)
	2	water (water cloud)
	3	mixed (supercooled/mixed-phase cloud)
	4	opaque ice (optically thick ice cloud)
	5	cirrus (optically thin ice cloud)
	6	multilayered (cirrus over lower cloud)
Cloud layer	1	low (0–2 km)
	2	mid (2–7 km)
	3	high (5–13 km)

3. Methods

Ground measurements of G_h with various sub-hourly temporal sampling intervals were used in this study. For UTSA, the sampling intervals considered were 5, 10, and 15 min and for ASF the sampling intervals were 1, 5, 10, and 15 min. For the UTSA data, the 10 and 15 min datasets (instantaneous data) were generated by decimating the original 5-min datasets so that the resulting grids match the times of the satellite imagery, as illustrated in Figure 2. For ASF, the 1, 5, and 10 min datasets were based on instantaneous data while the 15 min dataset was a combination of an instantaneous and a temporally averaged dataset. These four regularly gridded ASF datasets were generated in a similar way to the UTSA datasets by decimating the irregularly sampled high resolution (~0.16 min mean sampling rate) dataset using the nearest data to the desired regular grid which was chosen to match the times of the satellite imagery. When generating the 10-min gridded dataset one can choose to match the time of the satellite image 15 min before the hour or indistinctively the one on the hour, we chose the latter.

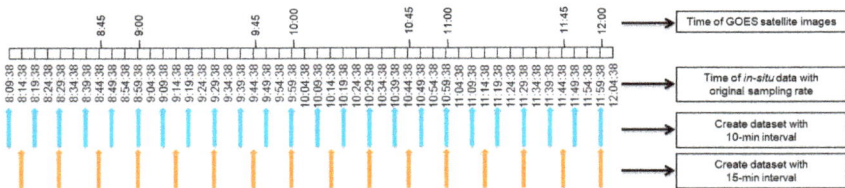

Figure 2. Graphical representation of the decimation procedure used to create UTSA datasets with 10-min and 15-min time intervals from the original 5-min dataset. The time of the GOES satellite images are also indicated in the figure.

To quantify the impact of clouds on solar irradiance, a clear sky index (CSI) [11,14,32] was used as follows:

$$\text{CSI} = \frac{G_h}{G_{hc}}, \tag{1}$$

where G_{hc} is the global horizontal clear-sky irradiance. Thus, CSI should be approximately equal to 1 under clear-sky conditions and typically smaller than 1 when clouds are present. In this paper, G_{hc} was calculated from the following model based on Reference [15]:

$$G_{hc} = a_1 \cdot I_0 \cdot \sin(h) \cdot \exp(-a_2 \cdot AM \cdot (f_{h1} + f_{h2} \cdot (T_L - 1))) \cdot \exp\left(0.01 \cdot AM^{1.8}\right), \tag{2}$$

where I_0 is the normal incidence extraterrestrial irradiance; h is the solar elevation angle, AM is the altitude corrected air mass [33], $a_1 = 5.09 \times 10^{-5} \times$ altitude + 0.868, $a_2 = 3.92 \times 10^{-5} \times$ altitude + 0.0387, $f_{h1} = \exp(-\text{altitude}/8000)$, $f_{h2} = \exp(-\text{altitude}/1250)$, and T_L is Linke turbidity available from the Solar Radiation Data website (www.soda-pro.com). For every decimated G_h dataset (i.e., 1, 5, 10, or 15 min sampling rate), a corresponding G_{hc} dataset was generated by calculating G_{hc} only for those times when G_h was available. The CSI was then calculated from Equation (1) for all sampling rates and averaged over a 60-min period centered on the time of each satellite image.

A variation of the CSI, but for evaluating the impact of clouds on energy rather than irradiance and for comparisons using the satellite retrieved cloudiness, is introduced here. The new index is referred to as the clear-sky energy index (CEI) and is defined as the ratio of two integrals. The numerator is the integral over one hour centered around the time of each satellite image of an array of n global-horizontal radiant energy values observed at a temporal sampling interval Δt and the denominator is the same integral but computed using the calculated global-horizontal clear-sky irradiance from Equation (2), as follows:

$$\text{CEI} = \frac{\sum_{k=2}^{n}(((G_h)_k + (G_h)_{k-1}) * \Delta t/2)}{\sum_{k=2}^{n}(((G_{hc})_k + (G_{hc})_{k-1}) * \Delta t/2)}. \tag{3}$$

To evaluate the impact of clouds on the temporal variability of solar irradiance, the following variability index (VI) was used [22,23] which was calculated from an equation similar to Equation (3), as follows:

$$VI = \frac{\sum_{k=2}^{n} \sqrt{((G_h)_k - (G_h)_{k-1})^2 + \Delta t^2}}{\sum_{k=2}^{n} \sqrt{((G_{hc})_k - (G_{hc})_{k-1})^2 + \Delta t^2}}.$$ (4)

As seen in Equation (4), the definition of VI involves adding variables with different units (i.e., energy flux for irradiance and time for the temporal sampling interval) which does not have any physical basis. To avoid this inconsistency, we proposed a modified version of the variability index (VI_{new}) defined as follows:

$$VI_{new} = \frac{\sum_{k=2}^{n} |(G_h)_k - (G_h)_{k-1}| / \Delta t}{\sum_{k=2}^{n} |(G_{hc})_k - (G_{hc})_{k-1}| / \Delta t}.$$ (5)

This variability index was thus based on the ratio of the sum of the changes of global horizontal irradiance (in absolute value) over each time Δt taken over one hour centered on the time of each satellite image to the same sum quantity except computed for clear sky. The numerator one of the absolute value of the changes of observed ground global horizontal irradiance over each Δt and the other integral same quantity but based on the global horizontal irradiance under clear sky. The VI_{new} in Equation (5) is thus a modified version of the VI proposed by Reno and Stein [23]. This definition is very similar to the previous one given in Equation (4) but avoids the inconsistency of adding values with different units.

All solar radiation indices (CSI, CEI, VI, and VI_{new}) were then paired with simultaneous satellite estimates of clouds for further analysis [23].

4. Results

4.1. Solar Attenuation under Different Cloud Conditions

Figure 3 presents the histograms of CSI at the UTSA (left panel) and ASF (right panel) study sites using all available ground data and their original sampling rates. The means of these CSI distributions are 0.74 and 0.87 for UTSA and ASF, respectively, and are indicated with red vertical dashed lines in Figure 3. The modes of CSI are, respectively, 0.96 and 1.02 for UTSA and ASF. In principle, the maximum value of CSI would be equal to 1 which represents clear-sky conditions, i.e., solar irradiance should not be attenuated when passing through a clear atmosphere. The fact that some CSI values in Figure 3 are greater than 1 may be associated with (1) a possible underestimation of the clear sky global radiation (G_{hc}) from Equation (2) and/or (2) a possible overestimation of the ground measurements (G_h) at both stations due to cloud enhancement [34,35]. Both CSI distributions peak at 1 and are left-skewed indicating that there were more clear-sky than cloudy conditions at both locations.

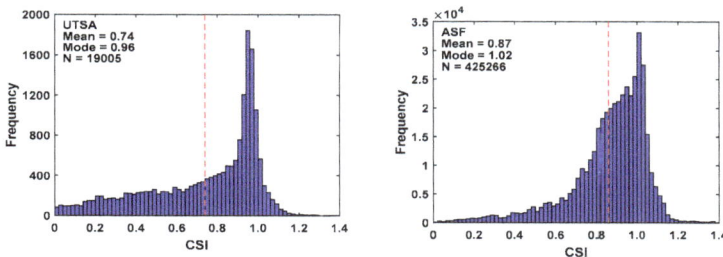

Figure 3. Histograms of the clear sky index (CSI) from all available ground data at their original sampling rates measured at the two study locations (UTSA and ASF), with the red dashed lines indicating the CSI mean.

The histograms of the averaged CSI values calculated using all available ground station data and the original sampling rates at the two study locations and for each cloud type are shown in Figure 4. Similar histograms, but for each cloud layer, are shown in Figure 5. Overall, two main conclusions can be drawn from these figures: (1) the distributions of CSI under clear-sky conditions in the upper panels of Figure 4 (>61% of all cases in both UTSA and ASF) peak at 1 and their average is greater than that for any cloud type or layer and (2) the mean of CSI is higher at ASF than at UTSA for all corresponding cloud types and layers. Among the various cloud types, water clouds have the larger CSI values, as shown in Figure 4. Values of CSI greater than 1 for clear-sky conditions reflect uncertainties in the CSI estimates since aerosols cannot enhance solar radiation. For cloudy conditions, CSI values greater than 1 are found more frequently for water clouds and less frequently for mixed and cirrus clouds suggesting that these cloud types are the most effective at producing cloud enhancement.

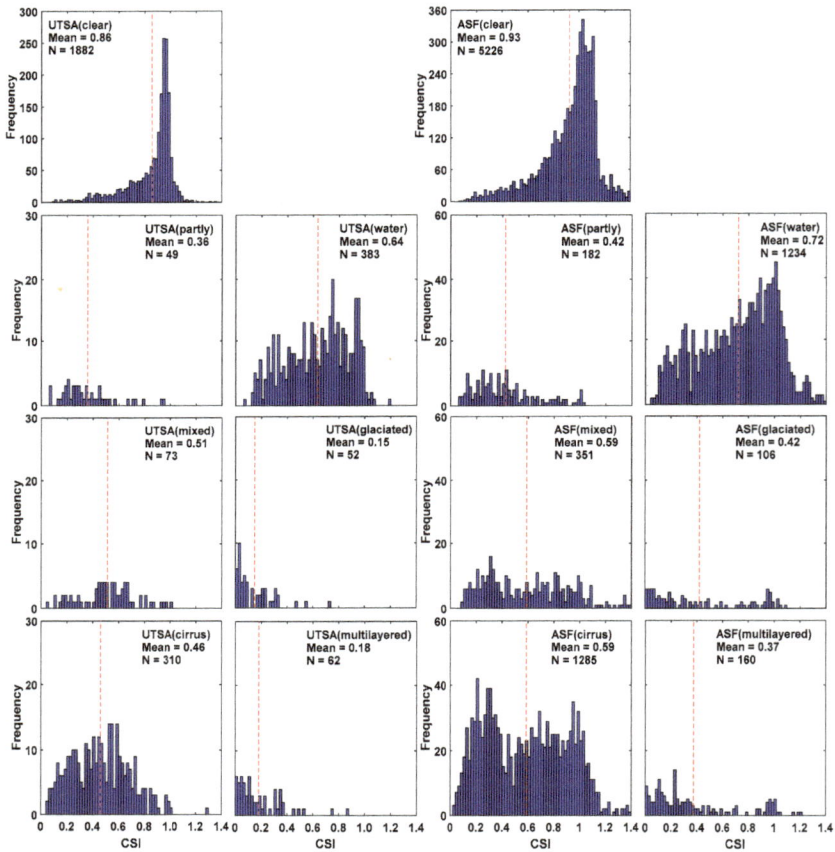

Figure 4. Histograms of the averaged CSI values from all available ground data at their original sampling rates for each cloud type at the UTSA station (**left two panel columns**) and ASF site (**right two panel columns**), with the red dashed line indicating the mean CSI for each cloud type.

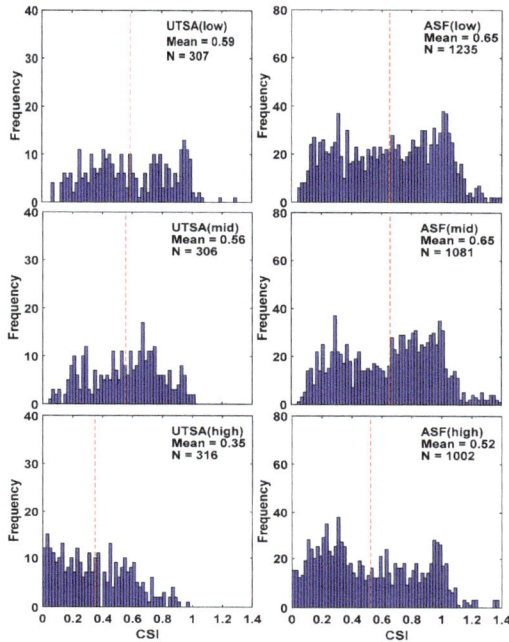

Figure 5. Same as Figure 4 but for individual cloud layers.

Among the three different cloud layers, CSI values are larger for low- and mid-level clouds than for high-level clouds, as shown in Figure 5. The water clouds are composed of liquid water droplets and the mixed clouds composed of supercooled water droplets or both ice and supercooled water. The water clouds absorb more visible and near-infrared radiation, resulting in lower CSI as compared to the clear-sky conditions [36–38]. Compared to water clouds, mixed clouds consist of water vapor, liquid droplets, and ice particles, and thus reflect more solar radiation and result in lower CSI. Cirrus clouds are higher-altitude, thinner clouds which are highly transparent to shortwave radiation. Partly cloudy conditions, in contrast, occur at the lower layer and are much thicker than cirrus clouds. The mean CSI, therefore, is higher under cirrus clouds than under partly cloudy conditions. The opaque ice clouds are composed of ice crystals or opaque clouds which are non-transmissive. The multilayer clouds consist of clouds from different layers. Both opaque ice and multilayer clouds reflect or absorb most of the solar energy, resulting in lower CSI than any other cloud type. It was also found that the distributions of CSI under clear and water cloud conditions are left skewed, while under cirrus clouds the distribution appears to be bimodal. Regarding cloud enhancement, the ASF data indicates that cloud enhancement occurs with similar frequency at all three levels.

Figure 6 shows the mean CSI and CEI with their standard deviation error bars under each cloud type and cloud layer for both stations. Similar to Figures 4 and 5, CSI averages vary more as a function of cloud types than cloud layers and are higher at ASF than at UTSA. CSI is larger under clear-sky conditions (type 0) followed by water clouds (type 2) and lower under opaque ice (type 4) and multilayered (type 6) clouds. For cloud layers, CSI is larger for low and mid clouds (layers 1 and 2) and smaller for high clouds (layer 3). As expected, the pattern of CEI is similar to that of CSI. One difference, however, is that CEI is not always larger at ASF.

Figure 7 shows the mean CSI and CEI with their ± 1 standard deviation error bars for various sky conditions as a function of solar zenith angles at the ASF station. The distributions of CSI and CEI means and standard deviations in Figure 7 show both similarities and differences. Both CSI and CEI means are larger for clear than for cloudy conditions and they do not vary much with solar zenith angles. For cloudy conditions, the larger values of CSI and CEI means are for water clouds and the lower values for opaque ice and multilayered clouds. For CSI under partly, water, mixed, and cirrus type clouds, the means are generally lower at high solar zenith angles except for opaque ice clouds which have the lowest value of CSI at the lowest solar zenith angle (0–15°). For CEI under water, mixed, opaque ice, and cirrus, the means show smaller values for intermediate solar zenith angles. For cloud layers, as shown in the lower panel of Figure 7, the mean CSI decreases with increasing solar zenith angle for zenith angles greater than 0–15°. This pattern is similar for CEI which is overall higher at low, compared to high, solar zenith angles.

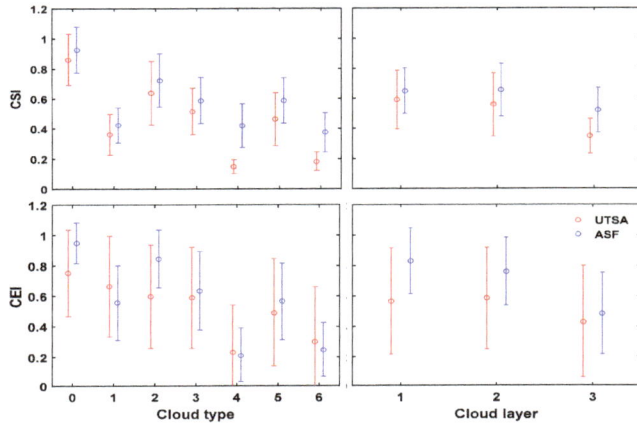

Figure 6. The mean CSI and clear-sky energy index (CEI) indices with their ±1 standard deviation error bars derived from all available data with the 5-min sampling interval as a function of cloud types (**left panels**) and cloud layers (**right panels**) (see Table 1) at the two sites.

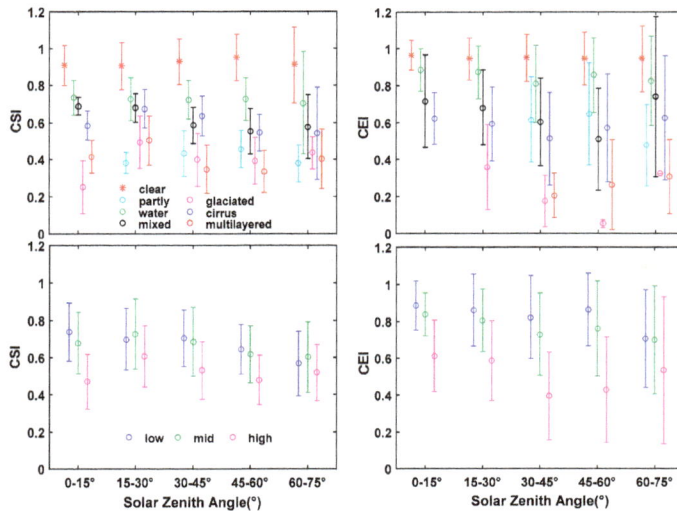

Figure 7. The averaged CSI (**left panel**) and CEI (**right panel**) indices (at 1-min time scale) for cloud types (**upper panel**) and cloud layers (**lower panel**) all plotted against solar zenith angle at the ASF station (not all cloud types have data over a 1-min interval at each solar zenith angle range).

With respect to CSI and CEI under clear-sky conditions, their values are nearly 1 and do not change much with zenith angle. For cloud types and layers, CSI may increase (e.g., opaque ice) or decrease (e.g., mixed) with increasing solar zenith angle. For CEI under cloudy conditions (both for cloud types and layers), the attenuation is generally higher at higher solar zenith angles (>45°).

4.2. Temporal Variability of Solar Attenuation under Different Cloud Conditions

Figure 8 shows the histograms of the temporal variability indices VI and VI_{new} calculated using all data available at the two stations with the 5-min sampling interval, indicating that these distributions are quite similar to each other. The means of VI are, respectively, 7.64 and 7.77 for UTSA and ASF stations, while the means of VI_{new} are, respectively, 8.40 and 8.96 for UTSA and ASF. The modes of VI and VI_{new} for the two stations are all equal to 1. Both means of VI and VI_{new} are somewhat higher at the ASF station compared to the UTSA station. This is consistent with the results of Xia et al. [27] which found that the measured G_h values at the UTSA station were overall biased low, due in part to improper calibration of the pyranometer at that site. The right-skewed distribution of the variability indices peaking close to 1 indicates that there were more cases with low temporal variability than those with high temporal variability in the study sites.

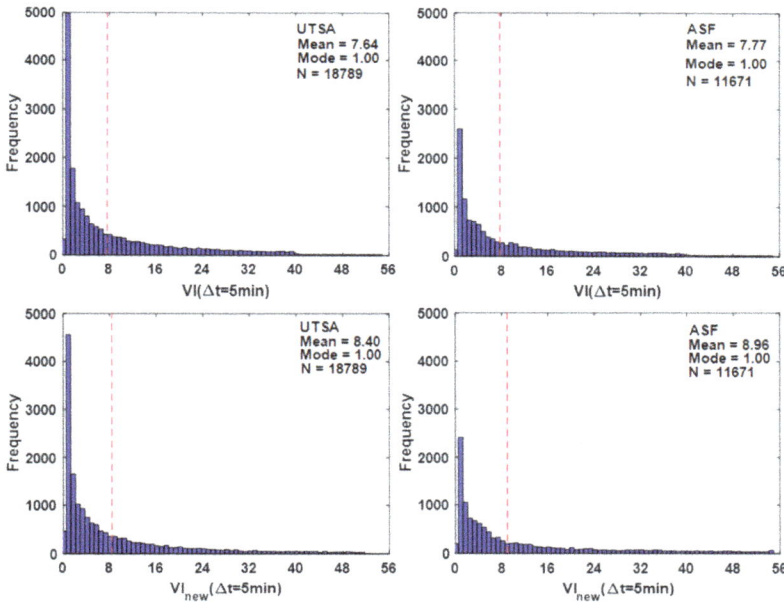

Figure 8. Histograms of the variability index (VI) (**upper panels**) and the modified version of the variability index (VI_{new}) (**lower panels**) at the 5-min time interval using all observations available at each of the two ground stations, UTSA (**left panels**) and ASF (**right panels**). The red dashed line in each panel indicates the average value of the index.

Figure 9 shows the means ±1 standard deviation error bars of the temporal variability indices (VI and VI_{new}) in terms of cloud types and layers under different time intervals (1-min, 5-min, 10-min, and 15-min). Clearly, the temporal variability as represented by both indices decreases with the increase of time interval. In addition, VI shows smaller values than VI_{new} under different cloud types and layers, especially at the 10-min and 15-min time intervals. It appears that the physically based VI_{new} is more sensitive to cloud type changes than VI for the four time intervals in this study and is thus recommended for future use. For cloud types, VI_{new} shows high temporal variability under water clouds (type 2) and low solar variability under clear sky (type 0), opaque ice (type 4), and multilayered clouds (type 6). For cloud layers, VI_{new} shows high solar variability under mid clouds and low solar variability under low clouds, which are reasonable results.

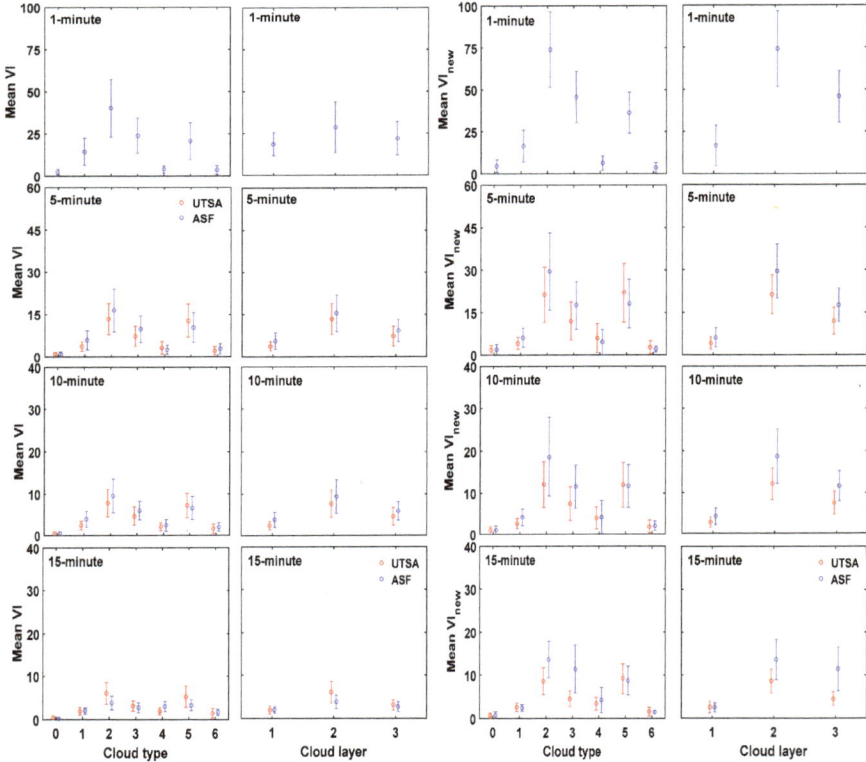

Figure 9. The mean variability index, VI (**left two panel columns**) and VI_{new} (**right two panel columns**) with their ±1 standard deviation error bars plotted against cloud types and cloud layers.

5. Discussion

Different approaches have been used in the literature to study solar variability and its impact on the electricity grid [22,23,26,39–41]. For a review that includes various definitions of solar variability indices, the reader is referred to Reference [24]. In this study, we followed an approach similar to Reno and Stein [23] who used global horizontal irradiance at two ground stations in Las Vegas, Nevada and cloud types from GOES satellites. Here, the relationship between ground observations of global horizontal irradiance at two stations in San Antonio, Texas and satellite-derived cloud types and cloud layers from GOES satellites was investigated using both the same and modified attenuation and attenuation variability indices used in Reference [23].

This study found similar distributions of CSI at the two ground stations, namely ASF and UTSA, but the mean of these distributions differed slightly in magnitude with the ASF means being somewhat larger than those at UTSA, as shown in Figures 3–5. These differences were attributed by Xia et al. [27] to the larger spectral band of the ASF pyranometer compared to the one at UTSA and to potential improper calibration of the pyranometer at the UTSA site. It is worth noticing that the CSI distributions in Figures 3–5 show some values that are greater than 1, which may be attributed to possible underestimation of G_{hc} and/or overestimation of G_h due cloud enhancement [34,42,43].

Assuming that the CSI values greater than 1 for cloudy conditions in the ASF data represent cloud enhancement, more cloud enhancement is produced by water clouds and in lesser degree by mixed phase and cirrus clouds. In addition, cloud enhancement occurs at all three levels.

In Figures 3–5, opaque ice and multilayered clouds have smaller CSI means because they contained ice crystals and opaque clouds resulting in high attenuation of solar radiation. The thin characteristics of cirrus clouds resulted in a lower ability to attenuate solar radiation as compared to opaque ice and multilayered clouds. Since there are more water clouds occurring at the lower layer and more opaque ice clouds at the high layer, the mean CSI is higher at the lower layer than at the higher layer.

Reno and Stein [23] found that at two locations in Las Vegas, Nevada the attenuation of solar radiation when arranged by cloud types according to CSI values from highest to lowest was clear, partly, water, cirrus, mixed, opaque ice, and multilayered. This organization is overall similar to the results of this study, except for partly-cloudy conditions. This study shows that the mean CSI for partly cloudy conditions is intermediate between mixed and opaque ice clouds, while for Reno and Stein [23] it is between clear sky and water clouds. These differences could be in part because their analysis was based on a coarser window of four by four GOES pixels, which could mix cloud types.

As expected, the attenuation indices CSI and CEI for clear-sky conditions, as shown in Figure 7, had larger values than for cloudy conditions and were nearly constant at all solar zenith angles due to the smaller attenuation of solar radiation in a cloud-free atmosphere compared to a cloudy atmosphere. Under cloudy-sky conditions, there was generally a decrease in CSI and CEI with increasing solar zenith angles. This effect was also expected since the longer the path that solar radiation travels through the cloudy atmosphere, the more the attenuation that would result.

The temporal variability of solar attenuation, as shown in Figure 9, is found to be dependent upon the temporal sampling interval with the mean variability indices (VI and VI_{new}) decreasing with the increase of the temporal sampling interval, consistent with Stein et al. [22]. The sampled ASF dataset at 15-min time interval could introduce errors due to the irregular temporal resolution. The variability indices of solar attenuation were lower under clear, partly, opaque ice, and multilayered clouds compared to those under water, mixed, and cirrus clouds. Of these last three, solar attenuation variability indices were higher for water clouds followed by mixed and cirrus clouds. All in all, these effects of clouds on attenuation variability are consistent with Reference [23].

6. Conclusions

Solar variability is considered a growing concern when it comes to the integration of the power from solar panel systems into the electric grid. Variation of solar irradiance at ground level results in variations of the harnessed solar power. The characterization of solar variability is thus very important for grid-connected solar photovoltaics and its impact on the power grid. A good understanding of short-term and long-term solar variability could contribute to grid reliability, power output forecast, and cost reduction.

In this paper, ground-based G_h observations and satellite-derived cloud properties were combined using attenuation (CSI, CEI) and attenuation variability (VI, VI_{new}) indices with the purpose of determining how solar radiation variability relates to cloud types and layers in San Antonio, Texas. As expected, it was found that on average solar radiation is attenuated the least under clear-sky conditions, followed by water clouds and cirrus clouds, and it is attenuated the most under opaque ice and multilayered clouds. Regarding cloud enhancement, the results suggest that it occurs more frequently under water clouds and less frequently for mixed phase and cirrus clouds. Cloud enhancement also appears to occur at all levels in the atmosphere. A new method was proposed for calculating the attenuation variability index (VI_{new}), which could be interpreted as the ratio of the rate of change of the observed global horizontal irradiance at some temporal sampling interval to the same quantity but computed from the global horizontal irradiance under clear sky. This definition makes more physical sense than the "length" approach posed by Reno and Stein [23], although gives similar results. The variability indices were found to decrease with increasing temporal sampling interval and were higher under water and mixed clouds and lower under opaque ice and multilayered clouds.

The results from this analysis illustrate how cloud classification from GOES satellites relates to solar attenuation and its temporal variability in San Antonio, Texas. It remains to be determined if these results could be extended to other regions. It remains also to be seen if similar results could be obtained using cloud products from other geostationary satellites like Meteosat or Himawari. The results of this study are overall consistent with the study of Reno and Stein [23] for Las Vegas, Nevada, but there are some differences. These differences, however, may be due to the different spatial resolution of the cloud datasets in the two studies rather than differences in optical properties of the clouds at the two sites.

The attenuation of solar radiation due to clouds has many uncertainties. This study, therefore, makes assumptions that are not able to be quantified, like cloud enhancement. An alternative way of studying the impact of clouds on solar irradiance would be using the optical properties of clouds in terms of optical thickness. The GOES dataset used here, if fact, uses the cloud optical thickness to parameterize the clouds. Some examples of studying the impact of clouds on solar irradiance using optical thickness include References [44–46].

Author Contributions: All authors contributed to the conceptualization; S.X. performed the formal analysis and original draft writing; A.M.M.-N. and H.X. contributed to the supervision; all authors contributed to the review and editing.

Acknowledgments: This work was supported in part by CPS Energy, Texas Sustainable Energy Research Institute (TSERI), and the University of Texas at San Antonio.

Conflicts of Interest: The authors declare no conflict of interest.

References

1. Atri, D.; Melott, A.L. Cosmic rays and terrestrial life: A brief review. *Astropart. Phys.* **2014**, *53*, 186–190. [CrossRef]

2. Hosenuzzaman, M.; Rahim, N.A.; Selvaraj, J.; Hasanuzzaman, M.; Malek, A.B.M.A.; Nahar, A. Global prospects, progress, policies, and environmental impact of solar photovoltaic power generation. *Renew. Sust. Energy Rev.* **2015**, *41*, 284–297. [CrossRef]

3. Reda, I.; Stoffel, T.; Myers, D. A method to calibrate a solar pyranometer for measuring reference diffuse irradiance. *Sol. Energy* **2003**, *74*, 103–112. [CrossRef]

4. Hodge, B.-M.; Milligan, M. Wind power forecasting error distributions over multiple timescales. In Proceedings of the 2011 IEEE Power and Energy Society General Meeting, Detroit, MI, USA, 24–29 July 2011; pp. 1–8.

5. Lappalainen, K.; Valkealahti, S. Output power variation of different PV array configurations during irradiance transitions caused by moving clouds. *Appl. Energy* **2017**, *190*, 902–910. [CrossRef]

6. Lave, M.; Kleissl, J. Solar variability of four sites across the state of Colorado. *Renew. Energy* **2010**, *35*, 2867–2873. [CrossRef]

7. Lave, M.; Kleissl, J.; Arias-Castro, E. High-frequency irradiance fluctuations and geographic smoothing. *Sol. Energy* **2012**, *86*, 2190–2199. [CrossRef]

8. Lave, M.; Kleissl, J.; Stein, J.S. A wavelet-based variability model (WVM) for solar PV power plants. *IEEE Trans. Sustain. Energy* **2013**, *4*, 501–509. [CrossRef]

9. Moumouni, Y.; Baghzouz, Y.; Boehm, R.F. Power "smoothing" of a commercial-size photovoltaic system by an energy storage system. In Proceedings of the 16th International Conference on Harmonics and Quality of Power (ICHQP), Bucharest, Romania, 25–28 May 2014; pp. 640–644.

10. Valero, F.P.J.; Minnis, P.; Pope, S.K.; Bucholtz, A.; Bush, B.C.; Doelling, D.R.; Smith, W.L.; Dong, X. Absorption of solar radiation by the atmosphere as determined using satellite, aircraft, and surface data during the Atmospheric Radiation Measurement Enhanced Shortwave Experiment (ARESE). *J. Geophys. Res. Atmos.* **2000**, *105*, 4743–4758. [CrossRef]

11. Beyer, H.G.; Costanzo, C.; Heinemann, D. Modifications of the Heliosat procedure for irradiance estimates from satellite images. *Sol. Energy* **1996**, *56*, 207–212. [CrossRef]

12. Marty, C.; Philipona, R. The clear-sky index to separate clear-sky from cloudy-sky situations in climate research. *Geophys. Res. Lett.* **2000**, *27*, 2649–2652. [CrossRef]

13. Smith, C.J.; Bright, J.M.; Crook, R. Cloud cover effect of clear-sky index distributions and differences between human and automatic cloud observations. *Sol. Energy* **2017**, *144*, 10–21. [CrossRef]

14. Mueller, R.W.; Dagestad, K.-F.; Ineichen, P.; Schroedter-Homscheidt, M.; Cros, S.; Dumortier, D.; Kuhlemann, R.; Olseth, J.A.; Piernavieja, G.; Reise, C. Rethinking satellite-based solar irradiance modelling: The SOLIS clear-sky module. *Remote Sens. Environ.* **2004**, *91*, 160–174. [CrossRef]

15. Reno, M.J.; Hansen, C.W.; Stein, J.S. *Global Horizontal Irradiance Clear Sky Models: Implementation and Analysis*; SANDIA Report SAND2012-2389; SANDIA: Albuquerque, NM, USA, 2012.

16. Ineichen, P.; Perez, R. A new airmass independent formulation for the Linke turbidity coefficient. *Sol. Energy* **2002**, *73*, 151–157. [CrossRef]

17. Perez, R.; Ineichen, P.; Moore, K.; Kmiecik, M.; Chain, C.; George, R.; Vignola, F. A new operational model for satellite-derived irradiances: Description and validation. *Sol. Energy* **2002**, *73*, 307–317. [CrossRef]

18. Gueymard, C. High performance model for clear-sky irradiance and illuminance. In Proceedings of the Solar 2004 Conference, Portland, OR, USA, January 2004; pp. 251–258.

19. Gueymard, C.A. Direct solar transmittance and irradiance predictions with broadband models. Part I: Detailed theoretical performance assessment. *Sol. Energy* **2003**, *74*, 355–379. [CrossRef]

20. Calif, R.; Soubdhan, T. On the use of the coefficient of variation to measure spatial and temporal correlation of global solar radiation. *Renew. Energy* **2016**, *88*, 192–199. [CrossRef]

21. Li, M.; Chu, Y.; Pedro, H.T.C.; Coimbra, C.F.M. Quantitative evaluation of the impact of cloud transmittance and cloud velocity on the accuracy of short-term DNI forecasts. *Renew. Energy* **2016**, *86*, 1362–1371. [CrossRef]

22. Stein, J.S.; Hansen, C.W.; Reno, M.J. The variability index: A new and novel metric for quantifying irradiance and PV output variability. In Proceedings of the World Renewable Energy Forum, Denver, CO, USA, 13–17 May 2012; pp. 13–17.

23. Reno, M.J.; Stein, J. Using Cloud Classification to Model Solar Variability. In Proceedings of the ASES National Solar Conference, Baltimore, MD, USA, 16–20 April 2013.

24. Schroedter-Homscheidt, M.; Kosmale, M.; Jung, S.; Kleissl, J. Classifying ground-measured 1 min temporal variability within hourly intervals for direct normal irradiances. *Meteorol. Z.* **2018**. [CrossRef]

25. Udelhofen, P.M.; Cess, R.D. Cloud cover variations over the United States: An influence of cosmic rays or solar variability? *Geophys. Res. Lett.* **2001**, *28*, 2617–2620. [CrossRef]

26. Nguyen, A.; Velay, M.; Schoene, J.; Zheglov, V.; Kurtz, B.; Murray, K.; Torre, B.; Kleissl, J. High PV penetration impacts on five local distribution networks using high resolution solar resource assessment with sky imager and quasi-steady state distribution system simulations. *Sol. Energy* **2016**, *132*, 221–235. [CrossRef]

27. Xia, S.; Mestas-Nuñez, A.M.; Xie, H.; Vega, R. An evaluation of satellite estimates of solar surface irradiance using ground observations in San Antonio, Texas, USA. *Remote Sens.* **2017**, *9*, 1268. [CrossRef]

28. Diak, G.R. Investigations of improvements to an operational GOES-satellite-data-based insolation system using pyranometer data from the US Climate Reference Network (USCRN). *Remote Sens. Environ.* **2017**, *195*, 79–95. [CrossRef]

29. Sengupta, M.; Habte, A.; Gotseff, P.; Weekley, A.; Lopez, A.; Molling, C.; Heidinger, A. A Physics-based GOES Satellite Product for Use in NREL's National Solar Radiation Database. In Proceedings of the European Photovoltaic Solar Energy Conference and Exhibition, Amsterdam, The Netherlands, 22–26 September 2014.

30. Pavolonis, M.J.; Heidinger, A.K. Daytime cloud overlap detection from AVHRR and VIIRS. *J. Appl. Meteorol.* **2004**, *43*, 762–778. [CrossRef]

31. Pavolonis, M.J.; Heidinger, A.K.; Uttal, T. Daytime global cloud typing from AVHRR and VIIRS: Algorithm description, validation, and comparisons. *J. Appl. Meteorol.* **2005**, *44*, 804–826. [CrossRef]

32. Skartveit, A.; Olseth, J.A. The probability density and autocorrelation of short-term global and beam irradiance. *Sol. Energy* **1992**, *49*, 477–487. [CrossRef]

33. Kasten, F.; Young, A.T. Revised optical air mass tables and approximation formula. *Appl. Opt.* **1989**, *28*, 4735–4738. [CrossRef]

34. Inman, R.H.; Chu, Y.; Coimbra, C.F.M. Cloud enhancement of global horizontal irradiance in California and Hawaii. *Sol. Energy* **2016**, *130*, 128–138. [CrossRef]

35. Tapakis, R.; Charalambides, A.G. Enhanced values of global irradiance due to the presence of clouds in Eastern Mediterranean. *Renew. Energy* **2014**, *62*, 459–467. [CrossRef]

36. Stephens, G.L. *Remote Sensing of the Lower Atmosphere*; Oxford University Press: New York, NY, USA, 1994; p. 544.

37. Matus, A.V.; L'Ecuyer, T.S. The role of cloud phase in Earth's radiation budget. *J. Geophys. Res. Atmos.* **2017**, *122*, 2559–2578. [CrossRef]

38. Tan, I.; Storelvmo, T.; Zelinka, M.D. Observational constraints on mixed-phase clouds imply higher climate sensitivity. *Science* **2016**, *352*, 224–227. [CrossRef]

39. Annathurai, V.; Gan, C.K.; Ghani, M.R.A.; Baharin, K.A. Impacts of solar variability on distribution networks performance. *Int. J. Appl. Eng. Res.* **2017**, *12*, 1151–1155.

40. Bright, J.M.; Smith, C.J.; Taylor, P.G.; Crook, R. Stochastic generation of synthetic minutely irradiance time series derived from mean hourly weather observation data. *Sol. Energy* **2015**, *115*, 229–242. [CrossRef]

41. Gan, C.K.; Lau, C.Y.; Baharin, K.A.; Pudjianto, D. Impact of the photovoltaic system variability on transformer tap changer operations in distribution networks. *CIRED Open Access Proc. J.* **2017**, 1818–1821. [CrossRef]

42. de Andrade, R.C.; Tiba, C. Extreme global solar irradiance due to cloud enhancement in northeastern Brazil. *Renew. Energy* **2016**, *86*, 1433–1441. [CrossRef]

43. Gueymard, C.A. Cloud and albedo enhancement impacts on solar irradiance using high-frequency measurements from thermopile and photodiode radiometers. Part 1: Impacts on global horizontal irradiance. *Sol. Energy* **2017**, *153*, 755–765. [CrossRef]

44. Rossow, W.B.; Schiffer, R.A. ISCCP cloud data products. *Bull. Am. Meteorol. Soc.* **1991**, *72*, 2–20. [CrossRef]

45. Wang, C.; Yang, P.; Baum, B.A.; Platnick, S.; Heidinger, A.K.; Hu, Y.; Holz, R.E. Retrieval of ice cloud optical thickness and effective particle size using a fast infrared radiative transfer model. *J. Appl. Meteorol. Climatol.* **2011**, *50*, 2283–2297. [CrossRef]

46. Krisna, T.C.; Wendisch, M.; Ehrlich, A.; Jäkel, E.; Werner, F.; Weigel, R.; Borrmann, S.; Mahnke, C.; Pöschl, U.; Andreae, M.O. Comparing airborne and satellite retrievals of cloud optical thickness and particle effective radius using a spectral radiance ratio technique: Two case studies for cirrus and deep convective clouds. *Atmos. Chem. Phys.* **2018**, *18*, 4439–4462. [CrossRef]

MDPI

St. Alban-Anlage 66

4052 Basel

Switzerland

Tel. +41 61 683 77 34

Fax +41 61 302 89 18

www.mdpi.com

Remote Sensing Editorial Office

E-mail: remotesensing@mdpi.com

www.mdpi.com/journal/remotesensing

www.ingramcontent.com/pod-product-compliance
Lightning Source LLC
Chambersburg PA
CBHW051839210326
41597CB00033B/5713